The Lure *of the* Land

A CENTURY OF EDUCATION AT HARPER ADAMS

Heather Williams

HARPER ADAMS UNIVERSITY COLLEGE

Main cover photograph by Gordon Dickins

ISBN: 1-870955-57-9

Cover and book design by The Graphic Terrace
Printed in Great Britain by Livesey Limited, Shrewsbury

Contents

Photo: Paul Humphreys/Alex Powell Studios

About the Author

Heather Williams has a degree in history from the University of Nottingham and was awarded a Ph.D. for her research on Nottingham artists. On arrival in Shropshire in the early 1980s she was a founder member and subsequently has been elected Chairman of the Newport and District History Society. During the 1980s she taught adult education classes throughout Shropshire. In the late 1980s she was asked to investigate the possibility of forming the archives at Harper Adams and today is Archivist of a wide-ranging and impressive collection of material stretching over one hundred years.

She believes that history should be enjoyable and accessible to all.

This book is dedicated to all the people
who have made Harper Adams what it is today.
Too numerous to name, their contribution
lies at the very heart of this history.

Acknowledgements

First and foremost I must acknowledge the immense help and support given to me in writing this book by the editorial team of Tony Harris and Colin Smith. Truly the book could not have been written without them. When ideas and enthusiasm ebbed, they were always there providing direction and encouragement. On a more personal note, when Life's storm clouds gathered, completion of the book was only possible through the friendship shown to me. I am greatly indebted to Josie Sheppard for her untiring selflessness and love and to Julie Brook for her patience, understanding and wisdom.

My mother, Joan Williams, did not live to see the completion of this book, but her faith in me and the love and pride which she always showed me remain with me and have undoubtedly played a very important part in shaping this book.

I have received immeasurable help from all at Harper Adams and my thanks are extended to them all. I must mention in particular: Alan Costley, Paul Custance, Mary Early, Robin Harris, Mike James, Wynne Jones, David Llewellyn, Richard Matson, Janet Parnaby, Liz Powell, John Roberts, Howard Robinson, Ian Robson, John Taylor and Janet Wilks.

Ralph Early gave invaluable help in suggesting the title for the book.

Basil Bayne has given me much encouragement and help with the sections on the NIPH.

Memories have formed a very important part of this book and I was shown considerable kindness by the following, when I arrived at their door, tape recorder in hand: Ian Baker, Shirley Barber, Ros Barsley, Angela

Billington, Mellor Boyns, Bob Bradley, Vera Bradley, Bill Burrell, Tony Chart, Steven Colman, Patrick Davies, Irene Doody, Maurice Eddowes, Ray Elkes, Tom Fordham, Dora and Tony Gale, Ian Gedye, Maurice Lancaster, Paula and Jack Large, Ellen Lintin, Dorothy Lomax, Peter Luscombe, Ken Mathias, Willard and Margaret Moore, Albert Morrey, Dorothy Neate, William Neate, Trevor and Christine Parnham, Ed and Liz Pridgeon, David Roberts, Albert Rowley, Elsie Tomkinson, Bob Tupper, Derek Turner, Frank Ward, Anthony Warner, Cecil Warner, Peter Watson Jones, John White, Jack Wickstead and Jim Wickstead. Written memories were gratefully received from Leonora Cottrell, Martin Cross, Arthur George and Gerard Poulter.

Jennie Preece did a brilliant task in transcribing the tapes. I thank Simon Hendrie of Maxell Europe Ltd for assisting in tape quality.

Staff at the Information Service and Shropshire Records and Research Centre at Shrewsbury have been of great assistance, in particular I must name Janet Williams of the former and Mary McKenzie, Tony Carr, Andrew Davidson, Kerry Dickins and Russell Evans of the latter.

Further assistance has been provided by George Baugh, Karl Cieciora, Bernard Lowry and Paul Stamper.

I am especially grateful to the photographic expertise of John Pick and Gordon Dickins.

My gratitude goes to The Earl of Selborne for agreeing to write the preface to the book.

I am also grateful to Dr Katherine Mercer Young for bringing her father, Wilfred Mercer's diary to my attention and to Mrs Mary Arden-Davis for information on her father, Reg Dalley. Tony Jeavons was diligent in finding and donating the lost wartime copies of Cat-A-Mountain.

Finally, I must thank Ann Parry for the index, Bridget Eckley for proof reading and Paul Brasenell of The Graphic Terrace for turning the draft history into the book that you see.

Wheelwright Cottage, July 2000

Preface by The Earl of Selborne

Most college histories are read only by those who have been closely connected with the institution. Dr Heather Williams has written a book which will be of great interest to a much wider audience. She guides us with a light touch through a century of agricultural education against a backcloth of the changing fortunes of the rural economy. Each generation of students has been required to meet very different challenges. The story of how Harper Adams has survived (at times with difficulty) and is now flourishing is one which will appeal to anyone with an interest in the history of British agriculture over the last century.

Harper Adams has always been jealous of its independence. From its inception as the first agricultural college to be established in this country by private bequest it has recognised that there are many advantages in being at one remove from Whitehall and from local government. The Governors have consistently been wary of any attempt from outside to exert undue influence. There have been periods in the College's history when this independence has been at risk. On several occasions crushing debts have forced the Governors to seek financial help from the government and the future of the College was threatened by ministerial edict at the start of the Second World War. As recently as the late 1980s Harper Adams faced enormous pressure to merge with other institutions, a challenge met by a firm resolution to remain an independent corporation. Subsequent events have demonstrated that this resolve to remain consistent to the College's traditions has been spectacularly vindicated.

The survival and ultimate success of this independent agricultural college has been due in large measure to the entrepreneurial skills and foresight of a handful of Principals at key periods in the College's history. Hedworth Foulkes, the first Principal, ensured that the teaching was sufficiently practical to win over the prejudices of farmers initially doubtful of the benefits that might be derived from agricultural education. Extra-mural courses run for the Shropshire County Council were soon well attended and much appreciated. The College's research activities were equally effective in breaking down the barriers between the College and the community as the research programme was kept highly relevant to the farming problems of the day. This emphasis on relevance to agriculture remains an enduring theme and as a strategy for survival and expansion this has paid off handsomely.

The second Principal, Dr Crowther was astute enough to realise that the harsh climate for farming created new opportunities. In the inter-war years, as the government put more resources into the Advisory Services, he ensured that the College benefited by building up a wide-ranging advisory expertise. Alongside the advisory officers there was also a whole range of research assistants.

After the war the government decided to divorce the advisory services from educational institutions in England and Wales (though not in Scotland). The loss of the advisory services was no fault of the College, but I suspect that an opportunity to shape the future of dairy production in this country was missed in 1932 when Bobby Boutflour's high yielding Friesian herd was dispersed.

One sad failure of nerve has to be noted. During the First World War and up to the end of the Second World War women enrolled in agricultural courses at the College. There was then a gap of twenty-five years when no women were admitted to study agriculture at Harper Adams. It was only because of the need to conform to the recommendation of the Pilkington Report that the Governors passed a special resolution in 1970 allowing for the admission of women to the Higher National Diploma course.

Harper Adams was initially slow to recognise the speed of change in Higher Education which arose from the Robbins Reports of 1963. This comprehensive review of Higher Education made special reference to the five independent agricultural colleges, including Harper Adams, and suggested that courses should be lengthened to three years with the possibility of offering degrees at a later date. The change of status in 1964 to a direct grant college and the introduction of the Higher National Diploma in 1969 were important developments in the College's history. However, it took another ten years or so for the Governors to embrace whole-heartedly the logical consequences of this new role in the national provision for Higher Education.

The last twenty-five years have seen these hesitations swept aside. The College put in place first a CNAA sandwich degree course in agricultural technology in 1981 and subsequently a range of other degree courses. Harper Adams courses now embrace land management, business studies, marketing, engineering and rural development alongside the core subject of agriculture. In 1996 Harper Adams was granted taught degree awarding status, an acknowledgement that the earlier decision to remain independent had been justified. The name Harper Adams University College adopted in 1998 is an appropriate recognition of the quality of the teaching and research and gives a clear message to the world of the role that Harper Adams now fulfils.

Harper Adams is now better placed than most other Higher Education institutions to respond to the many government initiatives which seek closer links between industry and academia. The tradition of practical teaching, advisory services and applied research, which has always been central to Harper Adams, now manifests itself through its research, consultancy and training programmes undertaken with the land-based industries.

I am confident that Harper Adams will never lose sight of the central role played by agriculture in the rural economy and that, however depressed European agriculture might be, the College will recognise yet again that with agricultural depression comes opportunities for both farmers and Harper Adams. The response of the College to the challenges of the last twenty-five years has been truly impressive. I have no doubt that its second century will see equally dramatic and equally impressive developments at Harper Adams University College.

Blackmoor Estates, July 2000

Chapter One —

Thomas Harper Adams, the Man and his Life 1817-92

It takes little to imagine the great pride with which the new Principal, Percy Hedworth Foulkes, and the Governors showed their numerous guests around the College at the official opening on 26 September 1901. The opening ceremony was 'the largest gathering of its kind in the neighbourhood for many years past'.[1] Seven hundred invitations had been sent.[2] Notable figures from the academic and agricultural worlds as well as Members of Parliament, Councillors and leading lights of the County attended the ceremony. There that day were the Earl and Countess of Bradford, Mr. H. J. Mackinder, Principal of Reading College, Professor Robert Wallace of Edinburgh University, known to Hedworth Foulkes from his university days, and guest of honour the Rt Hon. R. W. Hanbury, President of the Board of Agriculture.[1] The luncheon menu, prepared by the Royal Victoria Hotel, Newport,[3] consisted of salmon and mayonnaise, roast beef, pressed beef, partridges, roast chickens, roast ducks, rabbit pies, tongue, boiled ham, apple and plum tarts, cheese and salad. The band of the First Shropshire Artillery played[1] and a grand time was had by all.

Congratulatory speeches were made in a large marquee in front of the College. Praise was lavished on Thomas Harper Adams whose benevolence on his death in 1892 had created the College. However, little of real substance was said about the man. Strangely, it seems little was known, or was being said. Colonel Sambrooke Burne, a Governor and future Chairman of the Governing Body, had been one of the original three executors of Thomas Harper Adams' will, but, apparently, even he did not know the man or the circumstances of the will. In 1908 Lord Barnard, as a member of the Reay Committee investigating agricultural education, questioned Colonel Burne. The College, just seven years old, was naturally to come under the scrutiny of the Committee. Could Burne say what had led Thomas Harper Adams 'to make this magnificent foundation'? The sum mentioned to the Committee was £45,496. Burne lamely replied: *'No, I don't think I can. I have heard that he said that he got all his money out of the land, and he thought the best he could do was to leave it back to the land.'* Burne concluded by revealing that Harper Adams had taken little interest in agricultural education and had left his land

1.1 The College on completion, c. 1900.

'in a very queer state'.[4] We can sympathize with Lord Barnard's ensuing confusion. Indeed, until today, more questions could have been asked about the founder than answers given. However, with the recent discovery of documents and family papers the story can now be told.

By his death in 1892, Thomas Harper Adams was a man of considerable wealth, with land at Edgmond, Woodseaves, Chadwell, Newport and Church Aston. The value of his estate on his death was estimated at £63,000.[5] He was a gentleman farmer with a passion for buying land. Even illness, old age and the lack of an heir could not stem the dealing. It was in the blood. From the early eighteenth century, document upon document survives to show the Harper and Adams families amassing land through prudent business practices and well-chosen marriages within the tight circle of the Shropshire lesser gentry. Rumour had it that Harper Adams' peculiarities arose from inbreeding:

> *The families of both Thomas Harper Adams and his wife have intermarried considerably which partly accounts for the eccentricity of the two brothers [Thomas Harper Adams and his elder brother Joseph Harrison Adams] and the want of social intellect as compared with other men.*[6]

From at least the early eighteenth century the Adams Family was living in Tibberton. Thomas Harper Adams claimed kinship with the Adams Family of Sambrook, who in 1773 took the maternal name and arms of Anson and the estates of that family at Shugborough. Within two generations the family had received the Earldom of Lichfield.[7] However, neither modern research nor that of the College of Arms in the nineteenth century, when Harper Adams' brother, Joseph Harrison Adams, was applying for a coat of arms *(see photo page 39)*, have unearthed a link.[8] The Adams Family of Sambrook descended from a branch in Edgmond in the seventeenth century.[7] It is not improbable that the Tibberton Adams Family also shared Edgmond roots. The surname is a common one in Shropshire and lineages are difficult to trace when generations of parents have been doggedly determined to baptize their sons William or Thomas. However, such was the family pride in the Lichfield tradition that Harper Adams at one point thought of leaving his wealth to the Earls.[9] There is also the question of a link with William Adams who founded the Grammar School in Newport in 1656. No evidence survives and with William Adams being a man of considerably less wealth and kudos than the Lichfields, Thomas Harper Adams aired no such claim.

In 1804 the marriage settlement of Sarah Harper and John Adams, Harper Adams' parents, gave the Adams residence as Tibberton Hall.[10] The Hall was probably on the site of Hall Fields not far from the present church.[11] The parents were soon to move to the newly built house at the Ancellors, land along the Shrewsbury road between Edgmond and Tibberton.[12] Ancellor House is now the Principal's residence and the land forms the oldest part of the College Farm. This land had been acquired by the Adams Family from Robert Pigott in 1779,[13] but the house was not commenced until 1806. For £1,620 John Carline of Shrewsbury undertook to build 'a brick mansion' in the barn yard 'near to buildings there already'. The bricks were to be made from 'clay from the Barn Yard', burnt on the spot.[12] Additional costs were to increase the price by £200, when the invoice arrived in 1809. It was a house of some status with two best parlours and a best staircase.[14] An undated schedule of the farm lists 'waggon-stable, 2 barns, shead, feeding-stalls, cow-fold and stack-yard'. There were further stables, pig sties, 'cow-houses' and 'cart-houses'.[15] The spelling Ancellor, the form now

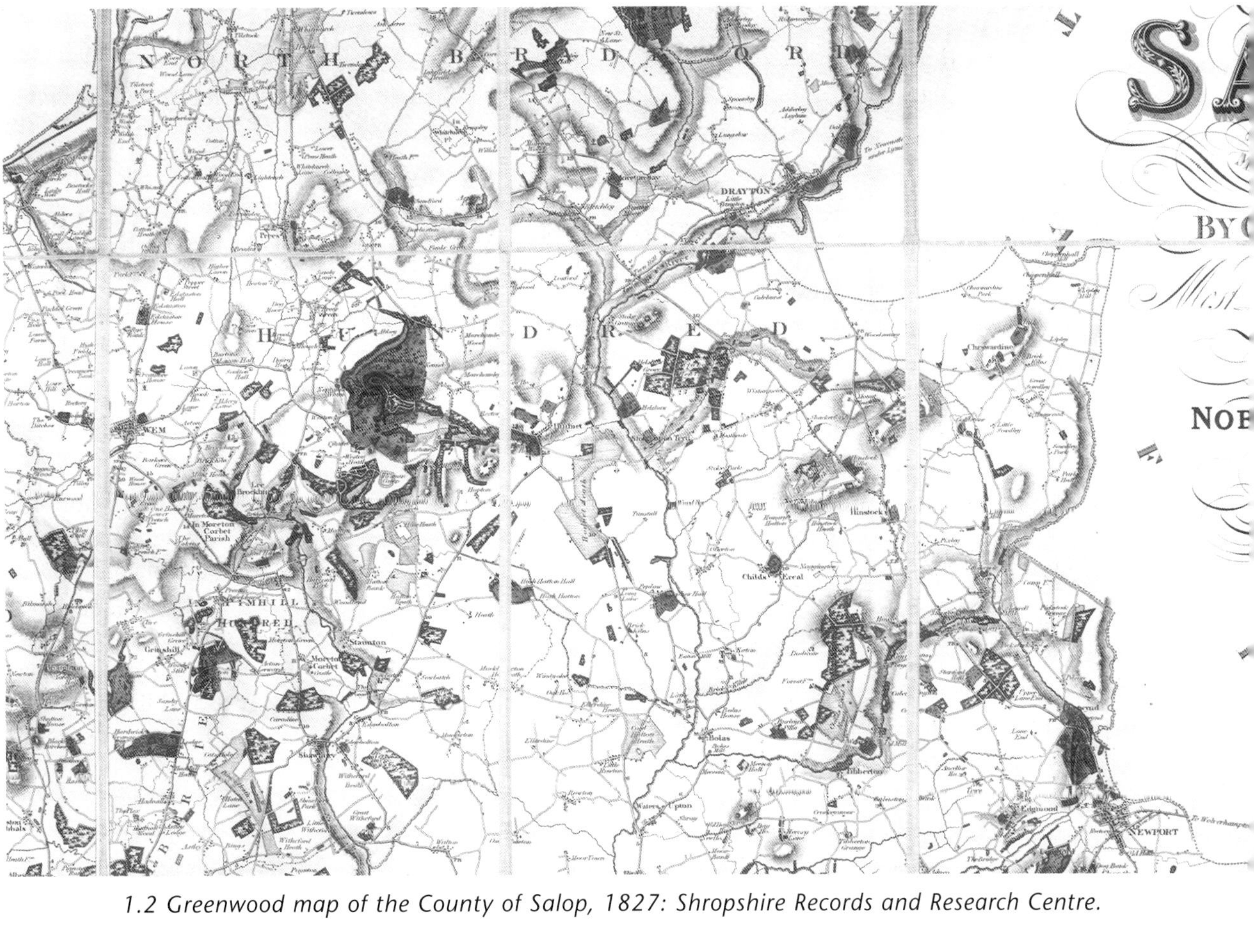

1.2 Greenwood map of the County of Salop, 1827: Shropshire Records and Research Centre.

current, varies from document to document over the years, for example, appearing as Ancellors in 1779, Ansiller in 1796, Ansellers in 1804, Anceller in 1812 and Hanclows in 1822.[16] The earliest reference to the name is Anslow in 1695 and it is this which could give a clue to its meaning.[17] Dr Margaret Gelling, an authority on place-names, suggests the word derives from the Old English 'law', or tumulus, combined with a personal name. Other local examples are Longslow and Munslow.[18]

It was at Ancellor House that Thomas, the youngest son, was born, probably in 1817.[19] Prior to John Adams' death in June 1832 he had settled his lands on his three surviving sons.[20] William received the Ancellor Estate, Joseph the Chatwell Estate and Thomas merely some land in Edgmond, as the Harper Estate at Woodseaves, bequeathed by his great-uncle, was to pass to him.[21] The facts reveal that Thomas Harper of Stanton upon Hine Heath was half brother to Sarah Harper, wife of Thomas Harper of Woodseaves.[22] The daughters of the marriage of Thomas and Sarah Harper of Woodseaves were Sarah Adams, the mother of Thomas Harper Adams, and Ann Dickin, the mother of Harper Adams' wife. The land at Woodseaves was owned by Thomas Harper of Stanton and it was this land which found its way to Thomas Harper Adams, his godson and great-nephew.[23]

Several stories survive as to a colourful history behind the bequest. The land was willed out of ordinary succession, evidently due to a family dispute. In one story Thomas Harper of Stanton has a son, whom he disinherits on account of his fiddle playing:

> *The son was an exceptionally good violinist much in demand at social functions. He accompanied the dancing at the Hunt Servants Ball which was held at High Hatton Hall. It was Saturday night and the dancing went on until the small hours of Sunday morning. Next morning his father enquired what time he came home and receiving the truthful reply seized the boy's violin and smashed it across his knees. The boy left home and went to London. He got a job playing the violin for a London orchestra.*[24]

After Harper Adams' death, the Newport Advertiser of 25 June 1892 wrote of a claim on the Harper portion of the estate made by a certain Mr Harper from London. The article claimed Mr Harper's ancestor had been disinherited by his father on account of his Methodism. Research into the early Methodist history of the Moreton Mill Methodist Church has revealed the role played by William Harper, a pioneer preacher in the area.[25] Indeed, as Harper Adams lay dying, one Edwin Harper visited him to request his rightful inheritance. Harper Adams would not accept the legitimacy of his claim.[26]

By 1841, the time of Harper Adams' marriage to his cousin Sarah Dickin of Prees, he was a man of some substance.[27] Brother William had died unmarried in February 1835 and the bulk of his property had passed to Harper Adams.[21] This contributed to his already considerable prospects. A childless marriage and a bitter family dispute were to ensure that eventually the majority of the inheritance of the Harper and Adams families would pass by the will of Thomas Harper Adams to the College.

By the date of his marriage Thomas Harper Adams had just one surviving brother, Joseph Harrison Adams. Although older than Harper Adams by three years,[28] Joseph was very much under his brother's influence.[29] In 1836 Harrison Adams married a local heiress, Mary Ann Blakemore. Her dowry amounted to around £15,000, a very substantial sum. Their only child, Mary Ann, was born in August 1837.[5] According to Mary Ann her mother was bullied by her father into drawing up a will in his favour, no account being taken of the child.[29] The mother died in May 1856.[5] Harper Adams, it is said, then worked upon his brother, 'a very eccentric, and weak-minded man', inducing him in 1858 to make a will whereby all his estate was to pass to him. Harper Adams' own will of 1863 made reciprocal arrangements.[29]

Circumstances worsened when in 1862 Mary Ann Adams married John Minor, a gentleman farmer, against the wishes of her father and her uncle. The brothers considered him of little worth and, when the couple fell on hard times, declared he had lost his money through 'dissipation, horse racing and betting'.[29] Harrison Adams gave his daughter no money and refused even to see her, 'she believes through the baneful influence of his brother Thomas Harper Adams'.[29] Sad letters, written each year around the time of Christmas, testify to Mary Ann's failure to obtain a reconciliation with either her father or uncle.[16] Thus, on Harrison Adams' death in September 1889, his estate passed to Harper Adams. Mary Ann contested the wills of her father and mother and in April 1890 won a settlement of an annual annuity of £200, based on a proportion of the land brought into the family by her late mother.[29]

On Harper Adams' death in June 1892 [30] his will stated that the family fortune was to be used 'for the purpose of teaching practical and theoretical Agriculture in England either by means of lectures or establishing a School of Agriculture in England' *(see photo page 40)*.[31] Thomas Harper Adams' widow, Sarah, was a woman of independent means and had neither sought nor been granted an interest in the estate beyond a life interest in the home, Edgmond House.[29] Not unnaturally, Mary Ann Minor sought to have the will set aside, arguing the will had not been duly executed as the testator was 'not of sound mind memory and understanding'.[32] Her solicitors feared the settlement of April 1890 would go against her, as it was impossible to reason Harper Adams' insanity at that point and then to fix a point at which he had become mad.[29] On their part the solicitors for the Trustees of the Harper Adams Estate must have feared the overwhelming justness of Mary Ann Minor's case as the sole surviving relation of both brothers. After months of legal dealings the most satisfying solution was to come to an agreement out of court and a compromise settlement was reached on 26 January 1893. Mary Ann Minor had her costs and all duty paid and received a lump sum of £11,000 and the property in Newport which in effect had been her mother's dowry.[32] In return she was to withdraw all opposition to the proving of the will. The remainder of the estate was now secure and the process could begin of setting up a foundation.

Some could argue that the dispute had worked primarily to the advantage of the solicitors, both locally and in London. However, the boxes of legal documents have provided an unexpected bonus in the form of witness statements as to Thomas Harper Adams' character and state of mind in the closing years of his life. It must be remembered that many of these statements were compiled in an attempt to prove that Harper Adams was unbalanced in old age and thus to overturn the will. Nonetheless a credible picture emerges of a very odd man, with no fondness for people or society, making an art of miserliness. Most importantly, this picture can be confirmed by the memories of local resident Vera Bradley, based on her mother's stories of Edgmond life in the 1880s.[33]

Laura Brisbourne, Vera Bradley's mother, referred to Thomas Harper Adams as 'Old Squealer Adams' on account of his high-pitched voice. Miss Bradley told how her mother talked of local children running behind his cart and calling him names:

> *It was an iron-wheeled cart, a farm cart. There was a bench along to sit on - all open. [...] The children all attacked him without reservation. [...] He didn't particularly get on well with other people. He never joined in anything. He was just very very much a recluse really. [...] He was a hoarder. He went about as if he hadn't got a penny. He was really a miser.*

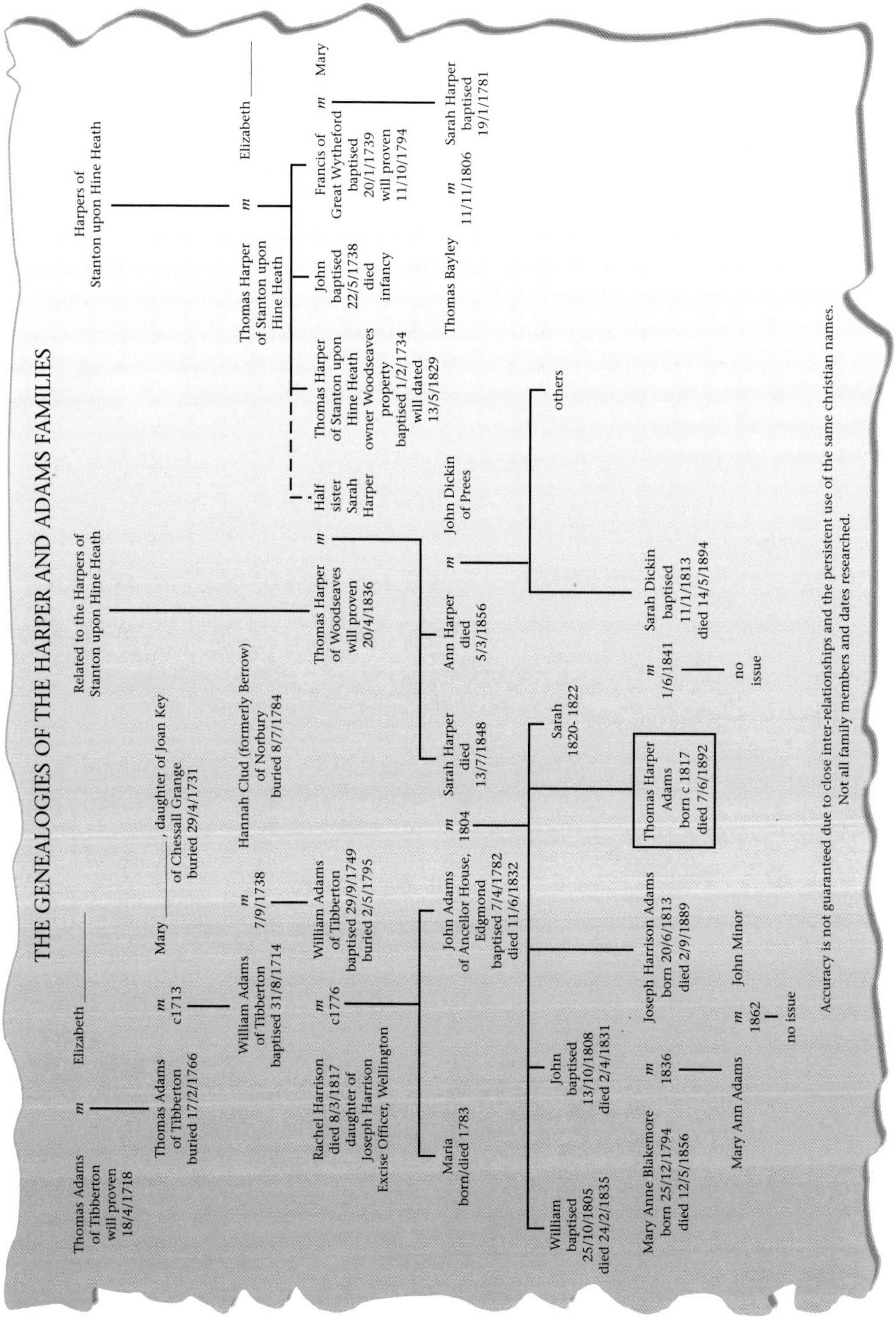

1.3 The family tree of Thomas Harper Adams

It is as a miser that Harper Adams frequently appears in the proofs of witness gathered for the dispute over inheritance. A journeyman tailor, John Morris of Edgmond, stated Harper Adams was 'an eccentric man and very selfish and very spiteful'. On one occasion he was called to shave him at Edgmond House and was paid 2d., when he originally had asked for 6d. Needless to say, Morris was expected to provide his own hot water.[34] William Scott, a labourer of Newport, told the following:

> *I heard some years ago that Thomas Harper Adams once sent some old clothes to the paper mills at Tibberton for sale, but as they would not give so much as he wanted he had them brought back and had them ploughed in the fields and the old people in the village picked them off the ground.*[34]

Others talked of Harper Adams giving away rabbits, a rare event in itself, only to demand the return of their skins.[34]

Amongst papers gathered for the brief for the defendant there is a reference to Harper Adams' dislike of people:

> *[He] lived in a very simple way keeping no company, never going to church, or taking part in any Parish Undertakings or assisting to promote any Charities or charitable object for the benefit of the poor of the Parish in which he lived so many years, in fact he was considered the pest of the Village and hated by every one.*[6]

This thumb-nail sketch is confirmed by an independent source. When news of his death reached the Newport & Market Drayton Advertiser, the Editor wrote:

> *We confess the announcement did not stir us to any deep emotion. To say that Mr. Adams had not made himself a popular man would be to put it in the mildest form. For years back he, and indeed his late brother too, the late John [Joseph] Harrison Adams, of Newport, appeared to cut themselves off from sympathy with almost all their neighbours, and from the share of public movements which is expected of men in their position. Not only did they fail to inaugurate any schemes for the general good, but their names were conspicuously absent from subscription lists, and as little was expected of them in private charity.*[9]

Such a damning obituary is very rare for the Victorians.

In defence of the man, from the late 1880s these anti-social and eccentric traits were undoubtedly accentuated by illness. About four to five years before his death Harper Adams had suffered a stroke which had resulted in facial paralysis.[29] By 1890 he was paralyzed down his right side and unable to use his right hand.[35] Dr Brereton noted down 'choreic movements in limb violent when not under influence of bromide ... locomotion becoming progressively worse'.[36] During the legal proceedings Ann Davey, his nurse, told of his jerking right arm and leg and his inability to feed and dress himself.[34] Modern medical conjecture attributes many of his symptoms to the stroke, namely his emotional displays and personality disorders, such as confusion and suspicion. The bromide would have been used to calm him. He may well have been showing signs of senile dementia in his seventies, which coupled with his already eccentric behaviour, led to rumours of madness.[37] It was these which were seized upon by Mary Ann Minor in her attempts to have the will set aside. As the dispute over the will heightened, Dr Thorpe of Newport, holder of a special certificate for examination in lunacy, was brought in to testify that he had visited Harper Adams on numerous occasions since February 1892 'and from the experience and observations while at the Lunatic Asylums has no hesitation in saying Mr. Adams was insane'.[38] Other doctors also testified that Harper Adams was not in a fit state of mind to make a will.[29] Mrs Adams denied this and, incidentally, all knowledge 'of his fear of being shot or blown up with dynamite'.[35]

Was Harper Adams deranged in his last years and incapable of sound judgement? Both sides in the dispute argued well and long. On reading all the evidence it is apparent that his eccentricity and disagreeable habits worsened considerably with age and infirmity. A week before his death a visitor found the gates to Edgmond House fastened with chain and padlock and the windows at the front of the house barricaded.[26] Harper Adams had grown increasingly paranoiac. He had always been anti-social, cantankerous and miserly and now as he was dying, he was suspicious that all were after his wealth. But, as his wife said: *'He was a very good man of business. [...] His memory was very good up to the time of his death.'* [35] He was undoubtedly seriously ill and at times agitated and nervous, but his determination that his niece should not inherit, although hard natured and unfair, was perfectly consistent with his character.

By 1890 events on the national and local stage had presented him with the ideal channel for the family wealth. A study of the times in which the will was drawn up reveals very clear evidence of a rational chain of thought behind Harper Adams' decision to leave the money for the purposes of agricultural education. After he inherited his brother's estate, the responsibility for the future of the fortune weighed heavily on him. The Cheswell farmer John Holland stated: *'On one occasion he told me he had an Estate left him "bothering" and he did not know what to do with it. [...] This was two or three years ago.'* [34] Significantly, George Adderley, a hairdresser of Newport, revealed that Thomas Harper Adams had said: *'There might be a good many surprised how I shall leave my money.'* While moments previous he had been saying that there were no good agricultural men about.[34] Sadly, no date is given for this. It seems that people confided in their hairdresser then as now.

The state of agriculture was not surprisingly a matter of concern to Harper Adams. By this time the agricultural depression was biting deep. Prices of cereals had been falling since the 1870s as a result of the influx of foreign grain. Land was falling out of cultivation and buildings were in need of repair. The landowner felt his own profits shrink, as a reduction in rents was often the only option to keep tenants. It is known, for example, that the valuers of Harper Adams' estate, acting after his death, placed a lower value on the land as they felt it would be prudent to reduce rents and retain the sitting tenants. In their opinion, they were paying too high a rent considering the dilapidation of the buildings.[5] This tale of neglect of the land is reinforced by the time and cost taken to improve the fields of the Ancellor Farm after they had been acquired by the College.[4] Everywhere the search was on to reduce production costs and for the innovator the answer lay in the introduction of new methods and ideas, allowing the developing science of agriculture to work hand in hand with the practice. Agricultural education seemed to offer a road to salvation, if only the farmer's natural suspicion could be abated. Harper Adams, although no innovator himself, was astute enough to recognize the path along which farming had to travel.

A letter from the local Member of Parliament, Colonel William Kenyon-Slaney, to Harper Adams, dated 12 December 1890, provides further evidence of Harper Adams' train of thought. It suggests Harper Adams had approached him for advice. Kenyon-Slaney wrote: *'I cannot help thinking that if your friend would consult any good family lawyer he would probably find that he has now full powers to found such an institution & to safeguard it satisfactorily against any of the peculiarities to what you refer.'*[39] Ever the sly one, Harper Adams had not owned up to the idea as his own and, characteristically, it seems he was trying to avoid paying for advice. So, in November 1890 Harper Adams was clearly contemplating an agricultural school.

Guest Information

Harper Adams
University

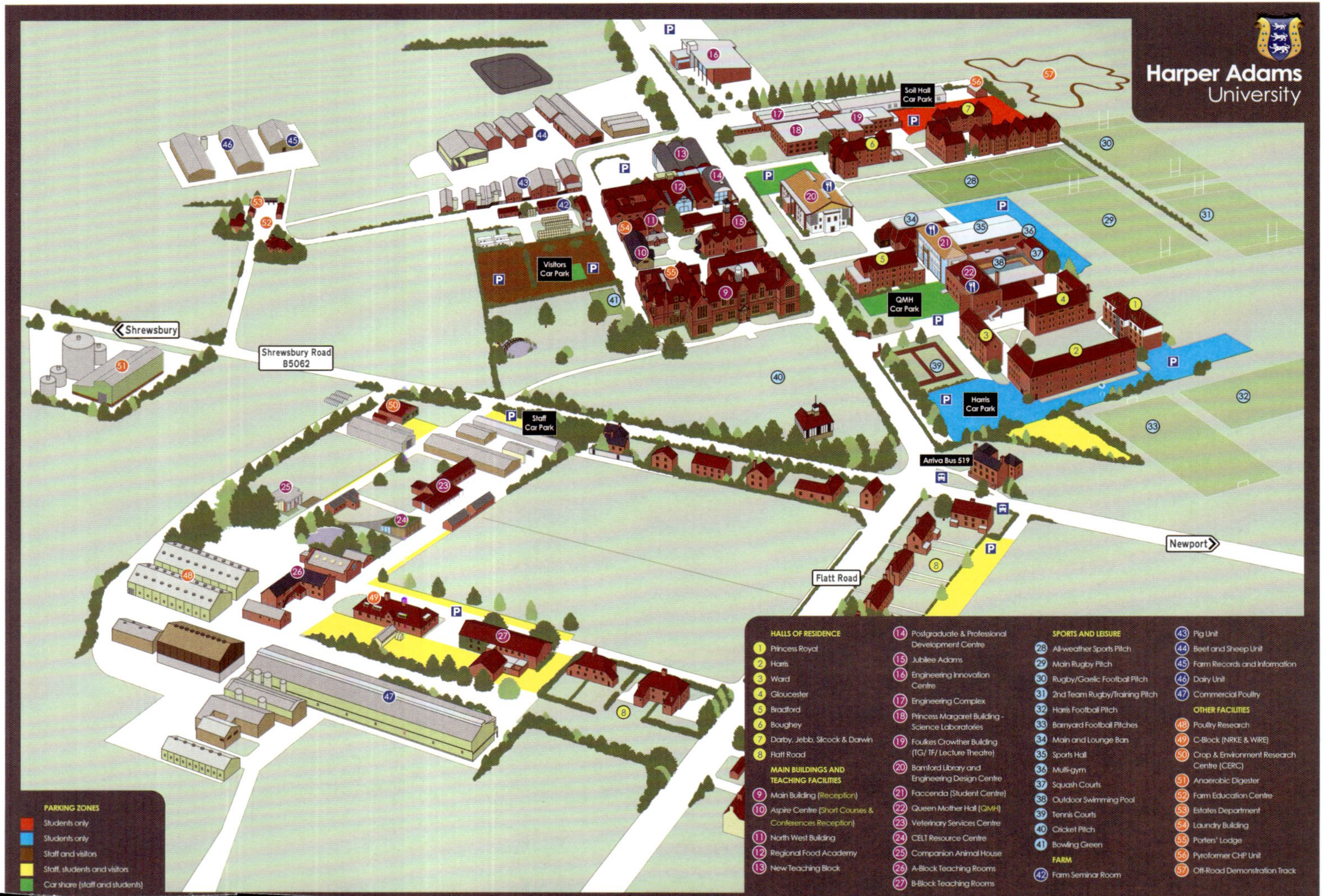
Harper Adams
University
Soil Hall Car Park
Visitors Car Park
QMH Car Park
Harris Car Park
Staff Car Park
Arriva Bus 519
Shrewsbury
Shrewsbury Road B5062
Newport
Flatt Road
HALLS OF RESIDENCE
1 Princess Royal
2 Harris
3 Ward
4 Gloucester
5 Bradford
6 Boughey
7 Darby, Jebb, Silcock & Darwin
8 Flatt Road
MAIN BUILDINGS AND TEACHING FACILITIES
9 Main Building (Reception)
10 Aspire Centre (Short Courses & Conferences Reception)
11 North West Building
12 Regional Food Academy
13 New Teaching Block
14 Postgraduate & Professional Development Centre
15 Jubilee Adams
16 Engineering Innovation Centre
17 Engineering Complex
18 Princess Margaret Building - Science Laboratories
19 Foulkes Crowther Building (TG/ TF/ Lecture Theatre)
20 Bamford Library and Engineering Design Centre
21 Faccenda (Student Centre)
22 Queen Mother Hall (QMH)
23 Veterinary Services Centre
24 CELT Resource Centre
25 Companion Animal House
26 A-Block Teaching Rooms
27 B-Block Teaching Rooms
SPORTS AND LEISURE
28 All-weather Sports Pitch
29 Main Rugby Pitch
30 Rugby/Gaelic Football Pitch
31 2nd Team Rugby/Training Pitch
32 Harris Football Pitch
33 Barnyard Football Pitches
34 Main and Lounge Bars
35 Sports Hall
36 Multi-gym
37 Squash Courts
38 Outdoor Swimming Pool
39 Tennis Courts
40 Cricket Pitch
41 Bowling Green
FARM
42 Farm Seminar Room
43 Pig Unit
44 Beef and Sheep Unit
45 Farm Records and Information
46 Dairy Unit
47 Commercial Poultry
OTHER FACILITIES
48 Poultry Research
49 C-Block (NRKE & WIRE)
50 Crop & Environment Research Centre (CERC)
51 Anaerobic Digester
52 Farm Education Centre
53 Estates Department
54 Laundry Building
55 Porters' Lodge
56 Pyroformer CHP Unit
57 Off-Road Demonstration Track
PARKING ZONES
Students only
Students only
Staff and visitors
Staff, students and visitors
Car share (staff and students)

Dear Guest

Welcome to Harper Adams University. We trust you will have a productive, enjoyable and safe visit. To the latter end would you please familiarise yourself with the contents of this booklet and in relation to the health and safety guidance, ensure that you act on its contents. Thank you.

History of Harper Adams

In 1892 Thomas Harper Adams bequeathed £38,000, the Adams homestead and Anceller Farm House to educate young gentlemen in the science of agriculture. This was sufficient to build the main college building which housed the Principal and his family, all of the students and some staff together with all of the classrooms, library and offices.

In 1901, 6 students enrolled on the first course. 17 years later, young ladies were admitted to the College. Today, the University farms over 500 hectares of its own and has approximately 2,700 students, and delivers professional development courses to over 3,000 industry delegates per year.

Our students all study degree subjects relating to the rural economy and the vast majority receiving a BSc (Hons) degree. During the middle of their course, students spend a year in industry, one of the factors contributing to the very high employment rate of Harper Adams graduates.

If you would like to discover more about the history of our University then please contact a member of the Short Course / Conference team who can supply you for a small charge, a copy of the Harper Adams Centenary Trail booklet or "Lure of the Land", the history of Harper Adams.

The Harper Adams Host

The Harper Adams staff member you are visiting, or in the case of Short Course Delegates your Course Manager, will, for courtesy and security reasons, be responsible for escorting you from and to reception and will assist you in case of emergency.
Conference guests, your conference organiser will be responsible for welcoming you and directing you around campus.

Security

You are requested to sign the Visitors Registration Book at Reception on your arrival. If you're attending a conference or short course you will be asked to sign a conference delegates list or Short Course attendance list.

Please do not enter any controlled area without authorisation.

Lost and Found

It is your responsibility to take care of your property and the University accepts no responsibility for any property brought to or left at the University. For general lost and found enquiries in the first instance should be directed to your Course Manager or Group Organiser or a member of the Short Course / Conference team. If the missing item is of great value or a credit card etc. this should be reported to the Short Course / Conference team during office hours and to the on site Security Officer at other times.

First Aid

There are a number of qualified First Aiders at Harper Adams and should you require First Aid during office hours you should contact reception by dialling 0 on any internal phone. In addition a Doctor or Nurse are available at allocated times during term time. Harper Adams reception staff will be able to advise you of the surgery times.

Outside of office hours please contact the Short Course Conference Office on 01952 815324 which is normally staffed until 5:00pm or Campus Security on 07980 061128. Alternatively an out of hours Doctors service is available on 111 (applicable for mobile or landline). In case of emergencies please dial 999 from a landline alternatively 112 from a mobile.

Fire Safety

In the event of a fire or other emergency a siren will sound. You must leave the building immediately by the nearest available exit, following the green and white EXIT pictograms (or instruction from Harper Adams fire marshals) to the nearest exit and appropriate fire assembly point. These are usually by car parks. In the event of smoke remember to keep as low as possible. Do not use the lift. You must not return to the building until advised it is safe to do so by a Fire Marshall or a senior member of Harper Adams staff.

If you become aware of fire, activate the nearest fire alarm call point and exit the building as above.

The fire alarms are tested every 2nd and 4th Wednesday afternoon in each month.

Health and Safety

Whilst the University takes responsibility for Health and Safety on its premises, you must do nothing to put yourself, other visitors, contractors or Harper Adams students or staff at risk. You must follow the Health & Safety instructions (Appendix 1) in this document and any additional instructions from your host.

Accidents / Near Misses

All accidents and near misses must be reported to your Course or Conference Organiser or to a member of the Short Course Conference team.

Children

Children are the responsibility of parents or carer's who bring them to these premises. They should always be accompanied by an adult indoors and outside. The children must be kept at a safe distance from ponds, animal enclosures and all other dangers on site. Please do not leave children unattended in residential accommodation.

Animals

All animals brought onto site must be authorised by the Conference and Short Course Office. No animals should be left unattended in vehicles. Animals permitted to be kept in residential accommodation should be housed in suitably designed containers. All animal waste must be disposed of correctly. All dogs must be kept on a lead except for training classes.

Disabled Facilities

Car Parking for disabled people is available on QMH Car Park, Bamford Library Car Park, Foulkes Crowther Building Car Park and limited spaces at rear of the Main Building.

Disabled toilets are located in the Main Building, Bamford Library, Queen Mother Hall, Foulkes Crowther and Faccenda.

No Smoking Policy

Harper Adams University operates a no smoking policy in all of the buildings and Visitors are requested to observe the law by not smoking within 5 metres of any entrance or window and to use the cigarette bins / smoking areas around campus.

Receptions

The University Reception is located in the Main Building. However if you are attending a short course or conference your reception will be in either the Aspire Centre or another location indicated in your joining instructions.

Contractors

In order to satisfy health and safety requirements concerning the provision of basic safety information to non-employees on campus, a contractor's pass system is in operation. It is applicable to companies and their staff who have been contracted by the University to carry out estimates, installations, servicing, or repairs to buildings, equipment services and work on infrastructure or utilities such as electricity, gas, water and drainage. Companies making delivery/ collections or agricultural contractors do not require a contractor's pass.

On arrival contractors must report to the Estates Foreman in order to sign in. They must also read the asbestos regulations/guidelines and any other buildings guidance for the intended area of operation before commencing work.

Toilets

Toilets are located on the ground floor in the Main Building, ground and first floor in the Foulkes Crowther Building, ground floor in the Bamford Library and Faccenda.

Disabled toilets are located in the Main Building, Foulkes Crowther Building, Bamford Library & Faccenda

Refreshments

There are vending machines located in the Foulkes Crowther Building, the Queen Mother Hall Refectory, NIPH B Block, Graze Café, Post Graduate centre and the Senior Common room. Snacks, light meals and beverages can be purchased from either the Kaldi coffee shop or Graze cafe.

Shop

There is a small shop that stocks snacks, newspapers, toiletries and a selection of general groceries located in the foyer of the Faccenda.

Public Telephones

Cash phones are situated outside the Queen Mother Hall, the rear of the Main Building and by the cross roads on the B5062.

Bedrooms

The domestic staff are pleased to assist you in any way possible. Your room and bathrooms are serviced every day and towels are changed every third day.

Beverages

Beverage making facilities are situated on each floor in each block in your hall of residence containing sachets of coffee, tea, herbal/fruit teas etc. Stocks are replenished daily. If your accommodation is in Darwin A & B, Darby, Jebb or Princess Royal hall each block of 6 or 8 bedrooms has a kitchen with the same provisions. Guests are requested to refrain from using the appliances in these kitchens.

Catering

If you have any special dietary requirements please inform the Short Course / Conference office or your Course Organiser who will then inform the catering department.

Messages

These will be given to your Course or Conference Organisers at meal times.

Sports Facilities

The Conference & Short Course office holds a limited amount of sports equipment that is available for guest's to borrow. Please liaise with your Course or Conference Organiser or a member of the Short Course / Conference Team to arrange this.

Damages/Repairs

All damages, repairs & any none daily cleaning must be reported to your Course or Conference Organiser or a member of the Short Course / Conference Team. Please note the University reserves the right to make an additional charge for any repairs / cleaning that would not be considered to be general wear & tear.

Useful Contacts

	INTERNAL	EXTERNAL
Short Course Office	5324 / 5300	01952 815324 / 815300
Porter/Security Guard	5321	01952 815321
Security Guard (Mobile)		07980 061128
Dentist		01952 811973
Doctor		01952 820400
Doctor out of hours		111
Emergency		999 / 112
Taxis		01952 501050

Vacating Rooms

Unless alternative arrangements have been made guests must vacate their rooms by 9.00 am. If you're key is not returned by 9.00am you may be charged an extra night's stay. Unless specific arrangements have been made with your organiser, please take your key to either the Aspire Centre Reception, Security Office or Queen Mother Hall Foyer, where A Key Return box will be situated.

Replacement keys are £30.00, replacement swipe cards or key fobs are £10.00 for any lost or damaged keys, key fobs or swipe cards.

Appendix 1

HEALTH AND SAFETY
BRIEF FOR COLLEGE VISITORS

Introduction

- The University acknowledges that it has a duty of care towards its visitors but they are reminded that they too have a duty of care for their own health & safety whilst on campus and should not expose themselves to any unnecessary risk.

- The purpose of this brief is to identify areas of the University where there is potential danger and advise on steps to minimise risks.

Roads & Tracks

Main Roads

- The University is bi-sected by the B5062. At times it is a busy road. Although traffic calming measures, including a 40mph speed limit, have been installed, care should always be taken when accessing the road by vehicles or crossing the road on foot.

Service Roads

- On campus there are a number of private service roads. Vehicle speeds should not exceed 5mph. To remind drivers of the speed limit and to ensure that there are no excesses, a number of 'sleeping policeman' ramps have been constructed on the service roads please take care when approaching these ramps.

Tracks & Paths

- Visitors are encouraged to walk the University's farmland, which is crossed by numerous paths and tracks. However they should not enter fields unless prior permission has been arranged. Care should be taken to leave gates as they are found, open or closed.

- Visitors should be aware that farm vehicles use these tracks and they are also liable to encounter livestock being transferred from field to field.

The Farm

Livestock

- The Farm buildings accommodate a number of livestock enterprises, including dairy and beef cattle, pigs, poultry and sheep. Visitors are not permitted to enter livestock areas without permission or unless they are accompanied by a member of staff. They will appreciate that, not only are they at risk from injury or infection, but that animals are exposed to similar risks.

Machinery & Equipment

- A number of University departments operate large pieces of machinery and sophisticated electronic equipment, which are potential risks. In no circumstances should any visitor attempt to use any item of machinery, plant or equipment without supervision or unless they have specific permission.

Ponds

- There are four small ponds on the University estate. However they are deep and constitute a danger to the careless or unwary. One is in fact the slurry lagoon, which is protected by a fence. Access to the ponds and lagoon is denied to visitors unless previously authorised and accompanied by a member of staff.

Recreational Facilities

- The University provides no supervision for recreational activity. Use of the facilities is at your own personal risk. Appropriate apparel should be worn when using the facilities to minimise the risk of injury.

The Swimming Pool

There are separate operating plans for the swimming pool and these will be provided on signing the document allowing access to this facility.

Squash Courts

- Injuries in Squash Courts are almost exclusively caused by players wearing unsuitable footwear, which cause's them to slip. Appropriate footwear, which minimises the chances of slipping, should be worn.

- Lighting for courts is activated by the insertion of a pre-purchased 'smart card' into a control box mechanism. If the lighting fails to activate, players should not attempt in any circumstances to interfere with the electronic control circuits. There is a danger of electric shock. Malfunctions should be reported.

Fire

- The risks and causes of fire are well documented. Primary causes in accommodation areas are the careless disposal of cigarettes, the use of unsafe electrical equipment or overloading of electrical circuits.

- Procedures in the event of a fire are displayed in each building and in each study bedroom. They advise you to sound the alarm, (if it has not already been sounded), and vacate the building by the nearest exit by following the green pictograms.

Electrical Appliances

- Other than electric kettles, no power appliances are permitted in rooms without the prior permission of the University. Visitors who bring appliances to the University must ensure that they are electrically safe. If a visitor has any concerns about the safety of an appliance the Estates Department can arrange for testing of items at the visitor's expense.

Medical

First Aid

- Each University department has members of staff who are trained in first aid. They are available during working hours.

Medical Treatment

- In the event of injury or illness, the University Medical Officer provides 24 hour cover. A doctor, from the Linden Hall Surgery, can be contacted by telephone: 01952 820400 (surgery hours) or 111 (out of surgery hours).

Notes

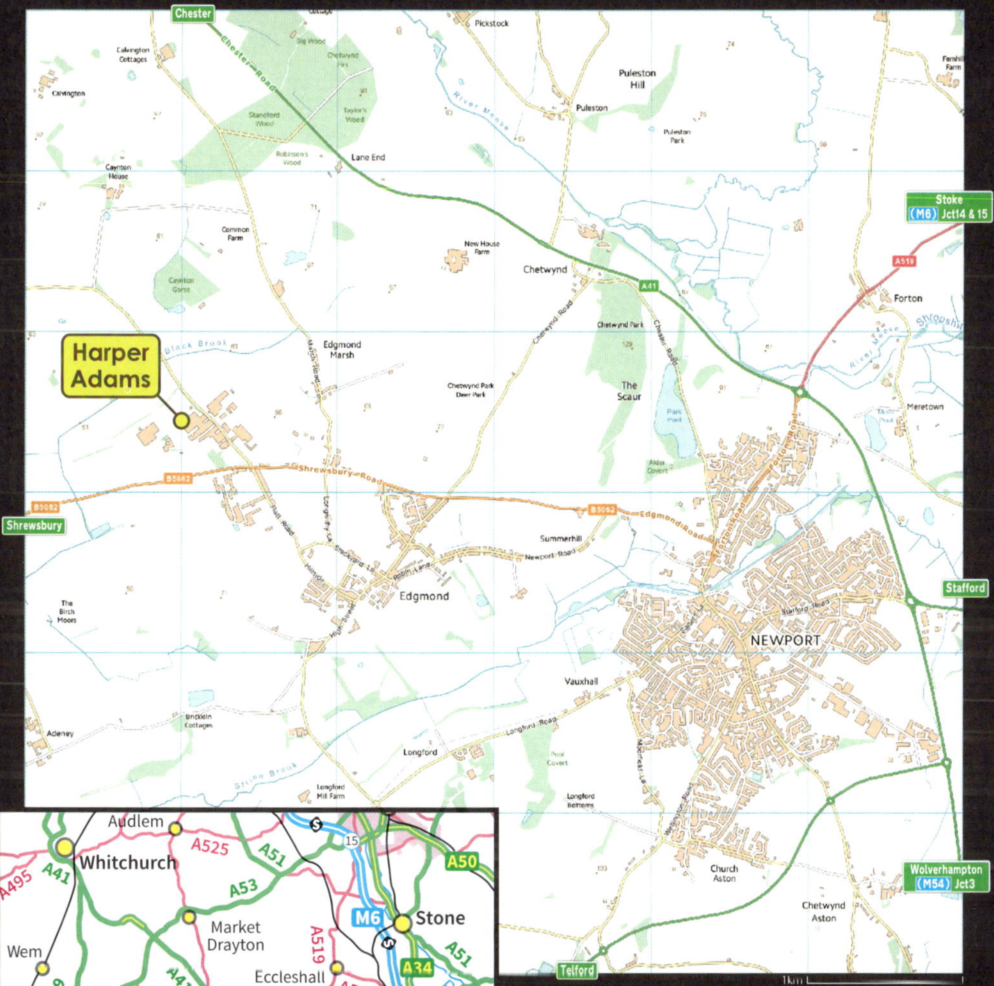

Contains Ordnance Survey data © Crown copyright and database right 2013

Harper Adams University is situated just outside Newport in Shropshire, off the B5062 from Newport to Shrewsbury (Postcode for SatNav: **TF10 8NB**). **www.harper-adams.ac.uk**

YOUR NEAREST...

Train station	miles
Telford Central	10
Stafford	15
Shrewsbury	16

Airport	miles
Birmingham	48
Manchester	58
Liverpool John Lennon	64
East Midlands	66

Harper Adams University, Newport, Shropshire TF10 8NB
General Enquiries: +44 (0)1952 820280 Fax: +44 (0)1952 814783

Harper Adams University

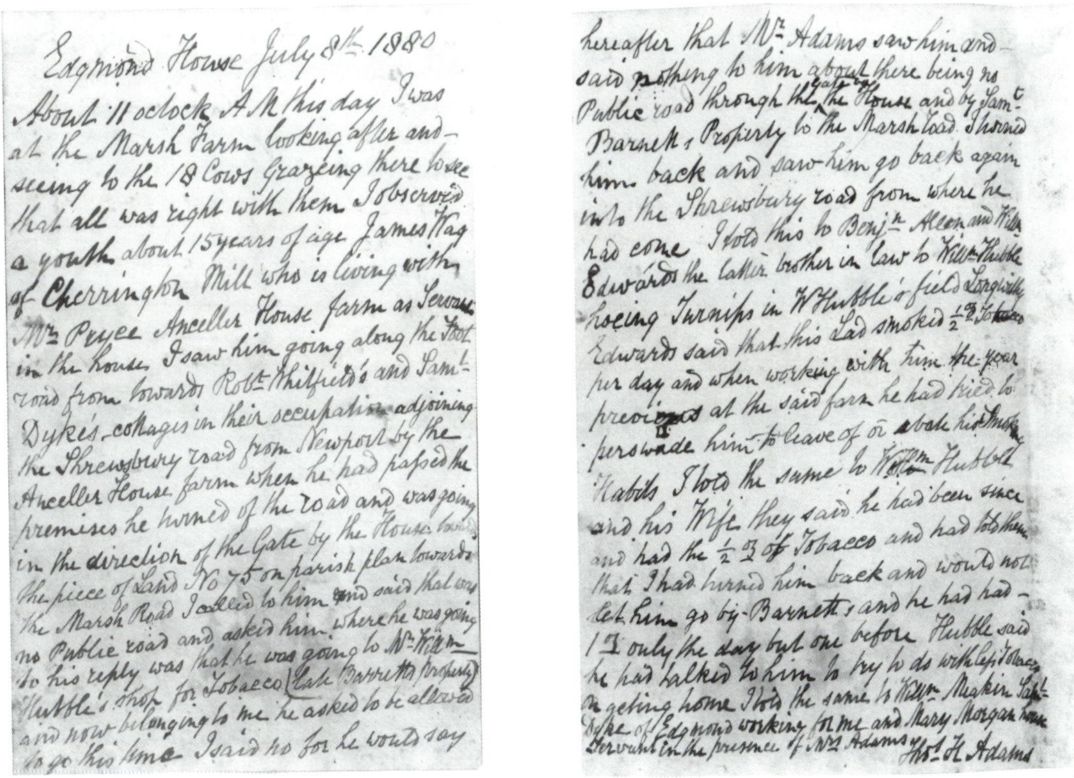

Edgmond House July 8th 1880
About 11 oclock A.M this day I was
at the Marsh Farm looking after and
seeing to the 18 Cows Grazeing there to see
that all was right with them I observed
a youth about 15 years of age James Hag
of Cherrington Mill who is living with
Mr Pryce Anceller House farm as Servant
in the house I saw him going along the Foot
road from towards Robt Whitfield's and Saml
Dyke's cottages in their occupation adjoining
the Shrewsbury road from Newport by the
Anceller House farm when he had passed the
premises he turned of the road and was going
in the direction of the Gate by the House towards
the piece of Land No 75 on parish plan towards
the Marsh Road I called to him and said that was
no Public road and asked him where he was going
to his reply was that he was going to Mr Willm
Hubble's shop for Tobacco (Late Barrett's Property)
and now belonging to me he asked to be allowed
to go this time I said no for he would say
hereafter that Mr Adams saw him and
said nothing to him about there being no
Public road through the Gate by the House and by Saml
Barnett's Property to the Marsh Road. I turned
him back and saw him go back again
into the Shrewsbury road from where he
had come I told this to Benjn Allen and Willm
Edwards the latter brother in law to Willm Hubble
hoeing Turnips in W Hubble's field Longwill
Edwards said that this Lad smoked ½ oz Tobacco
per day and when working with him the year
previous at the said farm he had tried to
perswade him to leave of or abate his smoking
Habits I told the same to Willm Hubble
and his Wife they said he had been since
and had the ½ oz of Tobacco and had told them
that I had turned him back and would not
let him go by Barnett's and he had had
1 oz only the day but one before Hubble said
he had talked to him to try to do with less tobacco
on geting home I told the same to Willm Maskin Saml
Dyke of Edgmond working for me and Mary Morgan house
Servant in the presence of Mrs Adams Thos H Adams

1.4 Statement in Thomas Harper Adams' own hand concerning a case of trespass, 8 July 1880: Martin Kaye Solicitors.

The all-important will which led to the establishment of the College is dated 28 October 1891.[31] There were, however, earlier wills. The first to refer to the cause of agricultural education is dated 31 December 1890.[40] With the will of 31 December the estate is left 'for the purpose of teaching practical and theoretical Agriculture in England either by means of lectures or establishing a School of Agriculture in England or such other means'. A revised will, dated 20 March 1891, provided substitute trustees.[40] The final will of 28 October came about as a result of fears that the pending Mortmain and Charitable Uses Act of 1891 might present complications.[29] The ancient law of Mortmain prevented corporations holding land without a licence from the Crown, thus preventing land being tied up in the dead hand (mortmain) of such artificial persons and safeguarding the royal feudal revenues.[41] It was this will which led directly to the creation of the College. It is therefore the months of November, with the letter to Kenyon-Slaney, and December 1890, with the will, which are worth studying. Significantly, very important developments were taking place in the history of agricultural education in Shropshire at this precise time.

Agricultural education had become an issue in Shropshire as early as January 1879, when at a meeting of the Shropshire Chamber of Agriculture, Mr Buckmaster of the Department of Science and Art had explained the scheme of the Department to encourage agricultural education.[42] The Department of Science and Art had been set up by the Government at South Kensington in 1853 as the main agency for technical education.[43] The Chamber resolved:

> *That the Chamber having heard the explanation of Mr. Buckmaster desire to express their approval of the Scheme he has brought under their notice and their sense of the necessity of providing a method of technical education for the sons of farmers and invite teachers in Shropshire to qualify themselves as Teachers of Agricultural Science where classes could probably be formed and that the Chamber of Agriculture Committee be a committee to promote the above subject.*[42]

The scheme, started in 1876, set up a special section for agriculture within the Science and Art Department and introduced a syllabus on the principles of agriculture.[44] It sought to stimulate the application of science to agriculture through education. By the 1880s, as in other parts of England, theoretical courses in Shropshire were being run by the Science and Art Department with Government grants for those enrolled.[45] The first classes had been established at Linley and Montford by the close of 1879.[46] Such courses were especially popular amongst elementary teachers.[44] This, however, was the only state support for anything approaching agricultural education. Neither was there encouragement for working farmers to take up training.[45]

Throughout these years developments in agricultural education were taking place on the national scene and the Minutes of the Shropshire Chamber show that the issues were not ignored locally. In February and March 1888, for example, members discussed the implications of the Paget Report and voted their approval.[47] The Departmental Commission on Agricultural and Dairy Schools of 1887, under the Chairmanship of Sir Richard Paget, was the first Government commission solely to consider agricultural education. The Report concluded there was insufficient provision for agricultural education and stressed the need for state aid. It recommended the establishment of a central school of agriculture maintained by the State, with five regional dairy schools, to which schools of agriculture could eventually be added. The Report also encouraged the setting up of local agricultural schools and, providing they were established by local effort, some funding would come from the Government. Ideally schools with experimental farms should exist in every county.[48]

As luck would have it, the machinery which could be used to implement this ambitious plan came about with the 1888 Local Government Act. This established a publicly elected local authority for every county. There remained the question of finance. In 1889 the Technical Instruction Act gave local authorities the responsibility for technical education, with special emphasis to be placed on agricultural education. A penny rate was introduced for funding. Most authorities made little haste to implement the measures.[49] That year agricultural education received a meagre £5,000 as the annual sum administered in grants by the newly established Board of Agriculture for the whole country. France in comparison allocated £160,000.[45] Improved funding came from a most unlikely source.

With growing pressure from the Temperance Movement the Government had set out to reduce the number of liquor licences. Additional duties on beer and spirits were imposed in the 1890 Budget to provide compensation for those who surrendered their licences.[45] However, public opinion balked at passing the £750,000 'whisky money' to the already prosperous brewers, who were the majority of licence holders. In a surprise move the Local Taxation Act was passed in summer 1890 with an amendment enabling the money to be applied annually for the purposes of agricultural, commercial and technical instruction.[50] At last the Government had given substantial financial recognition to this neglected area. Shropshire was quick to further the local cause of agricultural education.

Academic thought equates the agricultural counties with the worst response to the call for agricultural education, arguing that farming communities, favouring the practical, were slow to recognize the worth of the theoretical. Shropshire, however, was already clearly progressive when it came to agriculture, with men such as Jasper More, Member of Parliament for South Shropshire, to the fore. The Shropshire Chamber of Agriculture was established in August 1866, just one month after the founding of the Central Chamber of Agriculture in London. Significantly, More was the national Vice Chairman. Within two years there were sixty-seven chambers nationally, but Shrophire Chamber of Agriculture with a membership of 612 was the largest in the country.[51]

Moving forward in time, it was Shropshire again which was amongst the first counties to organize meetings in August and September 1908 to urge the creation of an association for farmers. This was in response to the demand from the butchers that farmers should give warranties with their fat stock and so indemnify butchers against loss through diseased carcases. The meetings led to the creation of the Shrewsbury and District Farmers Association in September. A mass meeting was held at Wellington at the end of the month with representatives from around twenty counties. Such enthusiasm led the Lincolnshire Farmers Union to suggest the formation of a national association of farmers. The National Farmers Union was formed in London on 10 December 1908, with Shropshire having played a significant part in its formation.[52]

In December 1890 the Shrewsbury Chronicle declared that Shropshire was the first county to devise a definite scheme for the allocation of its share of the 'whisky money', £6,543.[53] Indeed, Shropshire had been quick to act, appointing a sub-committee as early as August to investigate ways of spending the money.[54] It was October 1890 before Lancashire County Council, regarded as innovative, appointed a sub-committee to consider the best means to allocate grants for agriculture.[55] In early November 1890 the newly created Technical and Agricultural Education Sub-Committee of Shropshire County Council reported its suggestions to the Council. It stated it had set about ascertaining the wants of the County as regards agricultural education by advertising in the county papers, inviting ideas from all those eligible for a grant. Their report made the initial suggestion that two travelling instructors should be appointed, an idea which had been favoured by many including the West Midlands Agricultural Society and the Shropshire Chamber of Agriculture.[56] The instructors were appointed, but in the capacity of inspectors to ascertain the needs of the County.[53]

On 12 December 1890 a grand gathering of all interested parties was called at the Shirehall to present ideas on how to spend the money. Among ideas put forward were a travelling dairy school, as had been established by the Bath & West of England Agricultural Society, and the possibility of permanent schools and evening classes. James Whitaker, President of the Shropshire Chamber of Agriculture, presented the views of the Chamber. He spoke of sending lecturers into villages to give theoretical and practical instruction in dairy work, botany, entomology, veterinary work, chemistry and practical farming. He also advocated dairy schools and the establishment of one institution in the county 'where farmers might send their sons, when they had completed their education at the elementary school, and where the labourer might be taught all kinds of farm work'.[57]

All these activities in November and December 1890 were reported in great detail in the Shrewsbury Chronicle. Mrs Adams, describing her husband's activities in his later years, said:

'My husband took in the Stafford Advertiser, the Newport Advertiser and the Shrewsbury Chronicle.' [35] Miss Davey, his nurse, also spoke of his love of reading the newspapers.[58] Although ill and isolated from society, Harper Adams was aware of the events going on in Shropshire. Here, surely, was the source of inspiration for bequeathing his wealth to the cause of agricultural education. Further strength to this argument lies in the fact that Sir Thomas Boughey and Colonel Sambrooke Burne were two of the three executors chosen in the will of 31 December 1890, although there is no evidence that they were his friends. Both men were listed as attending the meeting of 12 December, Boughey in his capacity as a County Councillor and Burne as part of the Newport deputation.[57]

One important point of difference should be noted between the will of December 1890 and the final will of October 1891. The earlier will refers to 'establishing a School of Agriculture in England', while the latter expands on the idea, referring to establishing a school or schools of agriculture at Edgmond, Woodseaves or Chadwell 'or at some or one of such places or elsewhere in England'. However, it is the will of 20 March which first suggests a local location for the proposed school of agriculture, namely Edgmond. It is of course very important to the history of the College that a local site was chosen. Are there clues as to why? It could be argued that Harper Adams favoured a local site because of personal connections and the availability of land. But, events in Newport in the early part of 1891 point to a different reason. Winter 1891 saw the arrival of agricultural education on Harper Adams' own doorstep.

Late in January 1891 Mr Armytage, one of the two Shropshire County Council instructors appointed the autumn before, spoke to a gathering of interested parties in Newport to ask how they felt the money allocated for technical education, and for agricultural education in particular, should be spent in their area. Boughey and Burne were there and Burne chaired the meeting. Many ideas were suggested, although most attending felt need was greatest for instruction in dairying and cheese-making, which diminished the value of education in an arable area such as Newport. Boughey pointed out the smallness of the sum, just £6,543 for the entire county, and, similarly, he reminded the gathering that it was not known whether this would be an annual grant, or even if it represented a capital sum, with only the interest of £180 on offer. Inevitably a committee was formed to consider the matter further. All was reported in the Newport Advertiser and Harper Adams would have been aware of these events.[59]

At the first meeting in February the Committee put its support behind classes, declaring it could not recommend the establishment of a school of agriculture for the county as available funds were insufficient.[60] Interestingly, a leading member of the Committee was Charles Liddle,

1.5 Sir Thomas Fletcher Boughey, one of the executors of Thomas Harper Adams' will and the College's first Chairman of the Governors: Mate's Shropshire.

none other than Harper Adams' solicitor.
It would have been most surprising, considering the contents of the December will, if Liddle had not talked of these local developments to his client. Would it have taken a great deal for Liddle to sow the seeds of the idea of local support, when Harper Adams was already committed to the cause of education? Mary Ann Minor considered Liddle was far too influential in the affairs of her uncle [29] and rumours that Liddle was behind the idea of establishing a college persisted at least into the 1960s.[61]

Prominent among local developments was a series of dairy classes held in Newport Town Hall over two weeks in June 1891. These were part of the County Council scheme to provide technical education throughout the county. Spectators were invited at a cost of 3d. a day or 2s. for the whole course. The instructor, Mr C. R. Valentine of Ludlow, attended under the auspices of the County Council Technical Education Scheme.[62] These were the first agricultural education classes ever to be held in the Newport area. Although historically, these classes are directly tied to the County Council and not to the College, in spirit they must be regarded as the forerunners of education at the College. Interestingly, Mr Liddle's daughter, Constance, won the first prize for her buttermaking.[63] The local Committee's delight at the success of these dairy classes must have been short lived, when in July members received notification from the County Council of their share of the money allocated for technical education. The amount was £31 3s. This was to be shared between classes at Newport and Donnington.[64]

1.6 Colonel Sambrooke T. H. Burne, one of the executors of Thomas Harper Adams' will and the College's second Chairman of the Governors: Mate's Shropshire.

By the autumn of 1891 Harper Adams had made up his mind where his wealth should go after his death and all necessary legal details were about to be finalized in the will of 28 October. It is therefore of historical interest to quote from the Newport Advertiser of 3 October 1891. Reference is made to the forthcoming Oxford University Extension Movement lectures at the Grammar School:

> *There is demonstrative evidence that if the English farmer is to compete successfully in the future with his foreign neighbours in the world's mart he will need more than ever before to be fully equipped in every branch of knowledge bearing upon the cultivation of soil and the rearing of stock, and it is therefore hoped that all such opportunities as that at present offered will be eagerly seized upon, and that in any rate in this district farmers will be alive to the great advantages to be gained from such a course of lectures as is indicated in the syllabus of the County Council.*

Harper Adams' will was to extend 'such opportunities' beyond the wildest dream of the most enthusiastic farmer.

Ancellor Farm & House 1832

Shortly before his death, John Adams, father of Thomas Harper Adams, made a settlement of all his estates and possessions, dividing them between his three living sons. The settlement, dated 1 February 1832, included an inventory of the household effects of Ancellor House *(see photo page 41)* and the stock and implements of the farm. Rich in detail, it provides a fascinating insight into the life of the Adams Family. It is a lifestyle which is both affluent and unpretentious and one in which manual work is very much part of the equation. The following represent extracts from the document.

Kitchen
6 Kitchen chairs
Oak 2 leaf Table
Clock & Oak Case
Oak Kitchen table & forms
Brass & tin Candlesticks & 6 Spittons
5 Pair of Flat Irons
Tin Ware in Cupboard Forks etc in corner

Brewhouse
2 Coolers
3 mashtubs
Work ladder
2 Mash Staffs [used for mixing malt with hot water to form wort] and cleansing sive
2 laid [?] gawns [small bucket/tub holding a gallon] and 2 pails
Cheese press
4 Masling Kettles [brass or tinned-copper preserving pans]
Frying Pan chopper & spit
1 Iron pot and Saucepan
1 Copper Tea kettle
3 Half Hogsheads

Dairy
Cheese screen & frame
4 Tubs neading mitt [a four-sided wooden vessel used for kneading] and sigh [?]
3 Milking Cans and pail
22 Cheese vatts
Curdbreaker butter scales dolly
Chopping Block and tub churn
1 Bench wood bowl & 3
Cupboard pepper Mill & tinware

Salting House
1 Neading mitt 2 Tubs and 2 pails
1 Cooler and salting mitt [small vessel for household use]
Salt tub lined with Lead
9 Barrells and 5 benches
18 Brass milkpans
Barrell churn laid gawn & mitt

Small Parlour
Weather Glass
4 Prints framed and Glazed
Oak Dining Table
Two Oak Stand Tables
2 Arm and six elbow chairs
Waiter [?] and tea chest
Oak Cupboard

Pantry
Meat hastner warming Pan [hastener: stand/screen concentrating heat of fire on roasting joint]
Meat Cans and Tin ware
42 Pewter Plates 9 dishes 1 - Quart and tundish
Spinning Wheel and Cupboard

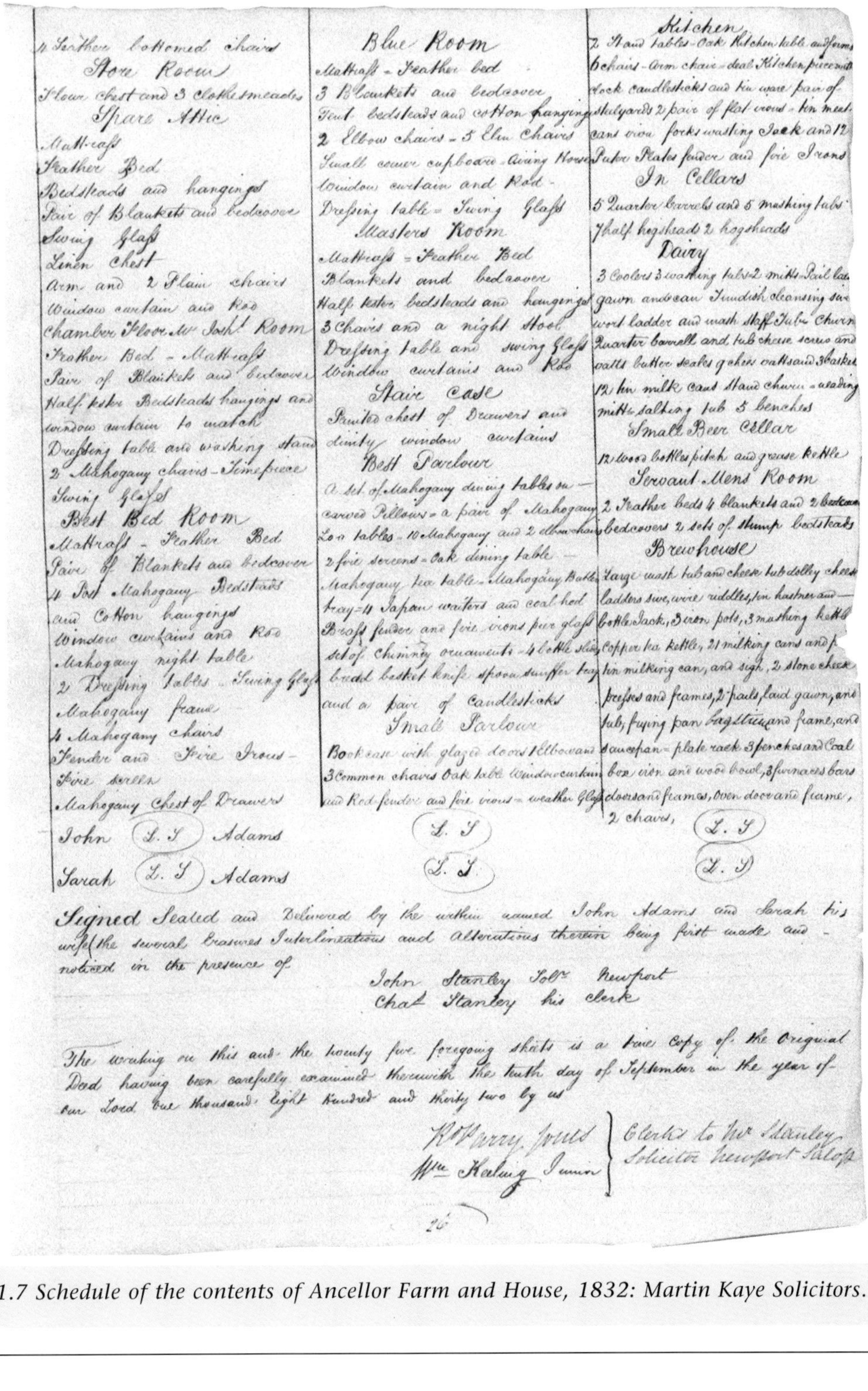
4 Leather bottomed chairs
Store Room
Flour chest and 3 Clothesmeaders
Spare Attic
Mattrass
Feather Bed
Bedsteads and hangings
Pair of Blankets and bedcover
Swing glass
Linen chest
Arm and 2 Plain chairs
Window curtain and Rod
Chamber Floor Mr Josh Room
Feather Bed - Mattrass
Pair of Blankets and bedcover
Half tester Bedsteads hangings and window curtain to match
Dressing table and washing stand
2 Mahogany chairs - Timepiece
Swing glass
Best Bed Room
Mattrass - Feather Bed
Pair of Blankets and bedcover
4 Post Mahogany Bedsteads and Cotton hangings
Window curtains and Rod
Mahogany night table
2 Dressing tables - Swing glass
Mahogany frame
4 Mahogany chairs
Fender and Fire Irons -
Fire screen
Mahogany Chest of Drawers
John (L. S.) Adams
Sarah (L. S.) Adams

Blue Room
Mattrass - Feather bed
3 Blankets and bedcover
Tent bedsteads and cotton hangings
2 Elbow chairs - 5 Elm Chairs
Small corner cupboard - Airing Horse
Window curtain and Rod -
Dressing table - Swing Glass
Masters Room
Mattrass - Feather Bed
Blankets and bedcover
Half tester bedsteads and hangings
3 Chairs and a night stool
Dressing table and swing glass
Window curtains and Rod
Stair Case
Painted chest of Drawers and dimity window curtains
Best Parlour
A set of Mahogany dining tables on carved Pillars - a pair of Mahogany Card tables - 10 Mahogany and 2 elbow chairs
2 fire screens - Oak dining table -
Mahogany tea table - Mahogany Bottle tray - 4 Japan waiters and coal hod
Brass fender and fire irons pier glass
set of Chimney ornaments - 4 bottle slides
bread basket knife spoon snuffer tray
and a pair of Candlesticks
Small Parlour
Bookcase with glazed doors Elbow and 3 Common chairs Oak table Window curtain and Rod fender and fire irons - weather Glass
(L. S.)
(L. S.)

Kitchen
2 Stand tables - Oak Kitchen table and forms
6 chairs - Arm chair - deal Kitchen piece with clock candlesticks and tin ware pair of Stilyards 2 pair of flat irons - tin meat cans iron forks washing Jack and 12 Pewter Plates fender and fire Irons
In Cellars
5 Quarter barrels and 5 mashing tubs
7 half hogsheads 2 hogsheads
Dairy
3 Coolers 3 washing tubs 2 milk Pail lead gawn and can Turndish cleansing sieve wort ladder and mash Staff Tubs Churn Quarter barrell and tub cheese screw and vatts butter scales q[illegible] oatts and 3 baskets 12 tin milk cans stand churn - [illegible] milk salting tub 5 benches
Small Beer Cellar
12 wood bottles pitch and grease kettle
Servant Mens Room
2 Feather beds 4 blankets and 2 bedcovers bedcovers 2 sets of stump bedsteads
Brewhouse
Large wash tub and cheese tub dolly cheese ladders sieve, wire riddles, tin hastner and bottle Jack, 3 iron pots, 3 mashing kettle Copper tea kettle, 21 milking cans and [illegible] tin milking can, and sigh, 2 stone cheese presses and frames, 2 pails, lead gawn, and tub, frying pan bag[illegible] and frame, and saucepan - plate rack 3 benches and Coal box iron and wood bowl, 3 furnaces bars doors and frames, oven door and frame, 2 chairs,
(L. S.)
(L. S.)

Signed Sealed and Delivered by the within named John Adams and Sarah his wife (the several Erasures Interlineations and alterations therein being first made and noticed in the presence of
John Stanley Solr Newport
Chas Stanley his clerk

The writing on this and the twenty five foregoing sheets is a true Copy of the Original Deed having been carefully examined therewith the tenth day of September in the year of our Lord One thousand eight hundred and thirty two by us
R Parry Jones } Clerks to Mr Stanley
Wm Keeling Junr } Solicitor Newport Salop
26

1.7 Schedule of the contents of Ancellor Farm and House, 1832: Martin Kaye Solicitors.

Dining Room
A mahogany Side board
A set of mahogany Tables circular ends
Mahogany Pillar and Claw table
10 Mahogany and 2 Elbow chairs
Mahogany Cellerett
Pier Glass 28 by 18
Mahogany butlers tray knife box and
3 waiters
4 Bottle Slides pair of candlesticks
cheese tray knife and bread basket &
1 pair of snuffer slides all Japan
2 Waiters and 2 teaboards

Drawing Room
A sofa
12 Mahogany chairs
Mahogany stand and copper urn
Mahogany Table and cover
pier glass & burnished gold frame
26 x 15 1/2

Room over Dining Room
2 Feather beds bolster and pillows
Bedsteads and hangings
3 Blanketts and 1 bedcover
2 Bedside Carpetts
3 Chamber Chairs and Stool
Mahogany night table etc etc
Mahogany Card table
Mahogany bason stand and ware
Swing Glass Mahogany Frame

Room over Drawing Room
Feather bed
4 Post bedsteads and dimity hangings [stout cotton fabric with raised striping or fancy figures]
2 Blankets and bedcover
Mahogany Dressing Table
Swing Glass Mahogany Frame
Mahogany wash hand bason and ware
3 Chairs and 2 Stools

Mr. Adam's Room
Feather bed
2 Blankets and bedcover
4 chairs
Dressing Table and Swing Glass
Deal Painted Cupboard
Window curtain and rod

Staircase
8 Day Clock
Oak Stand table
1 Trunk 2 Boxes 1 Chest and 2 Clothesmeads
8 Small chairs and 4 baskets
11 yards of Stair carpetting
19 Brass Rods and Hooks

Cheese and Store Room
1 Trunk and Cheese tub
Tin - and swift [A reel. Swifts: a machine for holding skeins of yarn for the purpose of winding them]

Chapter Two —

The Birth of the College 1892-1901

Thomas Harper Adams died on 7 June 1892 at his home, Edgmond House *(see photo page 42)*. The death certificate gives haemorrhage of the kidneys as the cause of death.[1] Liddle lost no time in notifying the chosen executors, Sir Thomas Fletcher Boughey of Aqualate, Sambrooke Thomas Higgins Burne of Loynton and Charles Cecil Cotes of Woodcote. Yet nearly a year was to pass before a start could be made to fulfil Harper Adams' wishes. Within days of the death, Mary Ann Minor had entered a caveat against the will and long months were taken up in legal wranglings. It was not until February 1893 that the will was proven and plans could now take effect.[2]

Information on events prior to the opening of the College in 1901 is scant. Cotes seems to have taken little part in the proceedings beyond irregular attendance at foundation meetings.[3] He died in 1898.[4] The instigators were primarily Boughey, Burne and Liddle. Boughey was a leading member of the landed gentry and a County Councillor. He was involved with many local matters, including, as its first President, the Newport and District Agricultural Society.[5] Burne had been educated at Magdalen College, Oxford,[6] and commanded the Newport Company of the Second Volunteer Battalion of the King's Shropshire Light Infantry.[7] In 1892, the year of Harper Adams' death, Burne held the post of President of the Newport and District Agricultural Society.[5] His biography says of his role in the creation of the College: *'He was appointed an original trustee by Mr. Harper Adams and the work of starting the College was left very much in his hands by his fellow trustees.'* [8] Liddle acted as a solicitor and later as Clerk to the Governors, but in 1899 illness forced him to retire in favour of his son Roland Pemberton Liddle. His early death in January 1900, at the age of sixty, was tragically to rob the College of one of its most capable founders.[9]

One of the first questions to have been asked must have been how much did the Trustees have at their disposal? Inevitably, on Harper Adams' death there had been wild speculation as to his wealth. The Newport Advertiser declared he left between £100,000 and £150,000.[10] A valuation prepared for the court case put his property and personal estate at around £63,000.[11] The probate figure of 24 February 1893 gives a sum of £18,763 for his personal estate after estate duty.[12] Clearly this is not the full value of his estate. It is probable, prior to the Land Transfer Tax of 1897, that the realty passed directly to the Trustees and was not a probate issue.[13] Figures in the Foundation Scheme of 1896 give railway stock and cash amounting to £37,078 and an additional 178 acres representing Ancellor Farm.[14] Colonel Burne stated to the Reay Committee that the net amount the Trustees handed over to the Governors was £45,496.[15] In the year 1894-95, just prior to the Scheme being published, the total sum of £78,000 was spent on agricultural education in Great Britain, including County Council expenditure and funds from the Board of Agriculture and the Department of Science and Art. The funds available to the Trustees can be put into perspective when it is noted that Harper Adams' bequest was equivalent to nearly sixty percent of that sum. Similarly, in 1891 an agricultural labourer could earn on average around £42 a year and a highly paid barrister could expect an annual income of around £1,342.[16]

There was no precedent for the Trustees to follow, as this was the first time an institution of this kind was to be set up by private bequest.[17] The will granted them a free hand to apply the income or the capital from the Harper Adams Estate in whatever way they thought fit in order to fulfil:

the purpose of teaching practical and theoretical agriculture to men and youths and a knowledge of dairying housekeeping and other subjects to women and girls which will tend to make them practical and useful women in their station of life in England either by means of lectures or by establishing a school or schools of agriculture at Edgmond or Woodseaves near Market Drayton in the County of Salop or at Chadwell in the County of Stafford or at some or one of such places or elsewhere in England.[2]

Hitherto much discussion has concentrated on the background to Harper Adams' proposal to establish a school of agriculture, as this was the actual outcome of the will. But, it should not be forgotten that Harper Adams provided an alternative proposal, namely to finance lectures, no doubt along the lines of those already organized by the County Council. In the final outcome, it is to the Trustees that the decision to establish a school must be attributed and it seems that this decision was made very soon after Harper Adams' death. The Newport Advertiser, just eleven days after his death, revealed: *'He has most agreeably surprised us all by leaving nearly the entire accumulation to found a College of Agriculture in this district.'*[10] Very early on too the Trustees resolved to build the school on the Adams homestead, Ancellor Farm. Thus, over the summer of 1893, land and property in Woodseaves, Chadwell, Church Aston, Chetwynd Aston, Newport and Edgmond were sold and the profits invested,[18] although it was not until July 1894 that Edgmond House was sold, following the death of Harper Adams' widow, Sarah.[19]

As early as the summer of 1893 documentation suggests the Trustees had been in touch with the Board of Agriculture. The Board declared that although there were 'several cases in France where local agricultural schools [were] endowed in this way, [it was] a novel departure on our part'.[19] The Board was of little help.

Similarly, the Trustees consulted the Charity Commissioners, but as Burne said before the Reay Committee: *'They responded by telling us that they looked to us as the experts in the matter, and hoped we should give them assistance.'* [15] Some encouragement was received from the Shropshire County Council which appointed its Chairman to act with the Trustees early in 1894.[18] Otherwise, it seems that it was very much the case of the Trustees feeling their own way. Principal Hedworth Foulkes was to say as much.[20]

Neither could the Trustees gain a great deal of assistance from kindred institutions. In the mid-1890s establishments offering agricultural education were few. Those which could relate to the requirements of Harper Adams' will, with its slant towards the practical as well as the theoretical and obviously directed towards the working community, could be counted on one hand. Burne was clear in his own mind about the market the College would be aiming for: *'From what I have read of the will, I think you will gather it was intended to meet the requirements of people of limited means.'* [15]
As has been stated, agricultural education was in its infancy. A Chair of Agriculture and Rural Economy had been established at Edinburgh in 1790, followed six years later by one at Oxford, although this was not filled until 1840. There was controversy as to whether agriculture was a suitable university discipline. It was not possible to take a degree in agriculture until the early twentieth century.[21] University education had scientific theory but less content of practical worth. The year 1838 saw the arrival of the Royal Agricultural Society of England, with a policy of promoting agriculture through the application of science in the form of shows, lectures and publications. In 1863 the Royal Agricultural Society formed a special education committee, introducing its own examinations in agriculture in 1868.[22] These were short lived but revived in the 1890s. A more positive move was the opening of the Royal Agricultural College in 1845. However, it was funded as a shareholding company, with priority of places going to those purchasing shares. Effectively the College catered for the sons or nominees of the nobility, the rich and the professional. The annual fees of £135 did little to broaden the entry band.[23] A college of wider appeal was Aspatria in Cumberland, opened in 1874 as a joint stock company with its origins in the Aspatria Co-Operative Society. With its lower fees, of around £50 a year for a boarder, it did aim to appeal to the small farmer.[24] However, Downton College in Wiltshire, opened in 1880 with annual fees of £100 to £150 for residents, set its sights on attracting the sons of the rich, whether land-owners, farmers or professional and business people. It was to be a short-lived venture, closing in 1906.[25] Mention should also be made of the British Dairy Institute which was opened by the British Dairy Farmers' Association in 1882.[26] The Paget Report of 1887 was clearly right in defining agricultural education as inadequate.

The early 1890s saw some improvements in the number of institutions teaching agriculture in England and Wales. The formation of the Board of Agriculture in 1889 was of great benefit to the cause of agricultural education, as the very subject was defined within its terms of reference. However, the annual budget of £5,000 to further the cause could not be considered generous. The Board recognized that the 'whisky money' was being used by many County Councils for local lectures and special classes and that much assistance could be obtained from linking these with agricultural departments in colleges or universities.[22] New departments of agriculture were springing up.[27] The first graduates in agricultural science at the University College of North Wales at Bangor appeared in 1889.[24] With the 1890s came departments at Leeds 1890/1, Aberystwyth 1891/2, Armstrong College at Newcastle upon Tyne 1891/2, Nottingham 1892, Cambridge 1892/3, Reading 1893 and the South Eastern Agricultural College at Wye 1894.[22] Agricultural education was at last taking off, but instruction in these institutions was essentially theoretical and as such, of little relevance to the Harper Adams will. What of the provision of agricultural education at a level more accessible to the ordinary man or woman?

Grants from the Board of Agriculture and the arrival of 'whisky money' in 1890 had gone some way towards that all-important step of state funding for agricultural education and, as has already been discussed, County Councils were taking the initiative and going out to the people, offering lectures, classes, demonstrations and the like. As a healthier climate developed for agricultural education, a number of counties recognised the need for a permanent teaching centre. Here were agricultural institutions of a character more relevant to the Trustees. It was the County Councils of Lancashire, Essex and Cheshire which took the lead. The Harris Institute, Preston, now part of the University of Central Lancashire, had been opened in 1828 and with County Council money it commenced agricultural courses in 1892.[28] Its sister institution, Hutton, the ancestor of Myerscough College, opened in 1894.[28] The County Technical Laboratories at Chelmsford, to become Writtle College, took over a permanent site in 1893, having opened two years previous.[29] In 1895 the College of Agriculture and Horticulture at Holmes Chapel, Cheshire, an ancestor of Reaseheath College, was inaugurated.[30] The Midland Dairy Institute, in 1905 to be renamed the Midland Agricultural and Dairy College, was established at Kingston on Soar, Derby, in 1895 by joint co-operation between the counties of Derby, Leicester and Nottingham.[31] The Cumberland and Westmorland Farm School, now Newton Rigg, joined the growing band in 1896.[24] Sparsholt College opened as a county farm school at its original site at Old Basing in 1899.[32]

It is known that the Trustees in their search for direction visited at least two institutions. Burne stated to the Reay Committee that visits were made to Holmes Chapel, Uckfield and Dauntsey, West Lavington, but dismissed the latter straight away as being a school for young boys and irrelevant.[15] Of the two there is no information to show how or if they provided guidance. A description of Holmes Chapel in 1898 suggests Burne would not have had the most constructive of days viewing 'well made and well drained paths', 'cucumbers looking fine and healthy' and students planting out lettuce and pricking out celery.[30] The reference to Uckfield relates to Uckfield College and High School, which had opened in 1881, preparing students for the universities and the professions. It was not until 1904 that it passed to the County Council and took the name of Uckfield Agricultural and Horticultural College.[33] What assistance an inspection would have been is anyone's guess. No other places were apparently visited. The County Laboratories at Chelmsford would have seemed an ideal choice, with classes in bee keeping, dairy work, poultry, smith's work and the practice of agriculture.[29] Chelmsford was not, however, visited. It can only be surmised that the Trustees, thrown in at the deep end, were possessed of little information.

One of the earliest stages in the history of the College must be the setting up of the Foundation Scheme on 29 June 1896.[14] It is a very thorough document. The proposed school, the term used throughout, was to be administered by eleven governors. Of the eight representative Governors, holding office for an elected term of five years, two were to be elected by Shropshire County Council, two by Staffordshire County Council, two were to be Justices of the Peace for the Petty Sessional Division of Newport, one was to be a representative of Victoria University, Manchester, and one of the Council of the Royal Agricultural Society. The three co-optative Governors were to be the original Trustees, holding office for life. On their deaths, future co-optative Governors were to hold office for eight years.

The presence of Shropshire and Staffordshire County Councils is not unexpected. The shared interest in agricultural education alone warranted inclusion. As with Shropshire, Staffordshire County Council had set up a Technical Instruction Committee in 1890, employing two peripatetic instructors the following year.[34] The issue of funding also made the County Councils natural associates. Shropshire County Council was an early and significant contributor to the College. At the County Council Meeting of November 1895 £2,500 was voted to the proposed college to pay for equipment. It was also agreed that there would be an annual grant of £1,000, of which £800 would go towards general purposes and £200 towards scholarships.[35] Staffordshire County Council, on its part, gave £1,000 to the building fund and also supported an annual grant and scholarships.[15] The relationship between the College and its neighbouring councils worked in many practical ways, as shall be seen, although this is not to deny that uneasy elements existed.

The County Councils ensured their presence was never forgotten. In the Foundation Scheme of 1896, for example, it is surely the County Councils behind the clause stating that, if student space was limited, preference would be given to residents of Shropshire and Staffordshire and amongst those to holders of county scholarships.[14] Of course, in this concern for the neediest of students, a concern also reflected in the low fees listed in the Scheme, it could be argued that acquiescence was rightly being shown to Harper Adams' wishes. Furthermore, Councils could always wave the righteous banner of protectors of the ratepayers. However, in this clause there does seem more than a hint of determination on the part of the Councils to flex their muscles. The Chairman of the Technical Education Committee of Shropshire County Council, speaking generally, put the issue very succinctly: *'It seemed to him the gratitude to the County Council was always actuated by the anticipation of future favours, and when those favours were granted, the Council was thrown over like a sucked orange.'* [36]

From the very beginning the College was wary of undue influence and was determined to maintain its independence. After all, it was unique in its origins and had yet to carve a niche for itself in the still very fluid structure of agricultural education. In 1907 Burne faced several questions from the Reay Committee, which was looking into the organisation of agricultural education and contemplating rationalization. Burne was asked: *'Do you think there would be any objection to your being placed under the guidance, I'll call it, of the County Council, that is to say, if the County Council should be induced to provide for and control this sort of instruction?'* Burne replied emphatically: *'You see, this property was left to establish a school or schools of agriculture in Salop or elsewhere in England; the money was not only given for Shropshire and Staffordshire. You must always remember that.'* Later he was asked: *'You wouldn't transfer the Trust in any way to them [Shrophire County Council] or the management of the Trust?'* Burne made his answer blatantly clear: *'I should say not'.* Burne felt Shropshire County Council had embraced the arrival of the College with perhaps too much eagerness: *'When they [Shropshire County Council] were in difficulty as to how to deal with agricultural education, this dropped down in their midst, and they greeted it as an excellent means of spending their "whisky money" on agriculture. [...] They had spent very little money on agricultural education until this came into existence.'*[15] Indeed the initial enthusiasm and speed with which Shropshire had taken up agricultural education had run into problems by the mid-1890s as classes were slow to recruit. In 1894 Shropshire County Council in fact decided more than a third of the annual Government grant should henceforth be used to relieve the rates.[37]

Later clauses within the Foundation Scheme covered administrative matters, the holding of property, details of instruction, entry requirements etc.[14] It was specified that the school was to satisfy the requirements of the Board of Agriculture. The threshold for enrolment was set at fourteen years. No allowance was made for women and girls, as required in the will, because in 1896 a scheme was agreed, financed mainly by Shropshire County Council, to establish the Technical School for Girls at Radbrook, Shrewsbury. Radbrook opened in 1901, offering classes in dairying and domestic science, and the College was to work closely with it.[38]

Finally, mention should be made of a short but nonetheless important clause in the Foundation Scheme. There was provision for 'a School farm for purposes of demonstration or practice in agricultural processes or otherwise in furtherance of the general objects of the Foundation' *(see photo page 44)*. Only sixteen years before, John Wrightson had provided a model by establishing a farm at his agricultural school at Downton. He maintained: *'Theory and practice must go hand in hand together if sound knowledge is to be attained.'* [25] When Harper Adams College opened in 1901, special pride was shown in the farm. At a time when only a few institutions had access to land, to own a farm and furthermore for it to be on the doorstep was a marketing point indeed.

With the Foundation Scheme agreed, the first meeting of Governors was held at the Royal Victoria Hotel in Newport on 17 December 1896.[3] Attending were Sir Thomas Fletcher Boughey, Sambrooke T. H. Burne, John Bowen-Jones, Henry Pooler and William Heath Lander. Educated at the Royal Agricultural College, Bowen-Jones was a prominent County Councillor with strong agricultural ties, including involvement with the Royal Agricultural Society and the Shropshire Chamber of Agriculture.[7] Pooler was a noteworthy agriculturalist, prominent as a founder and organizer of the Newport and District Agricultural Society. He had been educated at Oxford and farmed at Tibberton Manor. He was that rare mix of practical agriculturalist and educated man.[39] Another Tibberton man was William Heath Lander of The Grove. Like Bowen-Jones he had been educated at the Royal Agricultural College and, also like him, he was a County Councillor with a long interest in matters agricultural.[6]

Records for these early years are few. Even information on the building of the College is scarce. Burne is attributed a significant role: *'He seems to have been the leading spirit in the making of the College, which may be regarded largely as his handwork. Indeed, he used to design and draw the preliminary plans himself.'* [8] Burne, himself, attributed the day to day supervision of the building to Lander.[40] In the summer of 1896 the Trustees were inviting architects to submit designs in open competition.[18] The chosen architect was Harry Teather of Cardiff *(see photos pages 43 & 45)*. The contractor was G. I. Muirhead of Newport,[41] who is rumoured to have been bankrupted by the contract.[42] Much of the work was undertaken by local builders, namely the Talbots.[43] The College is said to have taken two and a half years to build,[44] with completion late in 1900.[45] It was to provide accommodation for fifty day boys and fifty boarders. Walls were 'faced with Ruabon red pressed bricks, relieved with Grinshill stone dressings, and the roofs covered with brown Broseley tiles'.[46] The farm buildings were designed and erected from the plans of Mr Vickers of Lilyhurst, surveyor to the Duke of Sutherland.[41] Burne put the actual cost of all the buildings at £22,260.[15] Crowther, the second Principal of Harper Adams, was to put the figure for building and equipping the College and the farm at around £35,000.[47]

With the College near completion in the summer of 1900 the Governors placed an

2.1 The builders of the College, c. 1900.

2.2 The earliest surviving photograph of staff and students, c. 1901. Front row from left: Principal Hedworth Foulkes sixth. Second row from left: William Neate, engineer, thirteenth: Private Collection.

advertisement in The Times inviting applications for the post of Principal.[18] The favoured candidate was to be under forty-five, married and in possession of an agricultural science qualification. He was to have control of the College and management of the farm. His salary was set at £200, with a capitation payment of between £2 to £5 per student, emphasizing the Principal's role as warden.[14] A brilliant choice was made in the person of Percy Hedworth Foulkes. Twenty-nine years of age, he was young, enthusiastic and full of ambition, not only for himself but for the College that he was to lead.[7] At a time when few were involved in agricultural education he came with impeccable credentials. He was born in Denbighshire, the son of a Welsh land agent [48] and had been educated at Bangor and Edinburgh.[7] He had graduated with a degree in agricultural science in 1892. From 1894 to 1900 he had lectured in farm and estate management at the University College, Reading.[7] He was possessed of excellent academic connections. His particular interest was in experiments, having initiated field trials with Gilchrist at Reading.[26] Hedworth Foulkes took up his appointment in October 1900,[7] moving to the Principal's residence within the College building, with his wife Beatrice, early in 1901.[45] Absolutely everything was new. The sense of excitement must have been as strong as the smell of wet paint.

The College Farm 1901

The Newport Advertiser of 27 April 1901 provided a detailed and impressive description: *'A fairly large-sized garden, planted with fruit trees of approved variety, lies between the College and the Dairy, the architectural features of which are in keeping with the principal building. The first room on entering the dairy is one in which the vessels are washed; and it is fitted with a rapid combustion stove for heating water for use in the Dairy. Adjoining this, there is a room for storing soft cheeses; and, from that we pass into the churning room, which is fitted with two separators. In the next room there are two whey tanks connected with pipes for running the whey into the piggeries, which are some distance away, across the farmyard. The ripening room is opposite; and the Farm Steward's office is adjacent thereto. The carpenter's shop and lathe-room adjoins the blacksmith's shop (with pent-house for shoeing purposes), a cartshed, and a store room. The farm buildings are on the other side of a wide, macadamised cart-road. The shippens are spacious and airy, and in close connection with the mixing house, so that the cattle can be fed with a minimum amount of trouble. Here, also, there is a machine for pulping. The hay is "chaffed" in an upper room, and falls through an aperture, on to the floor of the mixing house. The grinding mill stands in a corner, where it is fed, through a hopper, from the room above.*

There are three open yards with shelters all round, so that cattle can be protected from cold winds, and yet enjoy sufficient fresh air. There is a loose box at each end of the shippens; and a bull box, with an apparatus to facilitate feeding from the passage.
Off the said passage, there is accommodation for the engine which drives the pulping, chaffing, and grinding machinery, and also pumps the water from the outside tanks into cisterns, which feed the troughs in the yards. Here, also, there are weighing machines.
In the shippen with which we have been dealing, there is provision for 24 cows. In a smaller shippen, there is accommodation for eight cows. Outside, in the stackyard, there are two large Dutch barns, stocked with hay grown on the farm. In the granary, there are cake mills and chaff-cutting communication with the shippen below. [This was carried out by means of a five-horse-power vertical engine.] The piggeries consist of three yards, fitted with troughs so constructed as to allow of their being filled with food from the outside. Beyond the yards is a mixing house, with a tank, which receives the whey from the Dairy. The poultry are lodged in movable houses, the practice being to change the birds from field to field, as circumstances warrant. All the latest approved machinery, implements, and appliances are used; and the farm, throughout, is thoroughly up-to-date'.

One final important consideration was that all was designed to save labour. The whole of the waste liquid manure was, for example, conveyed by pipes into a tank in a field adjoining the farm. The tank had a chain pump erected over it so that the contents could be used.[a]

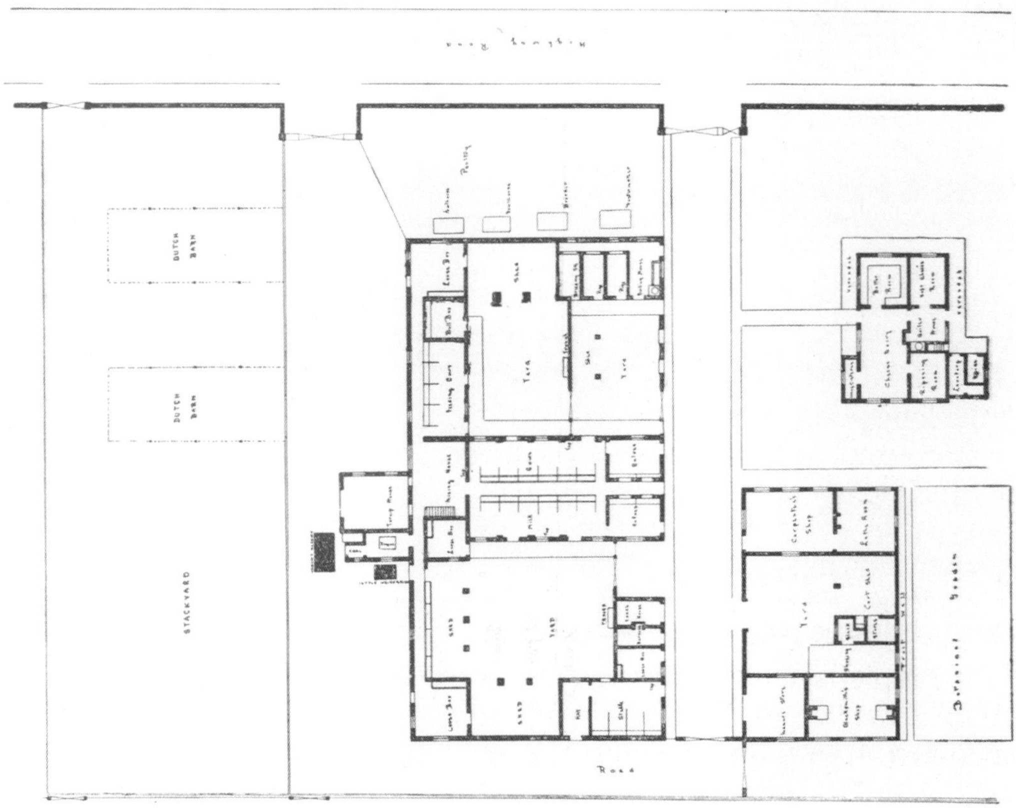

2.3 Ground plan of the farm buildings and workshops, c. 1901.

2.4 Farm buildings and dairy from the south, c. 1905.

Chapter Three —

Early Days 1901-07

All the long years of planning, persistence and patience came to fruition in 1901. It was no small achievement for the dreams of a dying man to be transformed into the bricks and mortar of reality. The momentous opening of Harper Adams College took place on 26 September 1901. What would the guests have seen as they were guided around? The College simply consisted of a main building for teaching and residential uses with a farm behind. The Adams homestead remained a separate complex, with Ancellor House soon to be used as the residence of the Farm Steward and for student accommodation.[1] The dilapidated farm buildings were eventually repaired and used as part of the soiling of milch cows experiment.[2] With the exception of the Ancellor, all was of the most modern design with little expense spared. Lander, Chairman of the Building Committee, attempted to excuse the costs involved: *'He had been somewhat extravagant in expenditure, but it had been his aim [...] to erect buildings complete, and in every way suitable for the purpose for which they were intended.'* [3] The Principal had a private three-storied house forming the west wing of the College. Here alone there were twenty-four rooms with domestic quarters occupying the top floor.

Douglas Scotland, student 1904-07 and a member of staff 1908-11, remembered:

> *The ground floor of the main building was arranged as follows: To the right of the entrance hall were Students' Common Room, Botanical and Veterinary Lecture Room and the Large Lecture Hall. To the left of the entrance hall were the Staff Room, Principal's Office and Dining Hall with Students' Cloak Room on the opposite side of the passage to the Staff Room and a Boot Room with lockers facing the foot of the main staircase. Upstairs there were four large dormitories divided by wooden partitions into cubicles with two smaller private rooms for staff ... there were, of course, bath rooms* [1] *(see photo page 45).*

Pride of place went to the Chemistry Department, separated from the main block by a covered way in an attempt to disperse any smells. It had its own lecture theatre which could be 'darkened for lantern demonstrations' and a large laboratory for thirty-two students, as well as ancillary rooms.[4] In addition to Scotland's list there was a kitchen, board room and library, biological museum with laboratory and a room devoted to machinery demonstrations in which students could take items to pieces and repair them.[5]

The premises were heated by hot water pipes and open fire-places. Emphasis throughout was on a healthy environment for the young students. Thus ventilation was 'on the natural system' and sanitary work was 'of the very best description, and comprises all the modern requirements'.[5] The first floor was fire-proofed.[6] Parents might have been less happy with the supply and quality of the drinking water which, although pumped from a specially constructed well, soon earned a reputation for foul taste.[7] Despite this the 1908 Prospectus boasted of 'a pure water supply of excellent quality'. The problem was not resolved until 1935 when cold water supplies for the College were first drawn from the borehole at the National Institute of Poultry Husbandry. The original supplies for the College were then utilized for hot water and raising steam[8] and the well stopped being used altogether in the late 1940s.[9] There were engine, boiler and accumulator rooms. Steam was required for cooking, for work in the specially constructed laundry and for heating the hot water for the pipes warming the building.[5] The passage of time is glimpsed when it is realized that the coal came by canal to Edgmond Wharf. Progress later saw it delivered from Newport by Sentinel steam wagon.[7] Electric lighting was generated by a dynamo driven by an oil engine.[5] Initially electric lighting was limited and candles were also used.[7] By 1922 the generating plant was working close to its capacity and Principal Crowther was contemplating a relaxation of restrictions on the use of spirit lamps for making the requisite student beverages.[10]

Of course, the College was far more than just state-of-the-art buildings. Staff were recruited from all the main centres for agricultural education[5] and a valuable contribution was made by non-academic staff as practical instructors. A prospectus of 1901 listed just six academic staff, including the Principal who taught agriculture and entomology. The Farm Steward, William Vaughan, was utilized as an instructor in practical agriculture. In addition the list included a farm bailiff, an engineer, a carpenter, a blacksmith and a matron.

As early as 1902 stalwarts of the College were making an appearance. That year additional staff were employed to undertake extension lectures for Shropshire and Staffordshire County Councils. Men such as W. T. Wilson, who was appointed the Veterinary Lecturer. He had worked at the Royal Agricultural College.[11] Soon known as Professor Wilson, although no such title was officially his, he was fond of drinking.[12] It has been said 'when he came from the College he called at every pub on the way to Newport'.[13] His favourite treatment for a sick horse was said to be half a bottle of scotch.[14] His convivial ways were coupled with a reputation as 'a racy raconteur'.[15] He remained at the College until his death only a few days before his intended retirement in 1928.[16] College Minutes ruefully lament that it would be impossible to get another good full-time veterinary lecturer for £275 a year.[17]

G. T. Malthouse arrived as the Horticultural and Botany Lecturer at the same time as Wilson. He had trained at the Royal Botanic Gardens, Edinburgh[12] and soon achieved national fame as a result of his pioneer research on potato scab. One of his first projects was to improve the layout of the gardens behind the College.[18] These consisted of vegetable and fruit plots. The old garden and orchard were kept at Ancellor, allowing the study of the old alongside the new.[19] There was also a botanical garden for the cultivation of specimen plants and a selection of agricultural grasses. An early prospectus concluded: *'The means of thorough ocular demonstration throughout the year are therefore ensured.'* [4] Much of his teaching was

3.1 1903-04 staff cnd students.

concerned with extra-mural classes around Shropshire. During the First World War the County Council set about providing its own lecturing staff and ceased to employ the College's personnel. Malthouse was thence employed as a horticultural adviser by the Council. From 1923 the College hired his services from the County Council on a part-time basis.[20] He retired in 1937,[21] but in the Second World War, with the shortage of staff, he returned to run the Biology Department and helped with horticulture.[18] Leonora Cottrell, nee Grice, student 1943-45, remembered Malthouse at this time as a man of mild disposition:

We found a special ally in the dear man who taught us the subject [botany]. [...] He was tall and thin and bearded. In our youthful minds he seemed quite ancient. [...] On the rambles which he organised for us, with the purpose of finding plant specimens and identifying them, he was always well ahead of the straggling mob of students - clambering over fences, nimbly leaping across water-logged ditches, in most enviable and energetic fashion. [...] I believe Mr Malthouse had a soft spot for the girl students. [...] I recall a day when one of us sat down on a log and cried, saying she didn't feel very well and could not go on. Mr Malthouse couldn't have been more sympathetic. He sat down beside her and took her hand, saying 'Never mind, my dear. I have daughters of my own you know'. He cut our expedition short and we all returned to College at a very moderate pace.

Malthouse died in 1952 at the age of eighty-two years.[18]

Despite staff such as Wilson and Malthouse, one of the first problems to confront the College was that of recruiting and keeping staff. Rates of pay were not good at the College and in an era when qualified staff were at a premium, frequent changes of staff were inevitable and getting a replacement was often difficult. Burne had grave concerns:

[The salaries] are too low for us either to get the best men or to retain the services of young men when we get them. I might mention in this connection that when we first started we engaged a dairyman at £65 a year, but he had not been with us very long before he got an appointment under the Irish Agricultural Department at £200 a year. We appointed a successor at the same salary, £65, and we increased him gradually up to £80 a year. He left us after about three years' service to go to Cirencester, where he received £100 a year, all found. Our next man we had to take on at the price we finished with the last, £80 a year. He only stayed with us three months, and he took a more lucrative post at Aspatria.[22]

Burne as Chairman of the Finance Committee had his finger on the pulse of the College. He must have known only too well the precarious financial knife edge on which the College balanced. The Foundation by means of stocks and securities gave the College an annual income of around £800. Shropshire County Council provided an annual grant of £800 and Staffordshire £350, although various stoppages for scholarships meant that the latter sum in reality fluctuated between £50 and £200.[22] In 1907 the Minutes record dire concern for the very survival of the College. An agreement had been reached between the two Councils to form a joint committee. Thus, from the session 1907-08 Staffordshire did allocate £350, but at the same time Shropshire reduced its grant to £350 to achieve parity.[23] Shropshire Councillors had long been complaining that the College only catered for a small number of Shropshire students and did not give proper returns in the form of lecture commitments to the County. In 1907 the figure for Shropshire students at the College was put at thirteen,[23] as compared with a total of sixty. Both Councils also offered scholarships, those of Shropshire amounting to £200 a year, although it should be remembered that not all scholarships were taken up. By 1907

Worcestershire and Lancashire were sending scholars [24] and by 1909 Warwickshire had followed suit.[25] It cannot be denied that local Councils were somewhat parsimonious when it is realized that in 1909 Wye College received £5,000 from Kent and Surrey [25] and Cheshire paid out £2,000 towards the costs of its county college and the experimental work associated with it.[26]

To complete the list of College income it should be noted that from the session 1902-03 the College received an annual grant of £300 from the Board of Agriculture. In 1907 it was raised to £400 but included much-needed money for research. A proportion of the income naturally came from fees, but these significantly were kept low, 'aiming for the sons of farmers of moderate means',[27] in accordance with the spirit of Harper Adams' will. Thus the Foundation document listed annual fees as £8 for tuition and £35 for boarding. From 1909 students resident outside Shropshire and Staffordshire paid higher fees of £60 for tuition, board and residence. In contrast, for 1903, annual fees at the Royal Agricultural College ran to £141 and at Aspatria students paid between £75 and £80.[28] From the very first it had been stated that the College was 'not managed for pecuniary profit'.[29] Quite simply, the income when tallied was insufficient. Hedworth Foulkes freely admitted financial difficulties and stressed the need for more money to the Reay Committee in 1907.[24] In 1909, at a meeting of Staffordshire farmers, he declared financially the College had 'practically come to the end of their tether'.[26] By this time the College had many commitments, in addition to regular teaching. Money, for example, was urgently needed to finance an ambitious building programme. That same year the College was advertising locally for tenders for a loan of £3,000 to cover costs for the erection of a new hostel.

With issues of funding to the fore, recruitment of students was naturally not overlooked. Counting heads was ever a crucial activity. The College had opened its doors to students on 17 April 1901, with the original entry numbering six.[30] The first full session began in September. In the early years students often enrolled at College without having visited, having chosen their education from 'a list of colleges in Whitaker's Almanack'.[30] In that first academic year numbers reached forty-nine. However, those on the standard two-year courses numbered just sixteen, the remainder of the students being on short courses, some for as little as two weeks. By 1907-08 a steady increase gave a total of sixty-five students, although again this figure included those on short courses. When the College opened there were 'no definite lines laid down to work upon'.[24] As the Principal explained to the Reay Committee, courses were developed 'as we found instruction was required by those coming to the college'. The Principal's remark illustrates not only the unorganized nature of agricultural education in these early years, but also the College's policy of ascertaining and meeting the needs of the market.

The early long-term students, after an entrance examination, were admitted to a two-year course, either the College Certificate or Diploma.[32] It should be said that length of course was very flexible in these first years.[24] The first, and most popular of the two, provided a thorough practical training in agriculture and included some science. It was intended for 'the farmer's son who has had little or no previous education in scientific subjects'. The College Diploma was of a more advanced and scientific character and intended for those with a good education, more often than not coming from public schools and intending careers in land agency, although a future in farming was not dismissed.[24]

The College Diploma was considered preparation for public examinations, primarily the National Diploma in Agriculture (NDA).[24] The first two Harper Adams students

3.2 1907-08 staff and students.

passed the initial stage of the NDA in May 1902. The origins of the NDA went back to 1895 when the Royal Agricultural Society of England instituted a Diploma in Dairying, with the first examination held at Reading in 1896. In 1897 the Royal Agricultural Society and the Highland and Agricultural Society of Scotland collaborated to form an examining board for a National Diploma in Dairying (NDD). The first examinations were held that autumn at Kilmarnock and Reading. Examinations for the NDA followed, the first taking place in Leeds in spring 1900. However, institutions also offered their own certificates and diplomas due to an ongoing dispute between the Royal Agricultural Society and the Agricultural Education Association as to who should control the national examinations, the Royal Agricultural Society or a new education council composed of all interested parties. Furthermore, both at Harper Adams and nationally, in these early years more students favoured college diplomas due to criticism of the lack of practical content of the NDA.[33] Hedworth Foulkes was firmly of the opinion that the NDA was too theoretical.[24]

A significant part of the College curriculum concentrated on short courses and until 1907 most students were enrolled on such. The nature of these was unique to Harper Adams.[34] Just as today, the appeal was obvious with a policy of tailoring the course to suit the client. A farmer's son who could not be spared for a full course could possibly attend for eight weeks from January to March when farm work was less pressing. He could also return to the College and extend his knowledge at a later date with a sixteen-week extended course.[24] Short courses were especially successful at Harper Adams: *'They have been taken up by young men of just the type we want to get hold of, young fellows from 20 to 26, who are living at home or on a farm and who come to us with a good knowledge of practical work and who know little or nothing of the theory of agriculture.'* [24] Short practical courses in dairying were also on offer but by special arrangement.[32] Dairying never achieved a high profile at the College, although a very small number of Harper Adams students did take the NDD. The first Harper Adams student was awarded the NDD in 1907.

The short courses were essentially practical in nature as would be expected for the clientele sought. Short-course student T. Jesse Wadlow wrote to his local newspaper:

> *I can testify to the great usefulness of the course to anyone who is or intends to be a practical farmer. Every lecture and demonstration was full of interest, just those things being put before one which cannot well be learnt on an ordinary farm, yet which every farmer wants to know - from the cause of bad butter to the most advantageous use of artificial manures, and, as regards practical work, from the laying of hedges, to the inoculation of calves.*[35]

However, the College prided itself on the practicality of all its teaching and its usefulness to the real world. In 1908 Hedworth Foulkes could state eighty percent of Harper Adams students for the years 1906 to 1908 went into farming. An important part of the syllabus was actual daily work on the College Farm. All the milking for example was done by students. Both the Certificate and Diploma courses required the passing of a practical examination, which included harnessing a horse, making butter, making a horse shoe, ploughing and driving a steam engine.[36] Interestingly, there is the suggestion that many students arrived with insufficient knowledge of farming. Hedworth Foulkes personally favoured some time being spent farming before arrival.[37] There was evidently no requirement for a year's practical work on a farm prior to enrolment, although by 1908 newly qualified students were clearly advised to gain practical experience before taking up farming on their own.[36]

What of the student profile? The early students were all male, as women were catered for at

Radbrook College. The average age was between eighteen and twenty, with farmers' sons on short courses averaging between twenty and twenty-six.[24] The reference in the Foundation Scheme to a minimum age of fourteen, suggests the Trustees envisaged boys of a younger age enrolling, although such boys would have left school. In 1918 the school-leaving age was raised to fourteen. In 1905 Hedworth Foulkes stated to the Shropshire Chamber of Agriculture that the minimum age had been raised to fifteen, but even so, few arrived that young. He stressed it was not the College's role to take up the secondary work of schools.[38] Yet, many of the students came ill-equipped with the necessary education. The Principal very much begrudged the time taken in teaching the students those subjects which they should have learnt already at school.[37]

The majority of students came from the larger farming families [22] and the professional classes.[24] For the session 1903-04 the breakdown was as follows: sixty percent sons of farmers, landowners, land agents or those connected with the land, thirty percent sons of professionals and ten percent sons of merchants. Nationally, this was very much the picture as well, with agricultural education providing the training which very often could not be obtained at home.[39] Harper Adams students did not come from the smaller and tenant farmer groups nor from the labouring classes, for which the College had been established. Neither the cheapness of fees nor the availability of scholarships could attract these sections. On average just a quarter, sometimes a mere sixteenth, of Harper Adams students, was in receipt of scholarships. In a time of agricultural depression the labour of a son of a small farmer could ill be spared for him to go to College, even if granted a scholarship.[24] Furthermore, prejudice against agricultural education was rife among farmers. Most felt farming could only be learnt at a father's elbow. As Burne put it: *'They [the farmers] doubted whether the scientific people could teach the practical people.'* [22] Even the Chairman of the Governors, Sir Thomas Boughey, had been known to disparage agricultural education: *'Everyone knew that farming was not carried on to a great profit by technical education. The farmers knew quite as much about their business, and many of their sons also, without this high technical education. They wanted a perfect climate for agriculture to thrive. [...] He was afraid what was called technical education would not put much more into the farmer's pockets as long as prices remained as at present.'* [40] With such ideas freely circulating, Boughey was addressing none other than the Technical Education Committee of Shropshire County Council in 1900, it would be essential to market the College well and to win its acceptance in the agricultural community, both at home and farther afield.

Hedworth Foulkes, more than any other, deserves the accolade of carving out a role for the College in the local farming community. Not content with that, he went on to develop it as a regional centre for the surrounding counties, as the Government rationalized agricultural education. His determination, energy, initiative and enthusiasm turned 'scoffers' into supporters.[22] Burne gave credit where it was due: *'What they [visitors to the College in 1906] saw today [...] was Mr Foulkes's creation and his creation alone.'* [41] Enthusiasm very much coloured the man. As the Bishop of Shrewsbury said: *'He has an enthusiasm which will carry him and carry the students who will come here to a high state of idealism and perfection.'* [42] By 1907 Burne could talk to the Reay Committee of strong farming support. There was a dawning realization that both the practical and the scientific had a part to play in training the farmer. But, this success was to have its price. Hedworth Foulkes was taken seriously ill in 1906, probably as a result of a nervous breakdown, and granted temporary leave of absence. The stress and workload were, however, to grow and,

3.3-3.4 Students at harvest work, c. 1905.

eventually, he and the College so close to his heart were to come to the brink of disaster. Before tragedy struck, there were years of considerable success. The College increased its number of students, expanded in size and took on greater and greater responsibilities as a regional centre for agricultural education. The message was clearly one of growing confidence and activity. Harper Adams had arrived and, despite significant and multiple problems, it was here to stay.

As early as 1904 the Board of Agriculture could state: *'[Harper Adams], though more limited in its aims than a university college, has succeeded in securing the position of an education centre for the agricultural work of a small group of contiguous counties.'* [43] The Government, for reasons of economy and efficiency, was urging greater joint action between County Councils in matters of agricultural education and Harper Adams' emerging pivotal role met with approval. From the date of the Foundation Scheme, as has been seen, Harper Adams had close ties with the County Councils of Shropshire and Staffordshire. The President of the Board of Agriculture at the opening of the College had made the point: *'Here was a local college for Shropshire and Staffordshire, the two counties working together, and the great advantage was that it must adapt itself to the needs of those two counties.'* [44] The County Councils gave grants and in return had representatives as Governors. In 1907 a Joint Committee of both Councils and the College was formed to enhance liaison between the three bodies.[22] At the very heart of the relationship was obviously the fact that in the early days most students came from the counties of Shropshire and Staffordshire. Between 1908 and 1909 forty-six came from Shropshire, twenty-eight from Staffordshire, twenty-seven from other counties in England, two each from Wales, Scotland and foreign countries and three from the Colonies. On the College's part, the relationship was valued, not only for the money and support it brought in, but for the opportunity it gave to get the College's name out and about in the community. The main areas of co-operation were extra-mural courses, research and advisory work.

In the matter of extra-mural courses some of the most successful were those run by Harper Adams for the Shropshire County Council in horticulture. Horticulture had had a successful history in the county since the 1890s,[45] and Malthouse continued that tradition when he began his peripatetic lectures late in 1902. Especially popular were his practical demonstrations and his visits to gardens to give advice. As he became known, he would be frequently asked to explain about experimentation at the College in matters of manure and soils. Barriers began to crumble and the College began to be perceived as 'a good thing'. By 1907 Malthouse was so popular that only about half the centres requesting him actually got him. A particularly successful class was held at Donnington, where Malthouse instructed miners and workers from Walker's factory.[46] Between 1902 and 1911 Malthouse gave 392 evening lectures and demonstrations in horticulture to an attendance of 13,151 and he made 693 visits to schools.

Veterinary and poultry classes were also available, although the latter were not offered to any great degree until 1911 when, at the request of Staffordshire County Council, Mr Rhodes, of the recently formed Poultry Department, began visits to seven centres. In February 1912 both Rhodes and Wilson took part in a new initiative led by Staffordshire, namely a specially organized day course for young farmers in Eccleshall. For all these extra-mural classes the College received payment from the County Councils, although it was not until the session 1907-08 that the College insisted, as financial difficulties mounted, that full costs were met. It should perhaps be mentioned that the relationship between the College and the Counties was not always one way. J. C. Rushbury, County Instructor in Agriculture for Staffordshire, was made an honorary member of the College staff in 1903 and undertook lectures for the students.

Also, in the area of education the College did important work providing nature study courses for school teachers in Shropshire and Staffordshire. This was an initiative on the part of the College in response to the Board of Education's call for increased emphasis on rural subjects within elementary schools.[47] Other colleges offered classes for teachers but along different lines.[22] Courses of a fortnight's duration were run very successfully every alternate year up to the First World War. By 1911 there had been five courses attended by 117 teachers on county scholarships. They were taught everything from insect life and the chemistry of common objects to the food of man and beast. The learning was sweetened with a picnic and cricket match.[48] Local teachers were vociferous in their praise of the instruction they received.[49] On his part the Principal considered courses for teachers of great value to the College, as teachers once won over to support could arrange talks in their areas, introduce students to courses and generally act as publicists for the College.[50]

Research was another area where co-operation between the College and Shropshire and Staffordshire worked to great mutual advantage. And it was with research that farmers could most easily be won over to support of the College.[51] The 1908 Guide to Field Experiments emphasized field trials and demonstrations had been seen to have made a 'marked effect upon agricultural education' in Germany and the United States. The results of experiments carried out by Harper Adams could be seen to have practical and financial advantages to farmers. Trials allowed the opportunity of seeing new varieties grown under similar conditions in one season in the same locality. In association with the County Councils a number of variety trials, demonstrations and experiments, principally to do with the manuring of crops,[22] were carried out at various centres and duplicated at the College. In Shropshire the County Council initially paid the College's expenses for trials around the county, although this stopped in 1905.[22] The College had sole control of the trials. In Staffordshire the College had expenses paid but less direct control, working in close association with Rushton. Trials had started in Staffordshire before the opening of the College.[24] College experiments were conducted either at the request of the County Councils or under the auspices of the Agricultural Education Association (AEA).[24] Since 1903 the AEA had masterminded experiments in centres around the country in an attempt at greater co-ordination.[52] The AEA schemes were not widely adopted,[53] but their acceptance at Harper Adams must have been down to the Principal. Hedworth Foulkes was a prominent member of the Association and one-time Secretary.[24]

The College also instigated its own research activities at home and these were equally effective in breaking down the barriers between the College and the community. Perhaps the earliest investigation to attract popular notice was work on the benefits of keeping cows on pasture during the autumn months. The experiment, begun in the autumn of 1901, caused widespread interest with its conclusion that it was beneficial to leave dairy cows out to late November or even late December. It resulted in a greater increase in live weight and lower costs as regards labour, food and bedding. Hedworth Foulkes, in his role of researcher, observed: *'The cows did not seek shelter from hedges even on cold nights.'* [54] One must sympathize with the Principal's dedication to research whatever the weather.

Up until 1908 Harper Adams' research activities may have been considered more praiseworthy

3.5 1908 staff with school teachers on the summer nature course at the College.

The Armorial Bearings granted to Joseph Harrison Adams and Thomas Harper Adams in 1857. Metal thread embroidery by Mrs Sylvia Harris, presented to the College in 1994.

The memorial to the Adams Family at Edgmond Church.

Ancellor House, the birth-place of Thomas Harper Adams and now the Principal's residence.

Edgmond House, the home of Thomas Harper Adams in his later years.

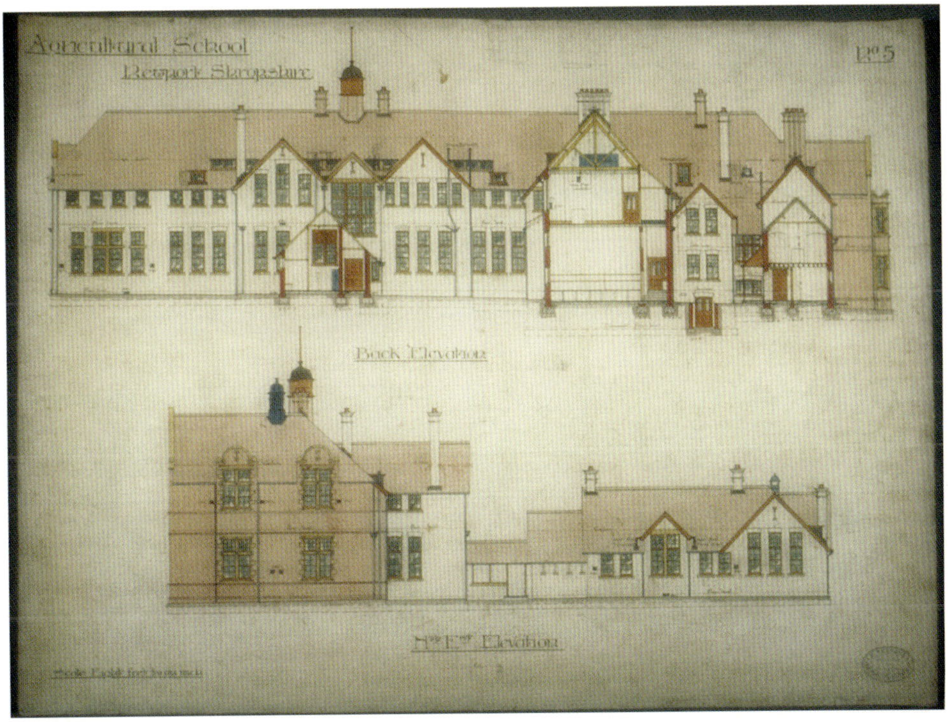

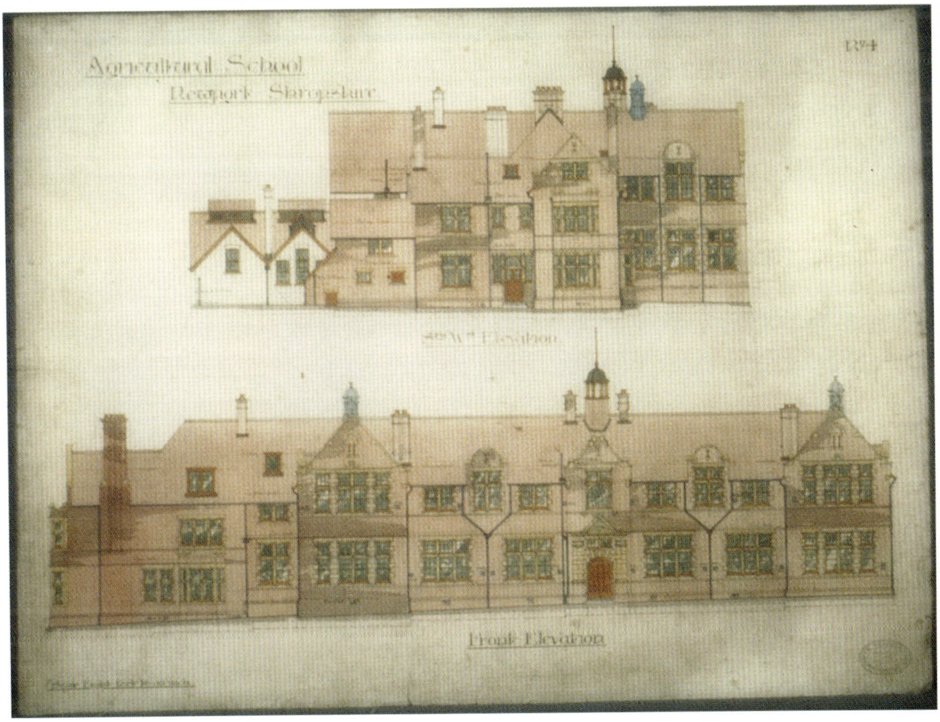

The architect Harry Teather's elevations for the College, late 1890s.

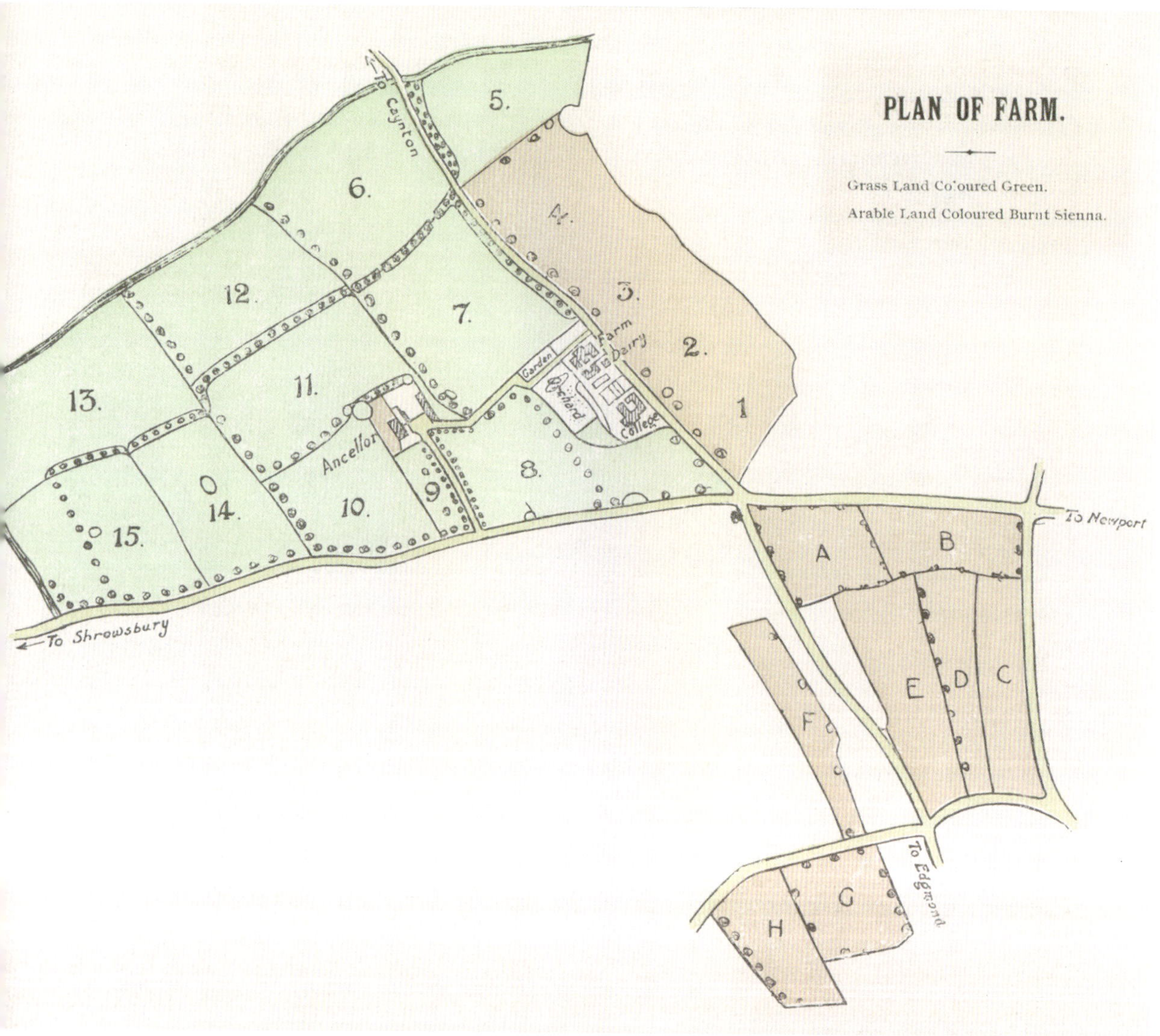

Plan of the College Farm, c. 1908. Grass coloured green and arable land brown.

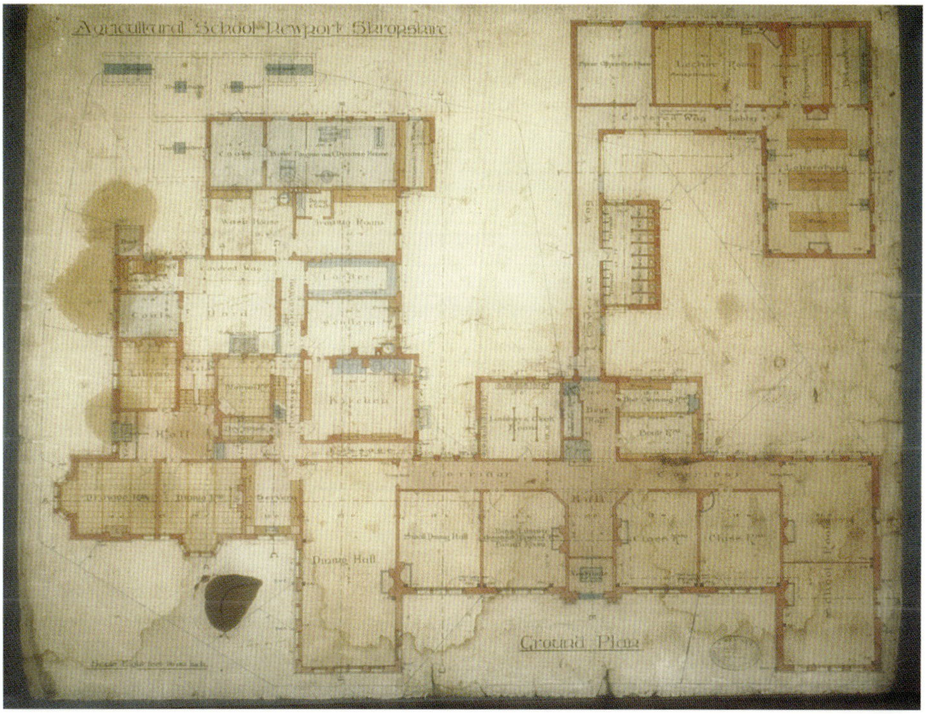

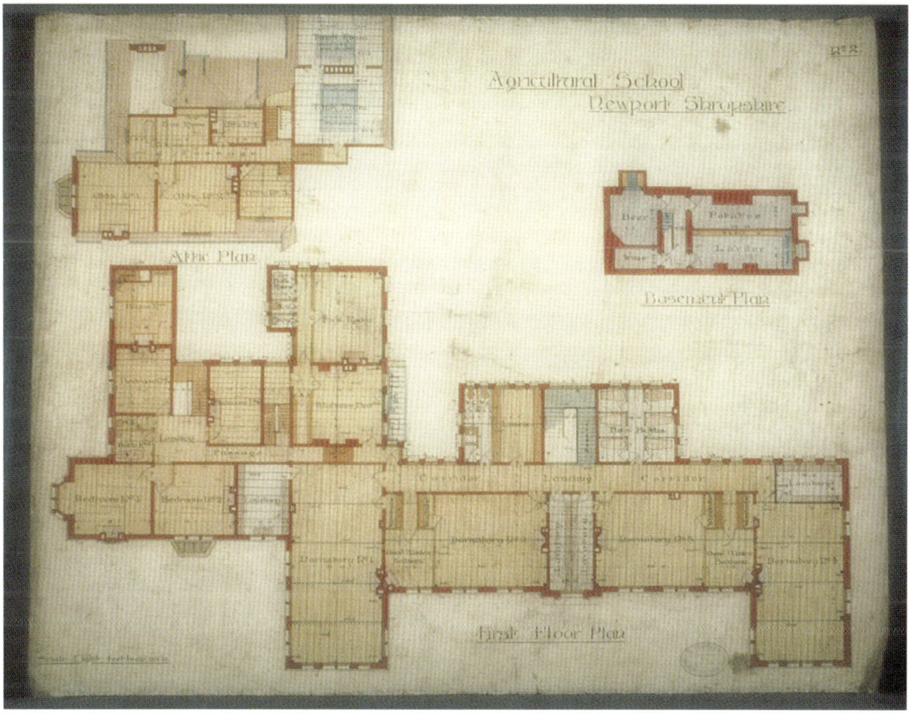

The architect Harry Teather's plans for the College, late 1890s.

College Diploma presented to Joseph Green, July 1919.

than noteworthy. From at least 1904 potatoes had formed an important part of the trials performed in association with Shropshire County Council.[55] However, in 1908 research of national importance was begun by Malthouse on black scab or wart disease of potatoes. Such was Malthouse's love of research, that in 1916, when research was severely restricted by war, he requested 'potato ground' in return for a Saturday afternoon's instruction at the Sankey Works. The Governors, however, gave him time in lieu.

The research into wart disease illustrates the College's willingness to meet the needs of the community and to break with any idea of academic isolation. Wart disease had been known in Shropshire since the 1890s [56] and the first samples of this disease had reached the College in 1901.[57] In 1908, at the request of the Rt Hon. Beville Stanier, Governor of the College, Malthouse investigated around fifty reported cases of wart disease.[56] In 1909 the trials of fungicides and wart-resistant varieties were widened in association with the Board of Agriculture.[54] From 1910 there was some funding from the Board of Agriculture, although this ended in 1913. As the Principal wrote: *'An intimate connection with growers and long experience of potato culture places Harper Adams in a good position for this work.'* These trials continued up to the time of the First World War.

Throughout, Malthouse published bulletins giving advice and reporting on the progress of the research. Requests for the publications came in from abroad and all parts of Britain. In 1910 Harper Adams took a further step in its role serving a community wrestling with wart disease. That year, for the first time, seed potatoes, provided by the Board of Agriculture and found to be wart resistant, were sold by the College, at a modest price and 'in lots to suit the cottager and allotment holder'. In 1912 Shropshire County Council arranged for J. W. Rylance, a member of staff at Harper Adams, to be appointed Inspector under the Wart Disease Order 1912. His task was not easy: *'Generally speaking occupiers seem quite unable to realise the virulence of this disease. [...] Frequently, the inspector had to be*

3.6 G. T. Malthouse with the horticultural class, c. 1908.

satisfied with [..]. scraps of information gleaned from the occupier's wife. In a number of cases the source of seed was not above suspicion, having been purchased from the [...] hawker's cart.' [58] By 1914 Rylance had travelled 1,376 miles by train, 1,466 miles by motorcycle and 386 miles by cycle and his contract as Inspector was for just four months of the year.[59]

Hedworth Foulkes and his staff realized the importance of getting the fruits of their labour circulating amongst the farmers. Research bulletins on all topics were readily published, official summaries were printed annually and regular articles appeared in the Newport Advertiser. The material was always popular and mostly distributed free of charge.[24] The staff also gave talks to farming groups and mounted displays and exhibitions at the numerous agricultural shows. In July 1903 the College had a stand at the Staffordshire Agricultural Show, giving veterinary demonstrations, seed testing and displaying, amongst other things, teaching apparatus, cheeses, diseased bones, carpenter's tools and methods of laying hedges. Field experiments with farmyard and artificial manures were illustrated by turves.[60] No show was too small or inconsequential. That same year the intrepid staff braved Hodnet Flower Show.[61]

Results were disseminated more often than not by farmers who, ever sceptical, were invited to visit Harper Adams and to see things for themselves. In the record year 1906, 500 farmers visited. These visits were very popular from the beginning. After the visits there would be the chance to buy souvenirs: *'Over £200 worth of seed has been sold off the farm in a season to farmers in the counties after their visits to the College.'* [24] Even if farmers did not visit, they could not resist casting a critical eye over growing crops: *'They [The College staff] have strips of oats, for instance, right across the field with the name written up large on a board, so that anyone going along the road in a motor or in a cart can see exactly what sort of oats are growing and can judge for himself of the condition.'* [51] It seems Harper Adams did not miss a trick in the war to win support. Yet elements of scepticism were slow to disperse. The donation to the College of radio-active manure by the Board of Agriculture in 1912 must have raised many a farmer's eyebrow. The theory was that it could stimulate plants to absorb manures previously ploughed in.[62] It was thus dutifully used and the mangels and swedes grown analyzed for the Board of Agriculture. They were shown to have no consistent differences from those grown on untreated land.

Harper Adams also had a very significant role to play in the community in an advisory capacity. From the beginning the College dealt with enquiries and gave help of the widest possible kind. In 1906 Shropshire farmers were enquiring whether they could hire a machine for charlock spraying. The College went out and bought a machine. A year later Wilson was asked by Shropshire County Council to investigate an outbreak of mortality amongst lambs in East Shropshire. The problem was worms.[25] In the session 1908-09 Wilson dealt with 154 enquiries concerning animal diseases.

The College also performed a useful analyzing service, dealing with everything from fertilisers and feeding stuffs to water. The Principal saw analyses as opening the door to requests for further information and thence to the realm of research.[24] Hedworth Foulkes never missed an opportunity to sell the College. Unfortunately, by 1907 pressure of teaching work and shortage of staff were leading the College to turn away many requests. In 1905 the College opened an official milk testing centre for Shropshire and Staffordshire at the request of the Board of Agriculture, which was establishing a nationwide scheme.[24] Periodic testing allowed farmers to know whether their milk reached the legal fat content as required by the Sale of Milk Regulations[63] and to determine which cows were worth

3.7 The Seed Testing Laboratory, c. 1908.

keeping and which to breed from. Farmers were charged 6d. a milk sample. A seed testing station had been set up two years earlier at the request of the Staffordshire Farmers' Association and Staffordshire County Council, with finance from the latter. For seeds, farmers were charged 1s. a sample. Again emphasizing the College's desire to be of service to the ordinary farmer, the fee was less than a quarter of that of other seed stations.[64] By 1911 the College had performed 411 analyses, 790 seed tests and 659 milk tests. On occasion their expertise saw the staff in unusual surroundings. C. D. Stewart, the Dairy Instructor, testified in court on behalf of a farmer in a milk standard case at Wellington in May 1902. He declared the legal standard was too high for April as the cows in question were in good condition and still unable to reach the target fat content. The case was dismissed.[65] 'To Harper Adams' must have been the toast in many a public house that night.

With the session 1908-09 student numbers made a sudden leap from sixty-five the previous session to 110. These figures included those attending the short courses, but from 1907 the majority was enrolled on the two-year courses. The reasons behind the growth are not apparent, but may be associated with the increased popularity of the College Certificate. In the period leading up to the First World War the College entered a phase of activity and expansion. It was on the crest of a wave, but, if one was to scratch beneath the surface, it would have become apparent that inherent problems had yet to be confronted. The perennial issue of funding remained and consequently in physical resources, in staff provision and in many associated ways the College was to become seriously overstretched. As yet these dangers were masked.

Student Life in the 1900s

The following extracts are from the diary of Wilfred Mercer, who was a student at Harper Adams from 1905 to 1907. He was an extremely able student and as well as studying for the College Diploma, he obtained the NDA and went on to study for a degree. He become Principal of Reaseheath and after the Second World War, Provincial Director of the West Midland Province of the National Agricultural Advisory Service. His diary, in the possession of his family, provides a remarkably detailed and unique picture of life at the College. It starts with his first day.

3.8 Surveying class, c. 1908.

Monday Oct. 9th 1905 - Got into Coll just before it was dark. Shoeb. [John Shoebotham, student 1903-?] showed us (2 other new chaps & myself) round the buildings, as well as the fading light would permit. A good many arrived after we did. [...] Hardly knew what to do with myself most of the time, but nobody took much notice of any of us new studts. [...] Had supper (beef) at 7.45 p.m. As soon as we were out, the waggon with the boxes on came and was unloaded. I got mine off & took them up to my cube [cubicle] - the worst in the Coll. I was told. It was very small & when both my boxes were in I hardly had room to stand.

Unpacked most of my things & then the man came to make my bed. Wandered about until 9 p.m. when we had to sign a book, the doors being locked at this time. As soon as this was over I went to my cube & got into bed.

3.9 Practical botany class, c. 1908.

Tried to go to sleep but did not manage it till after 10.30 when the electric light was turned off.

Thursday Oct. 12th 1905 - Put on my cord knickers this morning, which occupied some time. Breakfast this morning was porridge & frizzled bacon & butter & marmalade which Heath fetched from Tutors' table when they got up. This seems to be the regular thing. Went to Pract. Work (Shops) from 9 to 1. Helped the builders on the new building [extension to eastern wing]. Cut some fairly large mortices and helped put up some rafters. Dinner was boiled rabbit, potatoes & carrots & sago pudding & prunes. For Pract Chemy went a walk of 7-8 miles I should think & looked at a winding river & he (Mr [T. W.] Fagan) told us it was a rotten district for "Jology". Had a bad time with my diarrhoea but managed to hold out till we got back again. Tea - bread & butter & treacle. After tea went for a walk round Edgmond with Shoeb. Supper - bread & cheese & jam. Sat with a lot of others round Com. Room fire. Tried to read Pickwick but gave it up.

Tuesday Oct. 24th 1905 - In Practical Work we went into the mangold field & "hogged" mangolds till 9.30. Then did a bit of topping & then had to go & weigh little plots, in bags. Took us till about 12 & got very cold over it. They were manurial expts. After this we did a bit more topping. I immediately cut a piece off the end of my thumb. In the afternoon Mr Fagan went away so we had Pract. Work again. Loaded mangolds this afternoon. Went for a walk with Clark [Tom Clarke, student 1904-08] round the villages, after tea & bought a cake (6d) worked from 7 till 9. After Gatebook I messed about in various clubs till 10.30.

Friday Nov. 3rd 1905 - In Geology all we did was to examine specimens of various rocks & minerals - We all should have gone to the Duke of Sutherland's Nurseries etc in the afternoon, but a fair few did not go. Brake started at 2.0. It rained most of the way. A good few rode their bikes.
The Prinny drove in his trap. Had a look round the buildings attached to the saw-mills. The latter were very interesting - the various saws & machines for labour saving - afterwards we went to the nurseries & looked round. Got back about 5.30. Worked from 6-10.15.

Friday Dec. 1st 1905 - Instead of having Agriculture the second lecture this morning we had mathematics as the Prinny was away all day. For the same reason we had Chemistry instead of Entomology. From 11 to 1 when we should have had Geol. & Chem. we went a walk with Mr Fagan round Tibberton & Adney way. In the afternoon we finished surveying the field we started last Friday. They were thrashing all day & thrashed one of the divisions in one of the barns (oats) & then went away. Worked till 7, & after supper had a bath & changed my pants shirt & stockings, & thus got rid of what I think must have been a flea. At any rate it had caused me agonies during the day. Read Bleak House after supper, in bed till about 10.15 p.m. I separated [in the Dairy] in the morning but forgot all about it at night.

Chapter Four —

Activity and Expansion 1908-14

The chance to travel back in time to 3 July 1914 would be an opportunity not to be missed. One of the bloodiest wars in history was to break out in a month, but the bleak situation in Europe following the assassination of Archduke Ferdinand the previous week probably disturbed few that day in Shrewsbury. For five days the town was playing host to the Royal Agricultural Show and King George V was to visit that very day.[1] The whole town was 'a blaze of flickering, dazzling colour', with 'bunting in all the conventional forms, and of every conceivable colour'. Even the station yard was not overlooked: *'[It was] environed with Venetian masts, each surmounted with a royal crown, and graced with a pendant banner. A kindred mast stands in the centre of the yard, and from this radii of streamers are carried to the other masts, the effect being that of a tremulous roof of kaleidoscopic tints.'*[2] Those of the excited assembly at the Harper Adams College exhibition, gathered to witness the King signing the visitors' book, would surely have cheered themselves with the thought that any war, in the unlikely event that such would happen, could not but benefit British agriculture. Farming was still deep in depression, but the coming of war was indeed to stimulate agriculture as production was accelerated and prices rocketed.

The Harper Adams display formed part of the Agricultural Education Exhibition and 'was one of the principal centres of attraction on all the days of the show. [It] being intended to give an intelligible idea of the value of the alliance between practice and science in all matters appertaining to the farm and the garden'.[2] The Journal of the Royal Agricultural Society was full of praise: *'[It] was a most instructive, large, and well-staged exhibit, and the expressed approval and delight of many visitors must be a source of satisfaction to the Harper Adams authorities.'*

The models showing the effects of manuring grassland for milk were especially praised by the Journal as 'speaking of the country generally, that the grassland is not managed with the same degree of skill and attention usually devoted to arable land'. The bacteriological work of the College was illustrated by 'a unique collection of typical cultures with photo-micrographs below of the various organisms producing disease. [...] Another section of this new departure in agricultural instruction shows growths obtained from the air of an ordinary cowshed; others depicting growths

4.1-4.2 The College's exhibition at the Royal Agricultural Society's Show at Shrewsbury, July 1914.

from milk drawn from a washed and unwashed udder of a cow'.[2] Mr Malthouse's display of 'bottled fruits of divers varieties made a tempting display' and naturally he was not to be found without numerous photographs of wart disease.[2] However, the presence of over 300 varieties of potatoes growing in pots might have gilded the lily. Outside there was a garden with 'a rustic pavilion [...] set in a border of flowering plants'.[2] Beside this there were fruit trees and a vegetable plot with recommended methods of cultivation.

Mr Rhodes, the Poultry Instructor, brought along his chickens and even Mrs Hedworth Foulkes contributed with 'a charming little model of the College farm buildings, constructed to scale'.[2] The College exhibition was large, well planned and impressive. The message to the visiting public was clearly one of professionalism and confidence.

The six or seven years leading up to the success at Shrewsbury had been a story of achievement. Most noticeably, growth was in the air. Student numbers were rising and the site was undergoing development. For the session of 1910-11 student numbers had reached an all-time high. The figure was 131, although included in this would have been short-course students. In the next few years up to the war, Principal's Reports show numbers dropped but only slightly. The decrease was as a result of increased fees for students resident beyond Shropshire and Staffordshire.[3]

The College Certificate was still the most popular qualification.[4] The majority of the students did not go in for public examinations but favoured a training for practical farming. Those who took the Diploma course still favoured the College Diploma over the NDA. In 1911, for example, just four students gained the NDA, one with honours.[5] It had been customary to combine teaching the Diploma and Certificate courses at Harper Adams. However, in 1911 [6] changes in regulations and syllabus for the NDA meant that staff felt it would be impossible in future to do so as certificates of exemption required by the National Board in certain science subjects were of a more advanced standard than it was possible to reach with students on the Certificate course. From 1911 there was thus to be separate treatment for the two courses. The College also lengthened the Diploma course from two to three years, although those students with sufficient science before enrolment could still take the course in two years.[7] Nationally, as regards the NDA, there was no change. All papers could be taken at one time, or alternatively papers could be taken in two groups over a two-year period.[8] At Harper Adams it had been found that many students had insufficient knowledge of science to cope with a two-year Diploma course and had often spent an initial year studying on the Certificate course.[7] Clearly a longer time spent studying was the requirement. Three years of study for the Diploma remained the norm at Harper Adams until the session 1942-43, when the wartime need for qualified students led to a reduction of the course to two years.

The increase in student numbers, although reflecting the ongoing success of the College, created significant problems. The first was simply one of space. The College had been planned for fifty resident students with perhaps a total of one hundred for teaching purposes.[9] The fabric of the building had undergone minor changes from the beginning. As early as the session 1903-04 the two large dormitories had been subdivided into cubicles with wooden partitions to allow greater privacy for clientele who were older and richer than planned.[10] In 1905 the need for study bedrooms had led to alterations on the top floor of Ancellor House. The resulting nine rooms were immediately occupied. Accommodation there remained basic, however, with oil light[3] and no bathroom until 1914. During the First World War it was home to women training in land work and after the war it became the hostel for women. Ruth Adam, training to work on the land, described the basic life at Ancellor during the early years of the war: *'Only one WC and that downstairs. The pipes kept freezing up! We couldn't get baths and only a very small fire in the dining room-sitting room. We were always frozen! Hardly any blankets and no heating in bedrooms. Only one oil lamp downstairs, candle in bedroom.'* [11]

In December 1909 a detached hostel was finished to meet the requirements for private rooms, providing twenty-four study bedrooms

and a sitting room for students and two private rooms for staff. Each corridor had baths and lavatories and the whole was lit by electricity. Bathing facilities were, however, to continue to be much in demand. The Engineer reported that there was a chronic lack of baths in the establishment, especially for use by football teams: *'I have repeatedly known three students in one bath at the same time.'* He added with conviction: *'This state of affairs should not be tolerated in an up-to-date institution.'*
Today the hostel is part of Jubilee Adams, having previously been known to generations of students as Old Hostel and then renamed Adams Hall on the opening of the Jubilee extension in 1951. Until the revision of the Foundation Scheme in 1909, the Principal was directly in charge of accommodation, handling it as a private venture. Responsibility henceforth passed to the Governors. In 1913 greater privacy was granted to the Principal when his house was completely separated from the College building save for a corridor at ground-floor level.[12]

As well as the need to relieve pressure on the residential arrangements there was the need to improve and extend recreational facilities. In 1905 the eastern wing of the College building was extended to the north, filling the void between the lecture room and the chemistry laboratory. Sliding doors formed the division between the lecture room and a new student common room, thus allowing a larger space for community activities.[13]
In 1906 a billiard table was installed in the new student common room. By 1908 the room could also boast a piano and a miniature rifle range.[14] On the first floor above the common room, a small number of student rooms were created, later to be called the Warren.[15]

There was similarly the need to improve teaching facilities, especially in the sciences. In the summer of 1909 a boot room was converted into a biology laboratory, while the former biology laboratory became a staff laboratory for the Chemistry Department. Even so, the Board of Agriculture's Inspector in 1913 considered the biology laboratory too small and unsuitable for demonstrations and practical work. The balance and preparation rooms of the chemistry laboratory were converted into a student laboratory, providing space for eighteen more students. The costs of further laboratory accommodation

4.3 A study bedroom, c. 1908.

4.4 Recreation room, c. 1908.

for investigation work and the installation of a gas plant were met by Frank Bibby of Hardwicke Grange, Hadnall. The very latest 'petrol air gas installation' was acquired as a means of generating gas suitable for laboratory purposes. Mellor Boyns, appointed Head of the Chemistry Department in 1965, remembered the laboratory's distinctive appearance:
'I'd never seen anything quite like it. There were large bore pipes, one and a half to two inches, going up from the bench and all the way across the lab in all directions. [...] It looked just like a ship's boiler room.'

Naturally there was the question of paying for the building programme. No money came from the Board of Agriculture and the cost, £4,000 alone to cover the hostel and staff houses, had to be met by the College. There was also the matter of financing any improvement to the facilities needed to meet the growing demands on the College, whether teaching, research, investigations or enquiries.
As Liddle wrote to the Board of Agriculture in 1909: *'The time now appears to have arrived when it is necessary to extend if the Institution is to carry out the work for which it is intended.'* The headache of funding had not evaporated. Furthermore, the grant from the Board of Agriculture was not raised until 1910, when it reached £600, increasing to £1,000 in 1911.

With increased numbers of students also came the need for more staff. In 1909 the staff numbered ten, although not all were full time nor totally employed in an educational role. The staff consisted of the Principal, a lecturer in mensuration, surveying, book-keeping & agricultural engineering, a lecturer in chemistry, a junior assistant lecturer in chemistry, a dairy instructor, an instructor in practical agriculture, a part-time lecturer in veterinary science, a part-time lecturer in botany and horticulture, a farm steward and a new poultry instructor. The need to employ more staff was not unique to Harper Adams. The Reay Report made the specific point that many institutions needed to secure more highly qualified staff and to refrain from relegating important subjects to juniors.[16] Furthermore, the salaries of these staff did not reflect the growing workload. However, there was no money to finance either new staff or to improve salaries. Thus, in 1909, a dearth of funding led to the situation where the instructor in the newly created Poultry Department was employed on the understanding that he would expect no wages until the College was in receipt of additional grants for the work. His successor fared little better in 1910, receiving board and residence and an honorarium. The Governors' Minutes show that salaries did rise, albeit slowly. Interestingly, the Reay Committee had suggested in 1907 that employment of women might be one solution to Harper Adams' problems concerning the cost of staff. Burne reported that a lady dairy instructor had been employed on a temporary basis prior to June 1907, but there were 'certain objections to having a lady instructor there always'. He also added that ladies seemed to cost as much as men. This unknown lady enters the annals of the College's history as the first woman to be employed on the academic staff. In 1911 Miss E. Harris was employed as Dairy Instructor, although she was only employed for the duration of F. Pendlebury's absence due to illness. It was to be the exceptional circumstances of the First World War which brought women onto the academic staff on a more established footing.

Another area where financial limitations undoubtedly were felt was that of the College Farm. Hedworth Foulkes declared that with one exception it was the only institution to own a farm attached to it.[17] The College was indeed fortunate to own land right on its doorstep, but the need for more land could not be ignored as student numbers and research activities grew. By 1905 the College's original 178 acres had been increased by fifty acres of adjacent land, rented from Mrs Minor, Thomas Harper Adams' niece, at thirty shillings an acre.[18] The rented land was of a different character to the Farm and declared of great value for teaching and experiments. In 1914 the College began renting Flatt Farm, sixty-six acres in a state of some neglect, for a rent of £122 a year. Other demands on the budget were evident too. The jewel in the College's crown was its farm, but the Ancellor homestead amid the Agricultural Depression had suffered considerable neglect. Clearly, the College had to possess a showpiece farm if it was to function in its educational role and to win respect from the farming community. An arduous scheme of tidying up, repairing fences and roads and drainage was embarked upon.

The Report of the Board of Agriculture's Inspection of June 1913 provides an excellent description of the Farm at that date. It is evident that funding was a problem, but the Report illustrates the success of the balancing act imposed upon the College by financial limitations. The Farm was maintained as a mixed venture for teaching purposes. The soil was described as good 'and when in condition is capable of fattening a bullock per acre'.[19] It was worked 'on the usual four-course rotation', which traditionally consisited of roots, barley, clover and wheat. The livestock included six cart-horses, including 'shire horses of considerable merit', and one pony. The Farm was strong in dairying and beef production, with twenty-one milking cows, two feeding cows, one pedigree shorthorn bull, twenty-three calves, twenty-three yearling bullocks and heifers, twenty two-year bullocks and sixteen three-year bullocks and heifers. The bull 'Pride of Rosehaugh' was newly purchased and available for use by local farmers. The dairy herd was considered 'exceptionally fine', after an extensive programme of improvement. The original herd of 1901 had been a veritable mixture with Shorthorns, Kerrys, Welsh and Red Polls. Most animals were now home-bred Shorthorns.[20] The beef cattle were 'a credit to any holding'. The land was considered unsuitable for a breeding flock of sheep. Thus no permanent flock was kept, but a flying flock was used for breeding for a season.[14] In 1913 there were fifty-one Welsh improved ewes, forty-eight lambs and one shearling Ryeland ram. The Inspector did comment that footrot was prevalent. Pigs consisted of two large white sows, twenty suckling pigs, twelve store pigs and eight feeding pigs. The pigs were described as 'useful commercial stock'.

4.5 The farm buildings from the north, c. 1905.

4.6 Hard-cheese-making dairy, c. 1905.

The farm buildings did not receive the degree of praise accorded the livestock in the Inspector's Report. The only major building to be erected since 1901 was a Dutch barn in 1909. It is clear the economic plight of farming and the financial constraints of the College came into play here. The dairy was heavily criticized by the Inspector. Comments were made upon the small rooms, the need for better lighting and ventilation and the cleaning arrangements. Along with a steam jet cleaning system a totally new building was recommended. It is doubtful whether Mrs Foulkes would have welcomed new premises, as the door frame of the original dairy office bore the marks of the heights of her three sons. The dairy scales also proved of use weighing young children.[21] The cowshed was criticized for poor ventilation and bad design. As the building opened onto the dung yard, the cows inevitably brought in manure and the smooth concrete floor with inadequate grips was too slippery for the animals. The implements fared little better under inspection, being described as meagre and outdated. The list comprised of just a binder and a mower and both were well worn. The Inspector rightly felt that this was not good enough for a centre advocating new inventions. During the First World War the College became a centre for machinery demonstrations. Machinery was especially important as labour was in short supply and expensive to hire. Thus more modern machinery was brought in. The earliest reference to a tractor at the College dates from June 1916. This was probably the College Titan remembered by Albert Rowley: *'It was a bit of a monster actually. [...] The noise was terrific. It was horrifying. Terrifying things they were.'* Albert Rowley was the son of a College farm labourer. He was employed as a pantry boy at the College in 1926 and later joined the teaching staff of the National Institute of Poultry Husbandry. He retired in 1977, having achieved fifty years of service to the College, for which he was honoured with the MBE.

Overall the Inspector felt the Farm fitted admirably the education needs of the students and commented favourably upon the students' involvement with all farm activities, experiencing every type of work on a weekly rota. Students would spend about twenty hours a term milking and about five to eight hours a week on general farm work. Ideally he felt that students should be less involved with the common practical tasks. R. C. Andrew, enrolling as a student in 1902, would surely have agreed with the mundane nature of much of the work experience: *'I remember [Sam Lovett] skilfully working in the narrowest possible trench in land draining all the fields, and that was the outstanding job throughout our Course.'* [22]

There were growing experimental demands on the College Farm. The period just prior to the outbreak of war saw the birth of significant projects. As always there was the issue of funding, but some money came from the Government's Development Fund. In 1910 the Government set up the Development Commission in an attempt to boost the economic development of the nation. Government grants were made available for non-profit-making schemes, primarily in agriculture. A sum of £3,000,000 was earmarked for a range of projects from the strengthening of agricultural education with the establishment of farm institutes and the creation of a series of specialist research stations and advisory centres to more humble aims, such as the growing of new crops and improvement in livestock. The College received a number of grants from the Development Fund. As early as 1912, for example, Development money paid for a silo to be erected at Ancellor Farm for the Soiling Farm experiment.[23]

Money also came in from Industry, as in the case of sugar beet. The variety trials were as important as ever. In 1910, if not earlier, sugar beet was being grown at the College.[24] The Rt Hon. Sir Beville Stanier, a College Governor and already associated with the College in its investigations into wart disease of potatoes, was an early believer in growing sugar beet in this country. He grew the crop on his estate at Peplow. He was later to become involved with the building of the beet factory at Kelham, which was begun around 1913. Building was interrupted by the war and the factory was not completed until 1921.[25] The 1911 sugar beet trials at the College took place at the request of the Board of Agriculture.

There was a growing conviction that high quality roots could indeed be grown in this country.[26] The following year, at the request of the National Sugar Beet Association, and with grant aid from them, the comparative fodder values of dried beet imported from Germany and of home-grown mangels were investigated by Harper Adams.[27] The beet proved the better fodder. An experiment, at the request of the Board, was also undertaken to dry beet leaves. Considerable trouble was taken to obtain the dried leaves by means of a malthouse and kilns. Eighteen tons were reduced to just two after a process of constantly moving the wet matter. Then, after much cost and time, the cattle refused to eat the dried leaves and the lot had to go to litter. The College concluded that beet was not a profitable crop from the feeder's point of view.

Perhaps the most significant experiment to start at Harper Adams during the period was that of the Soiling Farm. Soiling farms had once been popular in Britain and were still to be seen on the Continent. The topic was of considerable interest in the United States.[28] However, little research had been done in this country on maintaining stock entirely on arable land, growing a succession of forage crops throughout the year to feed dairy cattle indoors even during the summer months. The spread of smallholdings and developments in intensive cultivation and crop varieties made it an area ripe for research. James Carter Brown, appointed Vice Principal in 1907, made the subject very much his own. His particular forte was the breeding of new varieties of fodder crops.[23] Fodder trials began in a small way in 1909 and experimentation with crops and a small number of cows took place in the years leading up to war. The experiment gained momentum in the season 1916-17 with the growing of a chosen selection of crops on a tested rotation and the erection of a wooden cowshed at the Ancellor. With the war had come the need to intensify production and, as at least three times as much stock could be maintained by soiling, interest in the experiment had grown. Similarly wartime emphasis on cereal crops and the ploughing up of pasture had given the work added weight.[28] After the war the system was to attract attention as a convenient way of setting up discharged soldiers and women on a small acreage of land.[11]

By 1919 the experiment had proved so successful that the Board of Agriculture accepted the College's scheme to establish the Soiling Farm as an independent demonstration unit. It became the first of a series of demonstration centres for milk production [29] and farmers were soon flocking from all over the country to see if the scheme could be adapted to their areas. To finance the project a significant capital grant of £4,000 was awarded by the Board of Agriculture in 1919. Delays in building a new cowshed due to high labour costs following the war meant that expansion of the herd beyond twelve cows was slow.[28] A worsening internal situation at the College resulted in the scheme never being fully implemented. With Brown's enforced departure as Vice Principal in 1922, one of the most important experiments in the early history of the College died a slow death.[2]

Against a background of growth as a college and achievement on the research front, developments in agricultural education were to propel Harper Adams onto a much wider stage in the years prior to the First World War. The Reay Report of 1908 allocated the College a very clear place in the structure of agricultural education. The Report spoke favourably of the great progress in agricultural education of the last twenty years and listed the twenty institutions in England and Wales in receipt of grants from the Board of Agriculture for 1906-07. The list ranged from universities with degree awarding status, such as the University of Leeds receiving £800, to the farm schools offering short courses, such as the Cumberland and Westmorland School, now Newton Rigg,

4.7 The Soiling Farm cowshed, c. 1919.

4.8 Interior of the Soiling Farm cowshed, showing stall divisions and open boarding of roof, c. 1919.

receiving £100. Harper Adams, teaching the NDA, was centrally placed in the list and in receipt of £400. The Reay Report went on to make recommendations for the future in the areas of education and advisory work. As Government plans developed, the question arose as to what Harper Adams' place in the evolving hierarchy would be.

The Board of Agriculture's scheme to promote agricultural education was ambitious and all embracing. It involved the creation of twelve provinces in England and Wales and encompassed education, research and advisory work. Within the provinces there would be chains running from County Council instructors and organizers through the farm institutes and colleges and thence to the universities.[31] Locally, a province was formed with Shropshire, Staffordshire and Warwickshire. The new partnership with Warwickshire was officially blessed in the revised Foundation Scheme of 1915. Warwickshire was granted two representatives on the Governing Body, as was the case with Shrophire and Staffordshire. Each county was also allowed, in addition, three Governors providing the County Councils made individual grants of £350 a year. Warwickshire funding in the form of a grant of £150 started in 1916.

Linked with the Government's thinking on the creation of provinces was the establishment of a network of farm institutes. At the time of the Reay Report just four institutions could be classed as farm institutes, in that they offered practical courses suitable for intending farmers with little secondary education. Among them were the forerunners of Writtle, Myerscough and Sparsholt.[32] The Governors considered Harper Adams' provision of short courses was adequate for Shropshire and Staffordshire and any plans to form a separate farm institute for the West Midland Province were therefore questionable.[31] A farm institute should be created in conjunction with the existing college. Fears over the creation of farm institutes were in fact premature. The First World War halted plans for a national system of farm institutes, only two being opened before the war, in 1913.[33] The Staffordshire Farm Institute at Rodbaston did not open until 1919. By that time Harper Adams was inundated with students and Rodbaston appeared no threat.

In matters of research the Board of Agriculture planned for each province to have a specialist research centre. Locally, Birmingham University was chosen with a specialization in zoology.[31] With the encouragement of research it was vital to set up a system to get results to farmers. A recommendation of the Development Commissioners dating from September 1911 led to the creation of a national advisory service.[33] Two advisers were to be appointed for each province[31] and provincial advisory councils were to be established as co-ordinating bodies.[34] The advisers would be attached to colleges, in the case of the West Midlands to Harper Adams, the College's already strong links with local farmers and the County Councils working in its favour.[31] The Development Fund would finance the provision of these specialist advisers, the institution providing the working accommodation. The colleges would, subject to the approval of the Board of Agriculture, select these advisers and be responsible for their work.[33] Their duties would cover everything from giving advice to local investigations and research and some element of teaching.

On the suggestion of the Board of Agriculture, the College and Birmingham University held a meeting in December 1911. The College welcomed the idea of assistance at a time when the number of enquiries was ever growing. The Governors declared that they would be glad to confer with Birmingham University *'upon the question of association in providing for the supply of technical advice to farmers, and for the investigation of local Agricultural problems'*. At the December Meeting it was

agreed that a joint board would be formed with the task of appointing expert advisers and of generally managing and overseeing the financial arrangements for the work.
As a consequence the Provincial Advisory Council for the West Midlands was formed at a meeting held at Birmingham University on 1 August 1912. It was composed of representatives from Harper Adams, Birmingham University and the three counties of the Province.[35] On 22 November 1912 the first meeting of the West Midland Provincial Advisory Council took place at the Shirehall in Shrewsbury.[36] Significantly, this was the first ever meeting to be held by any Provincial Advisory Council and as such deserves recognition of the initiative shown and the speed of response to the Board of Agriculture's directive.[37] Harper Adams was about to enter a very important phase in its history.

Early in 1913 the Government's plans were finalized with finances from the Development Fund.[38] Of the £50,000 allocated to research, £12,000 was earmarked for technical advice to farmers and was thus to be divided amongst the twelve advisory centres, Harper Adams being one. The Board of Agriculture advocated the appointment of a livestock officer for each of the ten English provinces with the task of promoting and administering the livestock improvement schemes financed by the Development Fund. The College's first advisory officer was appointed in December 1913 after consultation between the College and the Advisory Council. Conwy Bell, former Farm Steward, was appointed Livestock Officer with an annual salary of £400.
He was to work under the supervision of the Principal, visiting him at least once a fortnight. In 1918, as part of his work remit, Conwy Bell was instrumental in establishing the first milk recording society in the Province, that for the Newport Show area representing thirty-two herds and over 900 cows. On a lighter note, Conwy Bell's chief claim to fame seems to have been his inability to keep his expenses within prescribed limits. The Board of Agriculture declared in exasperation: *'No other officer exceeded the grant with expenses by such an amount.'*

No discussion of agricultural education during this period would be complete without consideration of the highly important developments which were taking place in poultry husbandry.

Significantly, these developments were centred on Harper Adams and were in time to win the College national recognition. Poultry husbandry had been taught from the beginning of the College, the poultry being kept in movable houses about the fields.[9] It was a subject within agricultural courses and it was also taught in extra-mural classes. However, poultry husbandry did not possess a high profile at the College nor on the national farming scene at this time. It was considered very much a marginal enterprise, associated essentially with the farmer's wife and the sale of eggs. Furthermore, in the scheme of things chickens fended for themselves.[33] Douglas Scotland, as a student in these early years, remembered: *'We had lectures on Poultry Keeping by C. D. Stewart, and our practical experience was derived from a few fowl houses, an incubator and brooder, and a Heath Robinson contraption for stuffing birds by forceful feeding - a disgusting affair. The stock was in no way outstanding. I remember when on Poultry Practical work we had the cleaning out of hen houses, the feeding of birds and the plucking and dressing of them for market.'* [39]

The earliest teaching of poultry husbandry began with the County Councils in the 1890s. More advanced courses for teachers and managers started with Edward Brown at Reading in 1894.[40] Brown was to become the kingpin of the poultry world. It was Brown, influenced by advances in the United States, who inspired the creation of the National Poultry Council in 1920 to oversee poultry research and education.[41] As the Poultry Industry rapidly developed there was a growing demand for qualified managers and poultrymen. Hedworth Foulkes, with his Reading background, was more prepared than most to recognize the opportunities which poultry husbandry could present a teaching establishment. Showing considerable entrepreneurial flair, he was personally behind the setting up of a specialist poultry department at the College in the autumn of 1909. This can be viewed as the first step towards the establishment of the National Institute of Poultry Husbandry (NIPH) at the College in 1926. Crowther, Hedworth Foulkes's successor as Principal, especially recognized 'the skill and tenacity with which he secured the College a position of leadership in poultry', praising the way he concentrated the slender resources at his disposal for developing 'sidelines' upon one branch.[13] By 1913 the Report of the Inspector for the Board of Agriculture could talk of an incubator room, store and hatching room, a fattening shed for table birds, six breeding pens and open-fronted houses and five colony houses scattered about the fields for laying hens. The Poultry Department was situated to the right-hand side of Ancellor Drive, approaching from the barn area.[39]

To Hedworth Foulkes is also due the credit for starting the egg laying trials at the College. Crowther did not underestimate their importance to the College: *'It is no exaggeration to say that no development has made so potent a mark upon the history and fortunes of the College as the Laying Trials. They have given us a public interested in our work far wider than any other institution can claim, and in this way have reacted favourably also upon our major activities in the sphere of agricultural education.'* [42] Within a few short years of the establishment of the Poultry Department, Harper Adams had moved on to the national poultry husbandry stage with the opening of the first trials in October 1912.[12] With the establishment of these national trials it could be claimed the science of utility poultry keeping began.[43] For the first time in Britain official records were available on the output of eggs per bird from all the breeds represented at these trials. Earlier competitions had been held in various locations since 1897 and the first twelve months laying competition had been held in 1907-08 under the management of the Secretary of the Utility Poultry Club at his farm in Rayne, Essex.[12] The trials at Harper Adams were the first to be held on a large scale, on a regular basis and with the benefit of continuity of site.

4.9 Students at work in the poultry pens, c. 1910.

The trials were initially run in association with the Utility Poultry Club, but from 1916 the College took over sole management of the trials.[41] They ran for twelve months at a time and started with one hundred pens, each pen having a house for six birds and trap nests. Poultry keepers from all over the country paid a fee and sent a pen of birds. The four-and-a-half-acre site was adjacent to a newly planted orchard, opposite the Main Building across the Caynton Road.[44] The considerable outlay was assisted by £500 from the Development Fund, but even then the bulk of the work erecting the plant had to be undertaken by the College for cheapness.[45] Henceforth the Development Fund provided annual maintenance grants, but the accounts had to be fine-tuned. By 1917 the College held the record in this country for the largest number of birds at one place under trial at the same time.[46] Never before had there been 1,200 birds together under trial. The Women's Prospectus

4.10 Early view of egg laying trials.

of 1918 claimed: *'These trials are the largest of their kind in the world.'* On the opening of the NIPH the trials retained their association with the College. The trials were held until 1942, when difficulties obtaining feed forced their closure. Each month the results of the laying trials were published, with the vast volume of clerical work falling to the College Secretary, L. Willoughby.[44] In 1915 over 2,000 reports were sent out by the College in addition to the 2,000 issued to Utility Poultry Club members. By 1916 demand for results was so high that a monthly report was issued as 'Poultry Competition Monthly'. In 1918 the Utility Poultry Journal superseded these monthly reports of the trials. Within a year 5,000 copies were sent out annually. It was especially popular with the military overseas. As before the reports were issued free of charge.

As interest in the egg laying trials grew, so more and more people sought to visit the College and view the trials for themselves. This led to the natural development of a Poultry Day, the first being held during August 1917, when poultry breeders from all over the country visited the trials.[47] This was to go down in history as the first poultry conference to be held at Harper Adams. At the Second Conference held in August 1918 a committee was appointed to act in liaison with the College to draw up a scheme to set the standard of points for the different utility breeds.
This decision did much to improve the standard of utility poultry. By 1919 the Conference had expanded to include outside speakers, demonstrations and an exhibition of stock birds. Around 250 visitors attended. With considerable pride Hedworth Foulkes wrote in the Principal's Report of 1919: *'The Poultry Department bids fair to become the mecca of the industry.'*

As Britain and the World stood poised to tumble into the anarchy of the First World War in the summer of 1914, Harper Adams College was similarly facing a defining moment in its history. Shortly, the College was to confront not the restrictions and limitations that a war can bring but the very reverse. Demands on the College grew phenomenally, adding to the already substantial strain on resources. Great things were accomplished at Harper Adams during the war years, but the very achievement of being one of the few agricultural institutions to remain open for the duration of the war came with a price. Within four years the College was fighting its own battle for survival.

The First College Regulations [a]

14.11 Group of students c. 1908. Front row from left: R. A. Dalley (second): Private Collection.

-Students are required to attend all lectures and demonstrations regularly and punctually, and permission must be obtained from the Principal before students may absent themselves.

-Students will be required to provide themselves with note and account books and drawing instruments. Scientific apparatus is provided, but any breakages must be paid for.

-Students must not give orders to or interfere in any way with the farm workmen.

-No student may carry or use firearms on the farm, and sporting, snaring, trapping game, or taking eggs of wild birds, is strictly forbidden.

-Students are forbidden to take any fruit or produce from the farm and gardens.

-Smoking is forbidden in the farm buildings, workshops, and stackyard.

-The continued breaking of College regulations will render a student liable to be sent down and to forfeit his fees. The Principal is authorised to remove students at any time who in his opinion are not endeavouring to profit by the instruction given, or whose influence on other students is injurious.

- Each student must be provided with the usual outfit, consisting of two pairs sheets, two pillow cases, towels, and serviettes.

- Meals will only be served in the Dining Hall at stated hours. [...] Students must be punctual at meals, a fine is imposed when meals have to be kept waiting.

- Doors are locked during the autumn and spring terms at 9.30 p.m., and in summer term at 10, and no student may be out after these hours without leave previously obtained from the Principal. The electric light is switched off at 10.30 p.m.

- Students are not allowed in their bedrooms between breakfast and one o'clock.

- Boots must be changed and left in the boot room, and students must confine themselves to one pair a day. Extra pairs will be charged for.

- Attendance at Church or Chapel is required, at least once a Sunday, unless an exemption is requested by the parent or guardian at the commencement of the course of study. All students, unless so exempted, are required to attend Morning Prayers immediately before breakfast.

And the Breaking of them:

Extracts from the Diary of Wilfred Mercer, student 1905-07.

- Saturday Dec. 9th 1905 - The most unfortunate thing of the whole day was the sacking of Watson.
He came out of exams v. early for some reason or other.
The Prinny saw him & ordered him in again & he refused so the Prinny told him he must leave the premises.

- Sunday Dec. 10th 1905 - At about 10 o'clock or so a paper was brought round for us to read & sign if we wanted. It was really a short letter to the Prinny telling him we should all stop working till he fetched back Watson. Everybody signed it. [No work was done on Monday and that evening Watson returned and the students were informed that work would continue as before.]

- Wednesday Dec. 13th 1905 - At 9 a.m. the Prinny put up a notice of lectures for this afternoon. [Normally, Wednesday afternoons were considered a free period.] [...] He said he would speak to the whole college at 2. Accordingly we all met in the Lecture hall at 2.0 & he talked all round the subject for about 1/4 hour. Everybody had been saying beforehand what ought to be said but no-one scarcely said anything to the point. He went out of the room leaving us in a worse and more unsettled state than before. He said the lectures this afternoon were not for a punishment, but to enable us to make up for the time we lost on Monday. [No one attended the lectures and that evening a list of fines was posted. Every student was to pay 1s by one o'clock the next day. The fines were not paid.]

- Friday Dec. 15th 1905 - [The ringleaders were sent down, without 'time to change even'. The Principal called for a meeting of the entire College.] He said he thought he had got the ringleaders & said he wished the matter to drop entirely & that the fines would be taken down & the chaps who had been sent down would come back next term if there was no more bother.
Most chaps were satisfied with this.

Chapter Five —

World War One 1914-18

War was declared on 4 August 1914. Early in 1916 the Governors asked Liddle as Clerk to investigate 'the question of insuring the College buildings against damage from air raids'. On 31 January 1916 the largest number of zeppelins yet seen over Britain, a total of seven, had taken part in a bombing raid, reaching as far inland as Staffordshire. In all fifty-four people were killed and sixty-seven injured that night.[1] It was not a zeppelin which was to threaten the College's survival but the dramatic fall in student numbers. For the session 1914-15 long-course students totalled just forty-one as enrolments declined and those in attendance left to enlist. Twenty-nine left to enlist during the 1914-15 session. Staff also left to join the Forces and were replaced by a succession of temporary appointments.

The munitions factories were especially enthusiastic about the College chemists. The number of clerical, estate and farm workers was similarly depleted, although the farm benefited from the employment of those over military age, internees, labour battalions and, towards the end of the war, military labour. Nevertheless, by autumn 1916 the threat of closure hung over the College. Student numbers on long courses for the previous session had reached just thirty-nine. Other institutions, such as the Royal Agricultural College, had been forced to close.

The reduction in staff numbers was exacerbated by the growing workload as a result of the war and the success of the College on the poultry husbandry scene. As early as the summer term of 1915 there was a crisis as a reduced clerical staff faced growing administration stemming from the popularity of the reports of the poultry laying trials. The illness of the Principal, the lynch pin of the College, caused added complications that term. Whether the arrival of the telephone in the spring of 1914 had helped or hindered the situation it is not known.

5.1 1916-17 staff and students.

Reduced student numbers meant reduced fees. Financial disaster threatened and with it the probability of closure. The initial loss in that first session of the war was over £2,000 in fees, board and residence. Economies were made with reductions in servants and 'working staff'. Even the positions of the remaining members of academic staff were under threat. Expenditure on repairs, experiments and outside demonstrations was kept to a minimum. Slowly the threat of financial ruin began to abate. In 1915 and 1916 special grants amounting to £1,750 were made through the Board of Agriculture from the Treasury on the recommendation of the Advisory Committee on University Grants. However, balancing the books must have seemed an impossible task as food, material and labour prices were all rising dramatically. There was little choice but to increase fees, although this was delayed as long as possible. The residential fees rose by two pounds and five shillings a term in 1918. However, an increase in tuition fees was avoided until 1919. That year tuition fees rose by £1 10s. a term to total £12 10s. for residents of Shropshire and Staffordshire, while others paid £22 10s. The benefits of an increase were not insignificant as from 1917 the number of students taking the long courses rose. In the session 1917-18 figures for long-course students reached seventy-three. With this the financial position improved.

Reciting figures for long courses does not give a true picture of education at Harper Adams during the First World War. In the session 1916-17 a startling 131 students were enrolled on short courses. Of great importance to the history of the College was the fact that 128 of these students were women. From the opening of the College, acceptance of women had not been seen as a good idea.[2] In fairness, it was still extremely rare to find women at agricultural colleges [3] and the College building was ill-equipped to mix the sexes. There was, after all, provision made for women at institutions such as Studley College, where women could study horticulture and dairying, and Swanley Horticultural College.[4] The war was, however, to bring dramatic changes to the way of thinking at Harper Adams.

At the outbreak of war Britain imported as much as half of its food requirements, but it was a mere handful of people, such as The Earl of Selborne, President of the Board of Agriculture, who appreciated the need to increase food production.[3] Some progress was made, but it was not until 1917 that major steps, such as the ploughing up campaign and the Corn Production Act with the introduction of guaranteed minimum prices and deficiency payments for some cereals, were taken. This period was very much a watershed in the history of the war. Among other things, the clear failure of the Allied offensive on the Somme in 1916, the appointment of Lloyd George as Prime Minister in December 1916, the entry of the United States into the war in April 1917 and the increasing impact of the German U-boat campaign all led to a new direction in thinking with greater organization and the subordination of everything to the needs of war. Home production of food received considerable impetus. Harper Adams, however, had been involved in the movement from the very early days of the war. One of the earliest problems to be confronted was the shortage of farm labour. In the Easter vacation of 1915, at the request of The Earl of Selborne, two short courses were arranged to train sixty women in the lighter branches of agriculture. The Governors had previously made it clear that such courses would be impossible in term time due to accommodation difficulties. Other institutions, such as Sparsholt, Swanley and the Midland Agricultural and Dairy College, had been asked to help, but Harper Adams acted quickly, thus becoming the first institution in the country to run such courses for women.[5]

The free courses were run in association with the Boards of Trade and Agriculture.[6] They ran for a fortnight each and covered everything,

5.2 Group of ladies training at the College for wartime farm work, 1915. Back row from left: Ida Ellis (first): Private Collection.

albeit briefly, from dairying to dock pulling. Women were invited to attend for interview at labour exchanges and interest was high with one hundred applying for the first thirty places.[7] At the end of the courses, with the exception of one or two, all students found employment, earning between ten and sixteen shillings a week.[8] To recruit, train and find the women work was no mean achievement and Hedworth Foulkes took considerable trouble with reams of correspondence as a consequence.[7] There was strong prejudice to overcome as few believed women could be good at farm work. There was also, as always, the obstacle of chaotic administration with layers of committees adding to delay and frustration. It was to be some time before order came out of chaos, for there was a myriad of voluntary bodies seeking to recruit women for the land, such as the Women's Legion and the Women's Farm and Garden Union.[3] In 1915 the Women's Farm Labour Committees appeared in the counties and became the recognized channels for recruitment.[9]
Some form of centralization arrived with the creation of the Women's Branch of the Food Production Department of the Board of Agriculture, but this was not until early 1917.

It was in March 1917 that the first recruits to a mobile force of women farm workers were enrolled. With the disastrous harvest of 1916 and the success of the U-boat campaign in sinking Allied shipping, the need to harness available labour became the more urgent. The idea was born of the Women's Institute working under the wing of the Women's Branch of the Food Production Department.[3] It was funded by the Government, administered by the Women's Branch of the Food Production Department and named the Women's Land Army. Harper Adams continued to play its part. During the session 1917-18 ninety-two members of the Land Army trained at the College. One local farmer, Edward Watson Jones, employed three of these women trained at the College. Initially he was aware of their limitations, but they soon settled to the work:

Neither the soldiers or the girls could replace experienced labour - man for man - but we were pleased to have them and put up with the fact that their inexperience was costly and it took two or three to do what used to be expected from

June 5, 1915. LADY'S PICTORIAL. 857

MAKING GIRL FARMERS.

THE BOARD OF TRADE'S EXPERIMENT IN THE MIDLANDS.

THE sound of horses' hoofs and the scrunching of cartwheels on a sandy road caught my ear as I stood hesitating at the cross-roads uncertain in which direction to go. I looked up to see a cart loaded high with manure coming towards me, drawn by a powerful cart-horse. Instead of a labourer, in clothes stained the colour of the soil, a tall girl walked at the horse's head. She was dressed in a long blue print overall which reached to the top of her heavy boots, and she walked with the free, swinging step of a typical English girl of the upper class.

FEEDING A CALF.

I was seeking the land of women farmers, so what need to inquire further? I followed the cart and its unconcerned and businesslike driver, till it passed by a piece of orchard ground where a number of girls in similar overalls were digging and planting. And across the road I saw a tall red brick building standing in its own grounds. It was the object of my visit, the Harper Adams Agricultural College, of Newport, a town in Salop, the College which, at the request of the Board of Agriculture, was providing short courses in farm-work for women that they might supply the labour that farmers have been unable to obtain since the war.

There were two courses of a fortnight's duration, and thirty women and girls were attending each. Board, lodging, and training were being given free to the students, all of whom had given an undertaking to engage in farm-work after their course of instruction was over, if they received suitable offers of employment.

The first group of girls had been at their work nearly a week when I visited the College. On my arrival I was put in charge of the farm-steward, who showed me the girls at work, and explained the method of instruction which was adopted.

The girls were divided into four groups, each group devoting three or four days in turn to the different kinds of work—farm, dairy, and poultry. Each of the departments is under an expert, a member of the staff, so that the girls have the best of teaching, and every opportunity of learning the proper way to do the work, which, as is briefly stated in the preliminary pamphlet, "can only be elementary."

The grounds of Harper Adams College occupy over 300 acres of cultivated and pasture land, where sheep, poultry, dairy, and fruit farming are all carried on.

As we left the College we passed a couple of girls bringing in the sheep from the pasture with the shepherd and his dog. Another group were digging, and away over the brow of the hill I could see them among the little hen-coops that mark the boundary of the poultry department. Everywhere dotted about the grounds a glimpse of coloured overalls and caps or sun-bonnets showed where the girl-farmers were at work.

"INSTEAD OF A LABOURER, IN CLOTHES STAINED THE COLOUR OF THE SOIL A TALL GIRL WALKED AT THE HORSE'S HEAD. . . ."

"You can see that all the work the girls do is regular routine farm-work. None of it is made for them," said the farm-steward, as in the course of our tour of inspection we stopped to look through the door of some outhouses, and he pointed to a heap of turnips and chaff cut

Our pictures show: (1) A scene in the Poultry Department. (2) Digging and hoeing are part of the day's routine. (3) Bringing in the sheep and lambs. (4) Feeding the pigs.

GIRL FARMERS AT THE HARPER ADAMS AGRICULTURAL COLLEGE.

5.3 Feature in the Ladies Pictorial of 5 June 1915, on training women at the College for farm work.

one previously. [...] The land girls got more experienced during the summer and Miss Fenn tried her hand at riding the tractor plough and soon became expert. These three girls [Marjorie Fenn and Madge and Winnie Wright] did good work and were very useful.[10]

Unfortunately, reports that the women were talking to their fellow workers, German prisoners of war, led to their being removed at once.[11]

Ultimately, the success resulting from the small beginnings at Harper Adams was down to the women themselves. The course was elementary and indeed many women, especially on the very first courses, were from urban centres and earning a living for the first time.[7] Yet, by 1916 farmers employing women were specifying that they wanted those trained at Harper Adams. Demand was greater than supply.[12] In their praise, the Farm Steward, Andrews, declared: *'They pick up things quicker than male students.'* [7] While Willoughby, the College Secretary, said: *'I think we can safely say these girls mean business.'* [7] In early April 1915 a reporter from Picture World had visited the first group at the College. He painted a memorable picture: *'Everywhere dotted about the grounds coloured overalls and caps or sun-bonnets showed where the girl farmers were at work.'* He interviewed an instructor who spoke of their love of hard work: *'We have a number of girls here who belong to the educated classes, and I anticipated some difficulty with them over the more unpleasant tasks, but I found that they seemed keener on those than on the easier and pleasanter ones. In fact the more strenuous the work the better they seem to like it.'* They evidently took their leisure seriously too, with a fancy dress ball one evening attended by Britannia, Farmer Giles and Mrs Hedworth Foulkes as Serbia.[13]

In 1916 the courses were repeated; this time each ran for a month. Recruits were now being sent by the Shropshire War Agricultural Committee and a number of women enrolled and paid their own fees. In order to provide for a total of 108 women students, courses henceforth ran into term time and accommodation was made available for thirty at a time at Ancellor House. That year, in July, requests from women for longer courses resulted in the Governors deciding 'to open the College to women on the same conditions as now provided for men the fees to be charged as in the case of male students'. The first five women enrolled on long courses covering all branches of agriculture that autumn. In 1919 Shumla Ridout, having passed in all but one subject, was listed among those students at Harper Adams who had obtained the NDA.[14]

Important demonstrations of women workers on the land took place at the College in July and August 1916 to coincide with the regular visits of farmers. By May 1916 the Government had realized reliance on volunteers for the Forces was insufficient and, as a result, conscription for all men between the ages of eighteen and forty-one was introduced. It was now increasingly important to get women working on the land and to overcome the farming community's suspicions of them. Thus the Government directed those organisations involved with women farm workers, such as the Women's War Agricultural Committees, to arrange demonstrations.[3] Women trained at the College and subsequently employed were seen doing all manner of work, such as working horses, spreading manure, hand-hoeing and milking cows.[6] A reporter from Country Life, while praising a woman's 'skilful handling of the plough' on hard dry ground, was even more impressed by the sight of a 'milking machine, which was entirely handled from starting the engine to washing the apparatus by the girl farm hands'.

In October 1917 Hedworth Foulkes discussed with the Women's Branch of the Food Production Department the possibility of running classes for women in tractor driving. The start of the Government's ploughing up campaign, an attempt to boost arable acreage by cultivating grassland, and the severe shortage of horses made the need for tractor drivers all the greater.[3] For the session 1917-18, fifty-five women trained as tractor drivers. Instruction began in November and was provided by two unnamed ladies, plus a Mr Breeze from Newport and the ever talented William Neate, the College Engineer.

The original course lasted six weeks and some of the first to qualify were sent to plough up the race course at Old Trafford.[15] Albert Rowley had lasting memories of the lady tractor drivers: *'The girls weren't put on one side to say "Christ, don't [involve] me". They were there part of it. [...] I looked on in awe. I couldn't believe they were allowed to drive these massive things and the noise was terrific. It was bloody terrifying it was. [...] You'd think they'd been doing it all their lives which was amazing. [...] They cut a good run.'*

There were also plans in 1915 for Harper Adams to undertake instruction of disabled soldiers and sailors. This was part of the initiative of The Earl of Selborne to promote the settlement of such upon the land. An initial scheme sought to train fifty such servicemen with the object of their gaining agricultural employment and of a selected capable minority receiving smallholdings.[16] Harper Adams and Holmes Chapel were chosen as the training centres. The costs would be met by the Board of Agriculture. Holmes Chapel was to close as a result of the war and there is no indication that this particular course ever took place at Harper Adams. By 1916 the instruction of women was well underway at the College and, in all probability, this worked against the influx of wounded servicemen in any numbers. In December 1916, however, there is mention of the Governors agreeing to reduce fees in the case of wounded soldiers from Shropshire and Staffordshire. Late in 1917 the War Pensions Committee enquired about the possibility of sending wounded soldiers for special courses, notably in poultry husbandry, an occupation considered eminently suitable. This time it is known that training did take place. Instruction was over three terms, but sadly, the nature of their injuries excluded many at the interview stage.

The College also involved itself with men in active service. In 1918 Wilson gave a series of lectures on the diseases of animals to the officers and men of the New Zealand Rifle Brigade at Brocton Camp. Veterinary lectures were given at the camp at Rugeley. Hedworth Foulkes proudly and no doubt accurately related that 'the attendance averaged 308.' [2] Poultry keeping talks were also given at war hospitals.

Harper Adams' war work was not restricted to the training of women and servicemen for farm work. Indeed, its range of work was vast. In the early stages of the war Mr Roebuck of the Biology Department undertook the gathering and drying of wild medicinal plants in view of the serious shortage of medicines. Prior to the war Germany had supplied many drugs.[17] Roebuck concluded that local organisations, including school-children and boy scouts, could probably collect the nation's needs. More conventional crops received due encouragement also. Within weeks of the outbreak of war, for example, 4,000 bulletins were written and sent out to the Three Counties of the Province on increased sowing of wheat.[18] Again, as the Provincial Centre, Harper Adams was involved with developing schemes for the reclamation of wasteland. Plans in association with Shropshire County Council to use the unemployed to cultivate the pit mounds of Donnington Wood collapsed as, it was commented, in a time of labour shortage there were no unemployed.

Hedworth Foulkes was personally involved with a committee to organize vegetable drying and fruit preserving in the Province. Needless to say this was far from the Principal's only committee work in the war effort.

In December 1917 the Governors had to refuse the Board of Agriculture's request that he be transferred to London in order to assist in the organisation of food stuffs. As part of the fruit preservation scheme two experimental factories were opened quickly in Warwickshire. In 1916 the College was asked to help with the saving of that year's large local damson crop. The College provided managerial and advisory staff free of charge. Primarily, the scheme was the responsibility of a Governor, Colonel Harold Sykes of Longford Hall, and the Department of Food Production paid all the costs.[19] Practically the whole of the work, some of it of a very heavy nature, was undertaken by around 120 women.[20] The work took place at Springfields Brewery in Newport, close to rail links. Steam was utilized from the Brewery to cook the fruit.[21] Pulping began in September 1916 and a year later production ran to around eighteen tons of pulped fruit a day. The pulp went to make jam for the Army.

Forestry came to the College as a result of war. With the war the need for wood increased. Rylance, the Assistant Horticultural Instructor, was engaged from Easter 1916 in the planting of conifer seedlings at the request of the Board of Agriculture. By the end of September 4,500,000 conifers had been planted. The Board of Agriculture stated: *'that the College probably possessed one of the largest and cleanest plantations of Conifer seedlings in the country this season.'* College trees were planted in Windsor Great Park. Forestry continued for a time after the war, but it never obtained a high profile on the academic side.

The College's all-important advisory work continued and increased throughout the war. The work of the newly appointed Livestock Officer of course raised the College's profile, but the routine work of the pre-war years continued. Especially important was the analysis of feeding stuffs as the dearth during the war meant that virtually anything was sold, its price very often being in excess of its real value.[20] Locally, crushed damson stones seemed a popular additive. Increasingly, much of the College's advisory work was made available to the public not through bulletins and reports but through daily and technical newspapers. The costly nature of printing at this time restricted the number of bulletins and reports that could be published. Sir Daniel Hall, Secretary of the Board of Agriculture, was full of praise for the College at the close of war:

> *There had been no occasion when the value of Colleges had been more thoroughly proved than during the present war. There were many things which came within the purview of farmers, things which might be of considerable assistance when they encountered difficulties, and no institution had done more for the agriculturalist than had been undertaken and carried out by the Harper Adams College.*

Numbers of visitors increased substantially during the war years. Many were attracted by the poultry laying trials, the importance of poultry increasing with the outbreak of war. Indeed, the dramatic increase in the popularity of poultry had led to significant expansion within the Poultry Department. Another facet of the work of the Poultry Department which grew at this time was the sale of day-old chicks. The scheme had started in the autumn of 1916, with Harper Adams as the centre for the West Midland Province, and had been initiated by the Board of Agriculture. By this means cottagers and smallholders could obtain poultry of improved strains at low prices, initially at six shillings a dozen. The demand far exceeded supply, for example in 1917, 2,598 chicks were ordered but only 1,326 chicks were sent. Added to these demands were orders to supply eggs for incubation, a total of 10,000 eggs being sent out. By the session of 1918-19 the demand for chicks and eggs was constant. At the same time the poultry trials and teaching needs were expanding and experiments were underway on alternative poultry feeding stuffs. Bracken fronds were tried successfully in the session 1917-18. All this led to an extremely busy department.

5.4 Students making poultry sheds during the First World War.

Visitors were also attracted to the experimental plots, which continued throughout the war. In 1917 the Principal spoke of the enormous numbers visiting the experimental plots, which he considered surprising in view of the increased costs of transport. On 10 August there was a record attendance of 1,100 Staffordshire farmers, the visit being an annual event at the College since early days. Hedworth Foulkes declared: *'[This] must be a record gathering for an Agricultural college.'* From late 1913 field experiments and trials had been co-ordinated in the Three Counties of the Province. This was a progressive and historic step, allowing identical experiments to be repeated in a farmer's home district.[22] At the meeting of the West Midland Provincial Advisory Council at which the agreement had been reached, it was stated: *'It is believed that this is the first attempt that has been made to correlate the work of a group of Autonomous Committees for County Agricultural Education.'* [23]

Perhaps, some of the most interesting events held at the College during wartime were the machinery demonstration days. The first was held in August 1917 as part of the routine policy of open days. The College tractor, an International Harvester Titan, a Burford-Cleveland with caterpillar tracks and an Avery tractor, the latter two being brought in for the occasion, were shown at work ploughing grassland. This was to give encouragement to the Government's ploughing up campaign. The press was most

5.5 Tractor demonstration at the College, 1918.

enthusiastic to point out that the Titan was driven by a young lady.[24] There were also several new implements on view, such as a tractor disc harrow and an automatic cabbage planting machine. Demonstrations continued throughout the following winter with a wider range of tractors sent from the Food Production Department. The tractors now included an International Harvester Mogul, a Samson Sieve Grip and a Fordson.[15] Varieties of crops were also being specially grown to illustrate their suitability to newly ploughed land.

An end came to war on 11 November 1918. Approximately 200 staff, students and ex-students had served in the Forces.[25] Of these, seventeen had died. Altogether students of the College received two Victoria Crosses, two Distinguished Service Orders, eight Military Crosses, one Distinguished Flying Cross and two Croix de Guerre.[26]
The College decided to remember its dead in a tangible way. In the autumn of 1921 an appeal went out for library funds in memory of the fallen.[27] Early attempts at setting up a library had met with a lukewarm response. In 1904 there had been a collection of books, chiefly fiction, held in the staff room, but these books were dispersed soon after their removal to the large lecture room in 1907.[28] In the session 1907-8 the Principal was first allocated a book fund amounting to ten pounds a year. Since 1909, at least, a reading room had been in existence, sited in a former dormitory on the first floor. From 1917 a small collection of reference books was in use. With the growth of the College there were hopes of improving the book collection, making it 'worthy of an agricultural college'.[29] By the spring of 1922, as a result of the appeal, an all-important but small library was a reality.[28]

The arrival of peace was clearly a time for the College to modernize and expand. Despite the drain of four years of war the College did attempt to seize the moment and launch itself into the new post-war era. However, as in the tale of the three pigs, the College's house had not been built on firm foundations. Serious problems had persisted, while others had arisen out of the upheavals of war and peace. In 1922 an official inquiry was established by the Ministry of Agriculture to investigate the running of Harper Adams. It must have looked as if the big bad wolf had arrived to blow the house down.

College Sport in the Early Days

Sport was strong at Harper Adams from the earliest of days, with half-day holidays on Wednesdays and Saturdays for such activity. The forerunner of the Student Union, established in 1908, was in fact the Sports or Athletic Club, formed as a result of a meeting held late in 1902. The first sports to be played were football, cricket and tennis. Rugby did not become a permanent feature at the College until around 1921, but by the mid-1920s its popularity was such that football was no longer the official game. Until 1905 football was played in front of the College, but that year a full-sized pitch was created in the field immediately west of Ancellor. Wilfred Mercer, student 1905-07, wrote in his diary for 21 October 1905: *'Did not intend playing footer but as they wanted to play North v. South I played. We played on the new field - only made this last week. It is below the cricket pitch and in the same field. It is 120 yards long. It was a good game, but the grass was rather long and also the field was bigger than any of us were used to. We all got rather tired. I think North won 2-0.'*
It was not until 1912 that the cricket pitch was first used in front of the College. In July 1909 the Cricket Pavilion was opened, financed by Annabelle, Lady Boughey, as a memorial to her late husband, the first Chairman of the Governors. By 1904 there was a tennis court and within a year another was added. The Hockey Club was started in 1909, although initially only away games were played as there was no decent pitch. From 1908 shooting was popular, with outdoor and indoor ranges, and there was an annual steeplechase.

The first sports day was held on 22 July 1908 with the awarding of several silver cups and the Victor Ludorum Silver Shield, donated by Mrs Hedworth Foulkes.
The Shield went jointly to F. Clive Osborne and William Arnold. The day included a 'novel obstacle or "pig driving" race'.
As the Newport Advertiser for 25 July revealed: *'Each competitor, armed with a toy whip adorned with a coloured ribbon, had to secure a pig wearing a small collar round its neck, and drive it from a given line round several hurdles, and finally between two hurdles set closely together. Endless amusement was caused by this novel race, and the confusion that was caused when all the pigs were released can be better imagined than described.'* Other more conventional sports seen that day were putting the weight, hurdles and the high jump.
The tug-o-war was between the Dormitories and the Private Rooms.
Honoured guests included Lady Boughey and the Duchess of Sutherland.

5.6 1905-06 College football team.

5.7 College tennis team, c. 1908: Private Collection.

Chapter Six —

Chaos and Crisis 1919-22

There is a moving account of the war in the Cat-A-Mountain written by an unnamed student.[1] He recalled fighting in July 1917 near a different Nieuport to that he was accustomed to:

[...] I was thinking of Newport, Shropshire - of dear old Harper-Adams where I would still have been had not Duty called me. My thoughts were soon brought back to earth tho', in the shape of a '5.9', which whistled over and burst a few yards away! [...] Nieuport, before the ravages of war had left their mark on its fair walls and surroundings, had been a quiet and peaceful Belgian town; but now, it was just ruins, ruins everywhere. Thank God our Newport and its surroundings were never dealt out a like fate. [...] Though old Harper-Adams has changed in many ways, and many of the innovations are not only strange, but a rude shock to me, yet it is the same Harper-Adams that I so often thought about, and I am inexpressibly thankful to be back.

Many of his colleagues whose studies had been interrupted by the war were equally grateful to be returning to Harper Adams. Others were new to the College and new to studying but were just as happy in the task before them. John Capron, enrolling in 1920, had 'the happiest of memories of College'. Having taken up a course offered to his brother, he 'jumped at this chance of learning farming quicker and in a more practical way'. He had been demobilized in 1919 and had soon grown dissatisfied with the lack of practical training at the Cambridge University School of Agriculture.[2]

Peace did not slow the pace of life at the College. The College bustled with activity as ever it had done. Students enrolled in ever larger numbers. The first full session after the Armistice, that of 1919-20, saw 146 students enrolled, while numbers for 1921-22 reached 155.[3] More students were rejected than accepted due to a lack of space.[4] In 1919 vacancies were booked as far ahead as 1922.[5] These post-war students were very different from their predecessors. Sixty-three of the 1919-20 intake had seen military service. For those who had served in the Forces, scholarships

6.1 1919-20 staff and students.

were readily available and fees were discounted. For those ex-service personnel who had not the time to devote to a full course, special one-year courses were arranged. As a consequence of the war many students were older than their former colleagues. The session 1920-21 saw an average age of twenty-one, an age higher than any previous year. Some students were physically disabled and a course for them in poultry and horticulture began in March 1920 under the direction of the Ministry of Agriculture. The last course for disabled servicemen closed in March 1921.[6] Other ex-service students bore psychological scars.

Most of the post-war students were seeking to enter practical farming and studied for the College Certificate or Diploma *(see photo page 46)*. Just five obtained the NDA in 1920, with three others passing in all but one subject. A year later the Principal could write that nationally the numbers entering for the NDA had increased, reaching a record 186. Significantly, that same year new NDA requirements were introduced, obliging candidates from 1923 onwards to submit evidence of practical knowledge with a complete year on a farm prior to study. The NFU had declared that the 'present system of granting Diplomas to students who are unfamiliar with practical farm work is extremely unsatisfactory'.

Another small but nonetheless significant group within the College was the female students, totalling twenty-five for the session 1919-20. With the end to the war special courses for women had finished. There had been praise for the College from the Director of the Women's Branch of the Food Production Department in connection with the training of women: *'The subsequent success of the students is strong testimony to the excellent work which has been done at the College.'* After the war, women continued to enrol on the regular courses at the College. Life would not be quite the same again. As one article in Cat-A-Mountain put it:

The female element about this College is exerting a subtle influence; with the result that one frequently hears remarks such as this - 'I wish I were a girl, I'd have shirts of pink georgette, collars of crepe-de-chine, and vieux rose ninon hankies!' This may be alright, but - well, I mean to say! This isn't the spirit to go back to the land with. Farming isn't all 'dress-me-up-to-make-me-look-pretty!' [7]

The next edition of the magazine was still finding women students fair game. This time women were blamed for everything from 'the awful accumulation of broken apparatus in the old chemical laboratories' to 'broken gateposts' and an abundance of kittens.[1] One lady student who arrived at the College in 1919 would, however, have given any gentleman a run for his money. Princess Helen Scherbatoff was the daughter of Prince Alexander Scherbatoff, President of the Imperial Society of Agriculture in Russia. She was a past student of Stebut School of Agriculture in Petrograd. Having fled the Bolsheviks, she *'is anxious to make use of the time she is in exile in England to learn all she can of agriculture in order to be able to help the tenants when she can return home'*.[8]

Another significant development was the noticeable numbers of students studying poultry husbandry. Many ex-servicemen were attracted to poultry keeping because of the small capital investment required.[9] For the session 1920-21 thirty-five students were enrolled, the majority on a year's course. The first examinations under the National Poultry Council Diploma Examination Board were held at the College in September 1921.[10] Candidates came from around the country, but two were wholly or partly trained at the College. Helen Crawford passed in all subjects and was to become the first Harper Adams student to receive the National Diploma in Poultry Husbandry (NDP). As the direct result of the third Poultry Conference to be held at the College in August 1919, a committee had been formed to look into ways of continuing the liaison between the various poultry societies and groups. This had led to the establishment of the National Poultry Council in early 1920. One of its first tasks was to inaugurate a national system of examinations in poultry husbandry, along the lines of those already existing for agriculture and dairying. Thus the NDP had come into existence.

Plans were also in the air to establish a national institute to further research and education in poultry husbandry. Hedworth Foulkes attended meetings of the Poultry Advisory Educational Committee and plans for such an institution were reported to the Governors as early as February 1919. By the early summer the Ministry of Agriculture was investigating sites and Hedworth Foulkes was instructed by the Governors to prepare a scheme for the College. By 1921 the area for poultry at the College was already considerable, covering ten acres.[11] Expansion now took place, namely the appearance of an army hut to act as staff offices and store room, two brooder houses and various smaller houses. The original poultry building was converted into incubator rooms. As well as the growth in the teaching of poultry husbandry, demands for stock birds, day-old chicks and eggs were on the increase. Stock was being sent out to places as far afield as Switzerland and Nigeria. Trials of new feeding stuffs continued and problems of incubation and rearing were investigated.[4] More land was also being taken up with the growing poultry trials. By 1920 two birds were refused for every one accepted.[12] The ongoing success of the College's poultry laying trials led to their becoming the model for the International Trials set up at The Hague for the World's Poultry Conference of 1921. The fame of the College was spreading throughout the world of poultry husbandry. Reward came with recognition of the College as the site for the National Institute of Poultry Husbandry in 1923.[13]

The future for poultry husbandry at the College was bright. Other areas gave reasons to be optimistic too. The involvement of a livestock officer with the College ended in 1919 when the Board of Agriculture modified the scheme. However, in 1920 two advisory posts were allocated to the College staff. The Board, by now the Ministry, gave a grant to cover salaries and expenses and the College provided the facilities. The liaison was considered to be of benefit to both parties: the students kept in touch with developments and the results of

6.2 Chicken-rearing ground showing three types of brooder house, early 1920s.

6.3 Large poultry house, c. 1921.

investigations and the advisory officers with the experiments and work of the College. Appointments would be made by the College but with Ministry approval. In March 1920 the College appointed A. Roebuck, already a member of staff, as the Advisory Officer for Agricultural Entomology.[4] It was not until 1923 that a second officer, an Advisory Mycologist, was appointed.[14] Roebuck dealt with enquiries, made visits to farms to give advice, gave talks and undertook investigations and published the findings.[3] An important part of his work was to co-operate with the Plant Pests Branch of the Ministry of Agriculture by providing a monthly report on the current pests and diseases in the Province. Farmers, it seems, were even failing to recognize 'such common pests as the leather-jacket'.[4] The partnership between the College and the Advisory Service was long and fruitful, only ending in 1946 with the Government's creation of the Provincial Headquarters of the new National Agricultural Advisory Service (NAAS), at Tettenhall, Wolverhampton.

The wide range of activities beyond the lecture room had continued throughout the war and now gained renewed vigour. The College continued to handle enquiries, to undertake analyses and tests and to issue bulletins. Extra-mural classes did not continue on the same scale after the war, as the County Councils established their own teaching staffs. Short series of veterinary lectures and individual talks did, however, remain a feature of College life.[4] Similarly, field trials continued and demonstrations and open days were increased and extended, remaining a popular feature of the College. During the session 1921-22 882 visitors came to the College. Significantly, 484 came for the poultry.[3] Machinery demonstrations now had a part to play in educating the visiting farmer. In the summer of 1921 visitors could see a Simar rotary cultivator and a Wood's motor-driven self-binder in action. In 1920 a new venture was launched with a special day being set aside for Women's Institutes to visit. That year alone 300 members visited.[4]

The College was thriving and the future looked good. The Farm was also caught up in this optimism, encouraged by the novel departure of Government protection. During the war the need for production had led to Government intervention to achieve efficiency, with controls over everything from prices and profits to wages and cultivation. The Corn Production Act of 1917 had given farmers guaranteed minimum prices for wheat and oats and the Government's commitment to agriculture was reaffirmed with confirmation of protection in the Agriculture Act of 1920. In 1920 the College signed a tenancy agreement for Village Farm, forty acres of arable land adjoining the College. New cows were added to the dairy herd.[5] New implements were bought, including a Bamford's hay-loader and a Clay and Wood's manure spreader, of which it was said 'both these are American inventions and still comparative novelties in this country'.[5] Other novelties to the College were, however, not so welcome, at least by those who had to work alongside them. Albert Rowley has the scars of a mule bite to this day: *'One day I thought "Well I don't know about you, you beggar." Her was quiet and suddenly you'd hear and that was her teeth coming up behind yer, you know and nailing yer.'* While Capron remembered:

> *Pinky Brown [Vice Principal Brown, Supervisor of the Soiling Farm, so nicknamed because of his description of one potato variety as 'pinky-brown'] had foisted two ex-Army mules on the Ancellor girls [women agricultural workers under Ruth Adam], and no doubt my three-year experience with these animals helped here. These were good and fairly quiet workers - but temperamental, and badly teamed, Sam being smaller than Pussyfoot. Specially unsuited to ploughing, where they went in jerks [...] and a dog-leg furrow often told the tale.*

Capron's disaffection with the mules was only matched by his growing affection for Miss Adam. They married in the early 1920s.

There is no denying the College reflected an image of success, but on analysis, there were deep-rooted problems, which in the rush of events had gone either unnoticed or had been ignored. The College had gone from achievement to achievement in the balmy years before 1914 and then had hurtled straight into the maelstrom of the ever growing demands of a country at war. With peace came no let up as the College continued to perform its diverse tasks but now with the additional demands of students arriving in numbers

6.4 Ruth Adam and mules cutting peas and oats at Ancellor, 1920: Private Collection.

hitherto unknown. The College just could not meet all the demands placed upon it indefinitely. The fact that the College continued to operate as well as it did, for the length of time that it did, is perhaps more remarkable than the eventual crisis which engulfed it.

Even from today's perspective it is hard to untangle events. What is known is that in April 1922 the situation was so bad that the Ministry of Agriculture sent in a committee of enquiry to investigate mismanagement at the College. Only a very few copies of the report were ever made. After consultation by the Governors all were accounted for by Liddle and all have disappeared. Liddle informed Hedworth Foulkes: *'I may say that it [the report] has not gone out of my keeping and that the three or four copies which I had made in London for circulation to the Governors have all been returned to me.'* [15] By September 1922 the Governors had acted: the College had a new Principal and of the original members of academic staff only three remained.[14] Hedworth Foulkes had been asked to resign and the members of staff were dismissed outright.[15]

A great deal was wrong with the College which was not evident at first glance. Perhaps, the most crucial problem was the lack of physical resources. Simply, the College was just not big enough to cope with the numbers of students now arriving. Much-needed expansion had been put on hold prior to the war as uncertainty about the precise nature of the new Provincial Advisory Council had made it unwise to act. However, with the rapid arrival of new students in 1919 it was imperative to build more facilities for teaching and accommodation. Work began in the summer of 1919 [16] and for the next three years work of some sort was in progress on the site. This fact alone must have disrupted College life. Buildings could not appear overnight and the post-war situation created its own delays due to shortages of labour and materials.[17] In 1919 a rail strike added to the confusion.[5] In the meantime the situation was so critical that ex-army huts were brought in for accommodation and teaching. Older students were boarded in houses in the area and in Chetwynd Villa, formerly on the site of Chetwynd Grove, Newport. The Villa, soon to earn a notorious place in Harper Adams' history, was rented as a hostel for twelve students. One student's solution to the accommodation crisis was to request permission to build a bungalow on the College Farm. The Governors turned the offer down.

Work was begun on new accommodation on the site to the west of the hostel of 1909. It was not until October 1921 that the New Hostel was first occupied, providing rooms for forty-three students.[3] The rooms were small, the standard size being just nine by six feet[19] and building work was not of the highest calibre. In 1968 the demolition of the hostel, renamed Boughey Hostel in 1951,[20] was a move welcomed by many. The ladies were not forgotten. Ancellor House, the female hostel, was renovated and electric light and a hot water supply were installed. But, even with expansion the accommodation on offer was insufficient.

It was not until the new hostel was opened that alterations to make way for more desperately needed teaching space could be started in the Main Building. The recreation room, created in the building work of 1905-06, now became an additional lecture room.[19] Other building work was also undertaken at the time to meet growing demands on the College, namely a new laundry, pump, engine and engineer's rooms and boiler house. A Cornish steam boiler was installed and the opportunity was taken to purchase a new electric lighting plant. The lighting and heating had been in dire need of improvement. Thus, that famous local landmark, the College chimney, made its first appearance in 1920. At its birth it was not so warmly received: *'Doubtless the new buildings, including the chimney stack, were designed with the object of providing a concrete illustration of Mr. Wibberley's famous book "Farming on Factory Lines".'* [21] It was demolished in 1968. Alterations to the kitchen were also included in the grand plan. Improvements included new stoves, additional cooking apparatus and a 'steam jacketed boiling pan'.[22] Catering for large numbers in the cramped surroundings of the early kitchen had been a nightmare.[5] This and post-war food shortages had led to student discontent over food.[23] Tea, at the time, consisted of half an inch of margarine and two teaspoons of jam, 'mostly stone', while breakfast was porridge and a small piece of bacon, ham, fish or scrambled egg made from egg powder with bread and margarine. Students were at a loss to understand the need for margarine and powdered egg at a college with a dairy herd and poultry.[5] Finally, the leisure requirements of the students were considered. New facilities were created around the quadrangle adjacent to the boiler house, primarily for recreational use, including that all-important game of billiards.[3] Martin Cross, student 1937-40, remembered 'a first world war asbestos building', which is now the black and white building named Tudor Lodge and home to the Crops Department.

6.5 The laundry, c. 1929.

6.6 Student common room in the quadrangle, c. 1935: A. J. Guy.

Financing the expansion was a significant problem. After years of dearth there was a debit balance of £2,178 in 1918-19. The question now was how could the biggest building programme yet undertaken by the College be funded? [24] The initial loan of £9,000 from the Ministry of Agriculture for building had to be raised in the summer of 1920 to £11,000. The College was to meet the costs pound for pound.[25] In the post-war situation inflation was spiralling upwards. The Minutes of the Students Union reveal a sovereign was worth just two-fifths of its pre-war value.[26] By January 1923 work was complete but total costs amounted to £28,386.[27] Liddle explained that despite careful supervision there had been

6.7 The College from the west, showing from left the carpenter's workshop, the boiler room, the laundry and Old Boughey Hostel, probably mid-1960s: Shropshire Star & Journal.

'unforeseen capital expenditure', adding with due aplomb that the original costs had after all been only provisional. With a note of inspired drama he declared that the costs had been 'incurred to meet a national emergency'. As the Ministry had only agreed to a loan of £11,000, he pointed out that the debt was a 'crushing burden upon the resources of the College' and 'unless it can be substantially reduced in the near future the most serious consequences to the efficiency of the work of the College are to be feared'. The Ministry grumbled, but, in June 1923, did agree to contribute half of the extra costs, although the total extra costs were accepted as £4,136 and not the £6,386 as stated by Liddle.[28] The remainder would have to be financed by the College by means of a bank loan.[24] However, it was made clear that a lesson had to be learnt: *'In authorising the grant, Their Lordships note that the attention of the Governors has already been called to the grave responsibility incurred by them in embarking upon additional expenditure without sanction, when it was clear that the cost could not be met out of the funds at the disposal of the College.'* [28] In 1922, at a time when the College was facing a severe situation, such financial problems must have made the alarm bells ring at the Ministry.

By 1923 the College had expanded and was far better equipped to take on the task before it, but of course, by that date the crisis had come to a head. The stresses and strains of managing excessive numbers with limited resources had told. As early as 1919 an article in the Cat-A-Mountain had expressed concern:

There is about a place, a subtle and indefinable something known as 'Tone' or 'Atmosphere'. We realize that there are varying degrees of this, and regret, that existing here is not as high and uniform as it might be. Without going into petty and annoying details, let it be said that there is abundant scope for improvement in a hundred and one different directions.[5]

Adding to the problem of management was the fact that the post-war students were a very diverse bunch. They were of wide-ranging ability and possessed varying degrees of commitment.[21] They were also of all ages and types, some fresh from school and others hardened by the experiences of the battlefield. Some were a dangerous mix of youth, wildness and wealth. Albert Rowley talked of the demobilized soldiers being sick of discipline after their experiences in the Army: *'They were cut and carved and downed to such a degree. [...] They sort of kicked discipline and all the rest that go with it. They'd just had enough. They were sick of it.'* Furthermore, the members of staff, more used to a public school approach to discipline, were probably of little use dealing with ex-servicemen. Rowley talked of staff haughtiness not going down well, describing some members of staff as 'a lot of puffed wind'. Much of the disruption was of a minor nature. In a letter to Hedworth Foulkes one parent lamented the loss of one of his son's football boots. Evidently, it had been stolen from his son's cubby hole, thrown at the cook and then burnt by the latter on the fire.[29] However, this underlying current of devilment was so easily to grow into serious indiscipline in the face of a mounting crisis.

By 1922 events have moved on considerably. Before students returned for the spring term in February 1922, all were required by the College authorities to sign a declaration stating that they would not take part in 'any demonstration or other act subversive of College discipline'.[30] This was rearguard action. The truth was that discipline had all but disintegrated and there were fears of nothing less than a riot. A month later the Minutes of the Students Union Committee read:

[The SUC] wishes to register a protest against the material means of restraint employed in one of the Hostels, namely the bars on the inside and outside of the windows. These bars are ridiculed by the public, who are thus compelled to form an unfavourable opinion of the students of the College, and do not appear compatible with the statement made at the commencement of term by the College Authorities, who expressed their confidence in the students.[31]

By this time it must have seemed like open warfare. By 1921 there were most serious signs that the College was unable to fulfil its educational role. Rowley confirmed the rumour of inadequate teaching. That summer there was an outcry at the poor results achieved by Harper Adams students in the NDA examinations. The College achieved a fifteen percent pass rate, whereas that of the West of Scotland Agriculture College was fifty-seven percent.[32] The Editor of the Cat-A-Mountain voiced the feelings of the students:

Disasters will happen we know, but when avoidable disasters occur, words are entirely inadequate. Help is required, active help from someone who is in a position to give it. [...] Of all the evils by which we are surrounded, perhaps this is the largest of them all; that such a subject should be mentioned in the 'Cat-A-Mountain' at the request of some Diploma Students who came here in order 'to be prepared for the higher examinations'. We merely say that evidently nothing in a prospectus should ever be taken literally or even seriously. [...] We venture to assert, however, that there is no more crying shame or more outstanding disgrace than that the College should have thought it fit and even possible to expect students to pass 'the higher examinations' with so few facilities to hand.[32]

The instruction had done little to prepare students for the examinations, especially in the subject of agriculture. The vivas, for

6.8 1920-21 staff and students.

example, seemed to 'hit upon the things we'd never even heard of, much less seen'. Hedworth Foulkes's address to students gathered at the opening of the autumn term suggested fault lay in the need for reform of the NDA, such as rearranging the subjects into more convenient examination groups.[33] More probably, fault lay in the disruption of facilities stemming from redevelopment. The practical side of instruction was something said to have been revived by Hedworth Foulkes's successor, which may suggest there had been shortfalls in this area.[14]

The College hurtled from one disaster to another in the autumn and winter of 1921-22. It is unlikely that improvement occurred. Continuing concerns over the education on offer undoubtedly added to the tension at the College. One student, Marjorie Godfrey, protested to the Principal that a delay of two weeks in starting the spring term meant that there was a mere eight weeks before NDA examinations began at the end of March.[34] The reason was given as 'reconstruction to buildings and the consequent re-arrangements of certain rooms and halls'.[35] Hedworth Foulkes declared 'being in the hands of the British workman it is impossible to plan far ahead'.[34] What both he and Miss Godfrey knew was that the official reason for the delay in starting term was not the true one. The prolonged absence of students from the College in fact ensured that a string of sensitive events, from the dismissal of the Vice Principal to an extremely acrimonious court case involving the College, took place with as little student involvement as possible.

Against this complex background the human element came into play, provoking a chain of events which was to leave the weakened edifice of the College tottering. The difficulties of 1922 are described by the Ministry of Agriculture as 'culminating in and consequent upon the complete breakdown in health of the late Principal'.[3] Hedworth Foulkes was at the heart of the crisis. The College revolved around the Principal. His valuable time was tied up with everything from the most mundane to his work on the national stage, notably advancing poultry husbandry education and his chairmanship of the Agricultural Education Association from 1919-22.[36] At this time the Principal was even expected to oversee a student's expenditure. The following letter from a father concerning 'reasonable expenses' for his son was sent to the Principal, although the matter was not in fact handled by him as he was absent due to illness. The concerns about a son away from home for the first time may well have been typical, although the desire for such a degree of distant parental control was hopefully not:

> *If he smashed his watch, I consider that he should have another one, as I look upon a watch as a necessity. I would not, however, give him a gold watch. I think a gun-metal watch from a good maker with a leather strap for a chain would meet the case. As regards clothes, I would not consider patent leather shoes, spats, extravagantly fancy ties and socks, top hats or morning suits to be reasonable, but I would like him to have a good stock of all necessary clothes, good substantial boots for farm work and substantial raincoat and overcoat, a couple of decent lounge suits outside of his working clothes, plenty of substantial underclothes, socks, ties and handkerchiefs etc. I would even go to the length of a dress suit with the usual accessories.*[37]

In hindsight it can be seen that the College should have reorganized its structure and spread the burden of administration. However, it was not until October 1921 that the first Bursar, Lt Colonel Traill, was appointed with responsibilities for 'the interior administration of the College and staff'. He possessed a distinguished Army background and was very much a military man. As one student remembered him in his later years: *'[He] was a very precise sort of man. He never would allow it to be too hot that he could take his jacket off. If he got very hot and it was very warm weather, we were going about with open necked shirts, he might undo the buttons of the sleeves of his jacket and just roll his cuffs back.'* [38] He retired in 1946.[39] By 1921 Hedworth Foulkes was suffering from severe stress.[15] He was a man of enthusiasm, energy and efficiency. As such he was the creator of the College, but his perfectionism and conscientiousness ill-equipped him to cope with the spiralling crisis. There is the possibility that Hedworth Foulkes could have recovered from the nervous exhaustion of the summer of 1921. He had done so before and the appearance of a Bursar and an end to the major building programme were both factors working in his favour. At this point, however, if life as Principal had hitherto been storm tossed, it was now about to be devastated by a hurricane.

The hurricane appeared in the form of Arthur Tyler, an assistant lecturer in poultry husbandry. Tyler was a new member of staff, employed in April 1920.[40] In April 1921 he became Warden of the hostel, Chetwynd Villa.[41] In October the Governors realized that, in view of the possibility that the College might become the National Institute of Poultry Husbandry, reorganization of the staff might become necessary. Thus Hedworth Foulkes, as a precautionary measure, was asked to give notice to the staff on the understanding that when the scheme was clarified there was the possibility of re-employment.[41] Such action obviously did little for College-staff relations at a time when the College was already beset with problems. Tyler was especially resentful and, it seems, somewhat unbalanced as a person. In mid-October he launched a blackmail threat against Hedworth Foulkes, seeking a withdrawal of his notice in return for his silence. Tyler made many wild accusations, some clearly petty such as 'issuing false advertisements in order to obtain higher prices for your eggs'. More seriously, he accused the Principal of 'compromising conduct with women students' and mismanagement of the College. The Principal reported all to the Governors who, after a hearing, considered the attack to be unwarranted.[41]

After the hearing the Governors considered Tyler's actions had made his continued employment untenable and he was dismissed immediately.[41] A petition signed by sixty-one students survives in the College Archives, protesting at Tyler's treatment and clearly showing a serious rift developing between the College and the student body. The petitioners declared: *'He has on all occasions taken every opportunity to help us in our studies, and at the same time has always commanded our respect and earned for himself a well-deserved popularity.'* Tyler, for his part, refused to vacate Chetwynd Villa, continued to demand his salary and disputed the right of the Governors to terminate his employment in such a manner.[41]

While Tyler continued to live at Chetwynd Villa, a very serious breach of discipline took place. In November 1921 a student was set upon by others and tarred all over, unimpeded by anyone. In effect the hostel was not wardened. Already of a delicate state of mental health as a result of wartime experiences, the incident and the cold bath which followed, hot water not being available, led to a fatal bout of albumenuria. The student died at home just after Christmas and the parents considered the College to blame for not supervising his health.[42] Tyler then left Chetwynd Villa.[41] Eight students from the hostel were sent home for three weeks as punishment [43] and the rental of Chetwynd Villa ceased.[44] It was this particular incident which led to the radical reorganization of early 1922, but the attempt to restore discipline and instil confidence was unsuccessful. By February 1922 the whole College was involved in the Tyler affair.

It seems that there was sympathy for Tyler not only amongst the students but also amongst some members of staff. Hedworth Foulkes no longer had the respect of all his

6.9 Arthur Tyler, the lecturer involved in the controversial case against the College.

staff, suggesting there may have been some truth in Tyler's accusations. Regardless of Tyler's accusations, clearly many of the staff realized the Principal had lost control. At this point Vice Principal Brown, possibly unwittingly, intensified the crisis. In the face of the Principal's refusal to present Tyler with a leaving present from the students and staff, [45] an action which was not surprising, Brown deputized at a ceremony on 9 December in his capacity as President of the Students Union.[46] There is evidence to suggest that relations between the Principal and the Vice Principal had not been amicable before this date.[15] Hedworth Foulkes and the Governors considered Brown's action to be one of gross disloyalty and proceeded to dismiss him in early January 1922.[15] If students and staff had been upset at the departure of Tyler, they were enraged at the action taken against Brown, who was a high profile researcher and an extremely popular member of staff. Brown was much admired for his patient and unassuming character, his involvement with student activities and his ability as a lecturer.[47] His wide knowledge meant that he was 'equally at home with Bimetallism or Gothic architecture as with his own hybrids [of wheat]'.[48] He was also the founder of the Student Union [45] and had been its President for fifteen years. The loss to the Student Union was described as 'irreparable and indescribable'. The Editor of the Cat-A-Mountain waxed lyrical: *'He has not left this place of happy memories, But taken all its heart and soul away.'* [35] Resentment simmered.

Meanwhile the Governors had set in motion a court case to evict Tyler which had resulted in a cross petition by him for recognition of wrongful dismissal and payment of his full quarter's salary.[15] The whole sordid affair and all manner of wild accusations were aired in the Shrewsbury Magistrates Court on 31 January, before a room crammed with students.[41] The matter of mismanagement was discussed in some detail, but the grounds for such an accusation were weak. A student, Helen Crawford, had enrolled at Harper Adams in January 1921. She had studied at University College, Reading, and at the Munster Institute in Cork, before the fighting in Ireland had disrupted her course.[49] That summer she gained the College Certificate with distinction and there was evidently much resentment. Accusations flew that the examination papers had been tampered with and that she had not taken all subjects and was therefore ineligible. Colonel Sykes, Chairman of the Governors, argued that no impropriety had taken place as her previous studies had been taken into account quite legitimately. However, there does seem to have been much confusion as to which subjects were required.[41] This particular incident, whether justifying criticism or not, did give rise to considerable resentment amongst the students. The Court's decision was reached in favour of the College with 'not the shadow of foundation for any of the charges' and the character of Tyler was brought into disrepute as 'a man with a

6.10 James Carter Brown, Vice Principal 1907-22.

burning grievance' who had made 'the grossest and vilest charges' and spread 'shameful and scandalous rumours'.[41] The resulting publicity was, however, bad. By early March Hedworth Foulkes's health had broken down completely [15] and in mid-March he left the College on a sabbatical. He was never to return as Principal.

Sir Daniel Hall, Permanent Secretary to the Ministry of Agriculture, became involved in early February, suggesting an investigation 'to enquire into the internal discipline and efficiency of the College and to make recommendations to the Governors with a view to securing its continual efficiency'.[15] With the departure of Hedworth Foulkes the Governors were in need of an interim principal and were virtually obliged to accept Hall's offer to send one. William Weir, one of the Ministry's Inspectors, thus became Acting Principal from March to September 1922. There is evidence that the indiscipline and unrest continued.[50] The Enquiry was held in the Easter vacation, as the Governors feared 'a strong undercurrent of feeling amongst the students and certain members of the staff' and 'anything which might lead to unseemly conduct ought [...] to be avoided'.[15]

The Enquiry's conclusion was initially revealed to the Governors on 12 April. Subsequently a report was written up but the copies produced were never widely circulated. Nothing less than sweeping changes were required for the very survival of the College. A committee of Governors and representatives from the Ministry was appointed to act. The Committee recommended the dismissal of eight members of the staff and the transfer from the College of the Advisory Entomologist, Roebuck. The Report also recommended the dismissal of Hedworth Foulkes. Significantly, the Governors objected to this, but failed in their opposition. However, as a compromise, it was decided to give Hedworth Foulkes the chance to hand in his resignation.[15]

There is sad correspondence surviving from August 1922 when Hedworth Foulkes seeks to return and Liddle, declaring his friendship, spells out to him that the matter of his resignation is not open for discussion. Hedworth Foulkes thus offered his resignation on the grounds of his ill health.[15] Hedworth Foulkes maintained the support and friendship of his colleagues. He went on to achieve a notable career in poultry education as Chairman of the National Poultry Diploma Examination Board, a post he held for thirty-three years. He died in 1965. The Governors marked his going with considerable regret and there was evidently considerable respect and friendship shown to him by the new Principal, Charles Crowther.[51] The first era of Harper Adams College had come to an abrupt close. The question was now, how difficult would be the road to recovery. To add zest to Crowther's task, a deep agricultural depression was now biting into student numbers. Both he and the College would have to show remarkable powers of resilience.

Electro-Culture Experiments at Harper Adams

Of all the experiments to take place at Harper Adams during the principalship of Hedworth Foulkes, perhaps the most weird and wonderful was that in electro-culture. Such experiments were widely performed, instigated and funded by the Electro-Culture Committee of the Ministry of Agriculture. The first to take place at the College were in May 1919.[a] The belief was that in electrifying the growing crop, in this instance oats, yield would increase. The procedure was described in the following way: *'Power from the College generating station drives a 100v dc. motor direct coupled to a 140v ac. generator.*

The power from this machine, after passing through a regulating resistance, is supplied to the primary winding of a transformer which raises the pressure to approximately 45000v. The current at this pressure is then again made uni-directional by means of a rotating rectifier and supplied to an aerial (7 feet high at the ends) stretched over the field. The wires of the aerial are 5 feet apart over half the electrified seeds plot and 10 feet apart over the remainder.' [b] As a child, Albert Rowley remembered the experiments: *'They had all gone stark raving mad [...], but it was all terrific stuff really. It excited people and a lot of these young farmers were interested. They were out here as thick as God knows what. Didn't know a thing. Didn't themselves learn a damn thing because they'd no idea what the hell they were talking about. [...] Dad used to get quite excited with all this menagerie of stuff and all these kinds of machines and hell knows what. [...] I thought somehow they were going to take off.'*

The results at the College showed only a small increase in yield. However, elsewhere, results were encouraging.[c] The reason for the apparent success of electro-culture soon became apparent. In the 1920s it was observed that wheat growing in the Fens grew considerably better under power lines. On investigation it was found that the copper of the cables had oxidized and washed onto the wheat, thus eradicating copper deficiency.

Chapter Seven —

Lean Days of Depression 1923-1939

The session of 1922-23 was crucial for the College's survival. As the new Principal, Dr Charles Crowther, later recalled: *'It was not without anxiety that the new staff opened the session on 7th October 1922, nor were the returning students without their unease as to what the new regime might have in store for them! [...] With staff and students on terms of armed neutrality, the one party determined to establish its authority, the other to protect its ancient privileges.'*[1] Yet, as Crowther looked back, it was the sense of excitement and challenge that he remembered. Discipline had to be re-enforced, education standards reimposed and confidence restored. Crowther's choice of staff was to be all important. Those of the 'old' academic staff were just three in number: Professor Wilson, Miss Crawford, the Dairy Instructress, and Rylance, the Horticultural Lecturer. Fortunately for the College, Crowther was an astute judge of people and of situations. He sought 'a young and active set of good mixers',[2] even if this meant flouting the conventional choice.

One such appointee was Douglas Hepworth Robinson. His memories captured the special character of these years:

> *We [the staff] began to wonder if Dr. Crowther had physical size as well as youth in mind when appointing his staff, for we were all on the six foot mark. Chalmers [W. S. Chalmers, Assistant Lecturer in Agriculture] delighted in a rough house, he and Malthouse had many tussles. I do not wish to give the impression that the staff were an unruly bunch, but in those days we were on a bit of a knife edge, somewhat isolated, and had to make our own amusements. There was a time when we all affected to smoke long-stemmed churchwarden clay pipes, an occupation which caused most of us, and in particular Dr. Crowther, to belch in a most unseemly manner.*[2]

Robinson was appointed Head of Botany and Zoology at the ripe old age of twenty-three, with just one year's experience of lecturing behind him. He went on to achieve respect as an author of academic books and papers and was awarded his doctorate in 1934. He was Editor of Agricultural Progress, the journal of the Agricultural Education

7.1 1922-23 staff and students.

Association, from 1931-45.[3] His nickname was 'Beetle' [4] and his particular talents lay in conjuring and ventriloquy, when he was assisted by 'Jimmy'.[5]

The new Vice Principal and Head of the Agriculture Department was T. J. Young. He possessed a practical farm background and experience as Principal of Holmes Chapel Farm Institute before its closure in 1916.[6] He was 'a mature, rotund person with a tremendous sense of humour and a profound detestation of the singing of the "Red Flag", then much relished by those with advanced socialist leanings. He was a great raconteur and kept the staff table very happy with scandalous tales about well known and high-up personalities in the farming world'.[2]
The new Head of Chemistry was a Shropshire farmer's son, William Morley Davies. He also took on the crucial role of Warden, gaining a reputation as 'strict but fair'.[7] He had worked as an industrial chemist before undertaking research at the School of Agriculture, Cambridge.[6] In 1924 he was to take on the additional role of Advisory Chemist at Harper Adams.[8] Students recognized that 'he was a brilliant chemist', but 'an awful lecturer'.[9] On the establishment of the National Agricultural Advisory Service in 1946, Morley Davies left Harper Adams to become the Provincial Director of the East Midland Province, but he returned to the College as a Governor in 1962.[7]

Morley Davies was assisted in the lecturing work by Thomas Oakley Wilson, the son of Professor Wilson. Money was so short that initially Wilson received a salary of 'less than £2 per week plus free lunches', which was paid from the petty cash.[10] Albert Rowley remembered 'Tommy Molecule' [11] as a strict man: *'He was dreadful, no playing around with him. He was the nibs that one and by golly everybody jumped six foot in the air when he came behind you. Oh my god, absolutely every word was weighed.'* Wilson was appointed Head of Chemistry in 1946 and worked at the College until his retirement in 1965. Dr Boyns, his successor, wrote of him: *'He was a character, a true conservative in all senses of the word. He didn't like change. This applied to his daily time-table, his dress and the organisation of his laboratory.'* [10] On his arrival at the College, Dr Boyns recalled: *'I found a jar of cyanide stuck on the shelf in his room. Later in an open cupboard I found a bottle of strychnine. These sort of things didn't worry Tom. I suppose nobody else dared go into his place.'* His first act of a morning was to light a Bunsen to keep his pipe going.[12]

The new Assistant Lecturer in Agriculture and Estate Management, W. T. Price, joined the staff from the Staffordshire Farm Institute. His fame was to come in 1946 when he returned to the College as Principal. He had served in the First World War as a fighter pilot, rising to the position of Captain. He had been lucky to have survived an encounter with the Red Baron, although badly injured in the fight.[13] He had been awarded the Military Cross. After the war, as a student at Reading, he was noted for 'the brightness of his socks and the length of his cigarette holders'.[2]
In November 1922 the new team was complete with the arrival of Herbert Howes from Wye. Howes had been Head of the Poultry Department at Wye and he now took over poultry at the College. On the establishment of the NIPH he took the post of Assistant Director. However, by his retirement in 1951 he was once again officially Head of the Poultry Department, after reorganization following the merger of the NIPH with the College in 1949.[14] His son, Peter, described his father's 'almost fanatical devotion' to work. Howes became a renowned expert in the Poultry Industry and was in great demand as a judge. He was an author of textbooks and bulletins and also wrote a weekly column in Poultry World for many years.[14] He regularly contributed to various other publications on subjects such as bees and gardening as well

as poultry. As a young boy in Edgmond, Patrick Davies remembered Howes' large eyebrows, watching them 'going up and down, expecting them to drop off any minute'.

Heading this diverse and mostly young and inexperienced team was Crowther, a man of forty-six with an established name in agricultural education and research. Crowther was a Yorkshire man. He had graduated from Oxford with first-class honours in chemistry and had then undertaken research at Leipzig, receiving his doctorate in 1901.[6] He became Lecturer in Agricultural Chemistry at the future University of Leeds in 1903. In 1913 he was made Professor [15] and his research interests were advanced by the establishment of the Institute for Research in Animal Nutrition at the University and his appointment as its Director.[16] In 1919, however, on the direction of the Board of Agriculture, the Institute was merged with its sister establishment at Cambridge. Not happy with the situation, Crowther moved into the commercial world,[6] but in 1922 the Olympia Agricultural Company went into liquidation.[16] Significantly, at this very time, the Governors of Harper Adams were in need of a new principal.

7.2 Principal Charles Crowther in the piggery mill.

The Principal's appearance was distinctive. His nickname, 'Little Middle White', did not solely derive from his passion for pigs.[11] Albert Rowley recalled: *'A very efficient little man. He looked what he was. He was a brilliant fella.'* Dr Harold Temperton, Head of the NIPH, spoke of him as the best fundamental chemist in the country.[17] His research achievements in animal nutrition bestowed glory on the College. On his retirement it was said that 'his close connection with the Ministry of Agriculture, combined with his rare ability to formulate or analyse projected schemes for the advancement of agricultural science, have placed him in a pre-eminent position in the sphere of agricultural education.' [1] In the early days of his principalship he was very much associated with all aspects of life at the College. He led by example: *'[At the End of Term Concert in 1923] much pleasure was felt when Dr. Crowther gave his songs.'* [18] He possessed humour and a sense of fun and knew all the students personally.[19]

Crowther was a firm disciplinarian but wise enough to accept the high spirits of youth. Rowley commented upon Crowther's short stature: *'He was such a little fella. He disappeared from sight when he was anywhere near the middle.'* However, he added that this appearance gave lie to the fact that he was a man who could command discipline. Arthur George, student 1933-36, referred to his 'raging temper'. Even minor misdemeanours, such as Alexander Reece's failure to wear a jacket at High Tea, could result in a summons to the Principal's office.[20] By the Second World War, rules as regards formal dress at mealtimes had been relaxed.[21] However, Crowther also knew when to turn a blind eye to student pranks. Such was the case with a memorable tea party on the lawn in front of the Main Building in the 1920s. Everything was 'laid out nice, [...] the solemnity was incredible' and 'there was water and water and water. All of it strange and peculiar stuff', including it seems ample supplies of urine. As Rowley concluded: *'Why some of them weren't poisoned, I'll never know.'* Crowther earned the respect and love of both the student body and the staff. As a family man with four children, he took an interest in the lives of his staff. A colleague declared on Crowther's retirement: *'To the staff he has been a real guide, counsellor and friend, assiduous in promoting the welfare of all, and his reliance on their ability and integrity has established a relationship unequalled in its friendliness in any similar establishment.'* [1]

That rare mix of strength and charm blended with the skills of both a diplomat and an efficient and intelligent administrator brought success. This is not to deny that the future of the College was to be a challenging one. The College soon ran into severe problems with student recruitment. During the session of 1922-23, 114 students were at the College, owing to the continuing presence of ex-servicemen.[1] However, for the session 1925-26 there were just forty-eight students.[1] From 1926 numbers at Harper Adams stabilized at around the seventy mark and remained so until the outbreak of war in 1939. However, this was not due to any improvement in the numbers studying agriculture, but stemmed from the popularity of poultry husbandry.[22] For the session 1930-31 numbers for the College were forty-two and for the NIPH forty.[23] As regards new enrolments the drop in numbers on the agricultural side between the sessions 1932-33 and 1934-35 amounted to fifty percent.

Crowther was only too aware of the reasons for declining numbers. First, Crowther admitted, there was the problem of the College's 'comparative failure to secure for our educational courses more than a small fraction of the sons of farmers'.[24] Although agricultural education was becoming more accepted amongst the farming community, as the difficult farming conditions of the 1920s and 1930s made many realize that science could help, Crowther made a vital

7.3 The College, 13 May 1927: Aerofilms Ltd.

distinction between the awakening of farmers to the subject and their conviction of its necessity.[24] Crowther did his best to encourage recruitment at the farm gate, going out of his way to convince visiting farmers of the practical nature of the College's courses.[25] In 1925 Crowther took to the air waves with a broadcast 'Training of the Young Farmer'.[26] Yet, if the message was getting across, it was not reflected in recruitment figures.

The root cause of recruitment problems was the depression in agriculture. Lean years may well have encouraged farmers to consider scientific knowledge as a way out of difficulties, but at the same time hardship deprived them of the means to finance education. Frank Ward, student 1925-27 and himself a member of a Shropshire farming family, was convinced that it was the lack of money rather than any shortfall in the College's reputation which hindered recruitment. At the end of the First World

War farmers were optimistic that the Government would back the need to make the country more self-sufficient in food production and would continue to assist their cause. However, in the summer of 1921, within months of the passing of the Agriculture Act of 1920, protectionism was withdrawn. As prices collapsed amid a flood of new and pre-war sources of supply, the Government had feared the outlay on deficiency payments which the Agriculture Act had entailed. Within two years the price of cereals was to halve. Prices improved slightly after 1926, only for the world market to hurtle into the Wall Street Crash of 1929. World recovery did begin around 1933 but, of course, improvement did not appear overnight. It was a depression in farming far worse than that encountered prior to the First World War. Hundreds of thousands of acres of land went out of production each year. Many left farming altogether and landlords found it impossible to rent vacant farms. Survival became a matter of reduced costs and improved management. All this did not bode well for recruitment of farming families to the College. Furthermore, it did little to assist recruitment of those from a non-farming background. Who would be foolish enough to contemplate entering such an industry?

As any good business man would, Crowther did not stop his analysis of falling trade there. In a report prepared for the Governors in 1934 he made a full study of the situation. Although accepting the importance of the NIPH in boosting recruitment numbers, he felt there were disadvantages to its presence. First, he feared that the College was becoming renowned for poultry, perhaps to the detriment of its name in agriculture. Secondly, Crowther felt that the large numbers of women on the poultry courses could alarm prospective parents not happy with co-education. Old myths were slow to die. A brother of a female poultry student in the 1940s jokingly accused the NIPH of being more husbandry than poultry.[27]

Crowther also considered the length of agricultural courses at the College made them expensive and thus less attractive than short courses on offer at farm institutes. The College had adhered to the Ministry of Agriculture's recommendation in the Memorandum on the Reconstruction of Agricultural Education in England and Wales of 1921 that the farm institutes and agricultural colleges were to offer different types of training. The former were to offer short one-year courses, with a minimum amount of science, while the courses of the latter were to be wide ranging and to run for a minimum of two years. As a consequence, Crowther felt, Harper Adams had suffered in the difficult times.

By the time that Crowther had prepared the report in 1934, there were hopes of improvement on the horizon. In 1931 the new National Government had established the principle of protecting British agriculture with a system of tariffs and quotas to regulate imports and of price subsidies and deficiency payments on selected home products. In this way the Government sought to tackle the root problem of over-production on the world market. The dumping of foreign food at very low prices was stopped and imports were regulated more efficiently. The most important subsidies were those introduced on sugar beet and beef. The Wheat Act of 1932 established deficiency payments to supplement actual market prices, in effect introducing a standard guaranteed price.

In addition to this the Government encouraged internal efficiency and self-help with the Agricultural Marketing Acts of 1931 and 1933. As many problems were attributed to marketing rather than production, producers were encouraged to set up marketing boards to regulate sales, to enforce grades and to improve packaging. The Act of 1933 permitted the restriction of imports of any commodity for which a marketing scheme for the equivalent home product was in force or in preparation. The most successful of the many boards which were set up was the Milk Marketing Board for England and Wales, established in 1933. Interestingly, Crowther attributed a sudden improvement in the employment of those students not intent on going farming to the appearance of marketing boards. As early as 1923 Crowther had recognized that many students were either unable or unwilling to take up practical farming and sought salaried posts.[24] In 1934 Crowther could write that the Milk Marketing Board alone had recruited seven Harper Adams students as Assistant Regional Officers. With these measures, confidence was gradually restored to agriculture. From 1934 the growth in agricultural output averaged from two to three percent a year. Over time, recruitment at the College could not help but benefit. Indeed, on the outbreak of the Second World War, as the Government ordered Harper Adams to discontinue its agricultural courses, Crowther declared with frustration: *'We had in sight the best entry for agricultural courses for many years - and had to scatter it to the "four winds"!'* [1]

The College Farm during the Crowther period, not surprisingly, reflected the national farming scene. It was in a far from sound state. Matters would have been far worse if it had not been for the capable skills of 'Big' Bill Thompson. Thompson possessed a degree and a NDA, had knowledge of advisory work and practical experience of farming in Lincolnshire. In 1927 he was appointed Farm Director and Lecturer in Agriculture, later becoming Head of the Department. He remained at the College for thirty-seven years.[28] As Tony Gale, student 1944-46, remarked: *'He was an interesting chap. Not purely science but he got a bit of mud on his boots as well.'* He was considered a brilliant lecturer by Leonora Grice, student 1943-45: *'As soon*

7.4 'Big' Bill Thompson explaining to students how not to damage the corn when spudding thistles in wheat, 1936: Sport & General.

as he swept into the lecture room it was as if the air became electrified: his facial expressions, his gestures, his dramatic delivery were those of an actor playing to the gallery.[...] He made us feel that farming was the most satisfying way of life possible. That living close to the soil, whether it be "a four-horse clay" and "grey as a rat's back", or whether it be "a lovely, friable loam", was the best, the only, worthwhile thing you could do.'

By the late 1920s the Farm amounted to 275 acres.[29] In June 1923 Flatt Farm, previously rented by the College, had been purchased.[30] If the deepening gloom of the national farming scene did not deter Thompson, the fact that the College Farm was only just emerging from a bad state of neglect must have at least given him pause for thought.[25] The crisis and confusion within the College had spread to the Farm. Slowly improvement came about. In 1934 the Principal was able to announce the prospect that the Farm would once again become a revenue-producing asset. By 1938 the Farm could boast improved and cleaner land, better stock and a substantially improved dairy herd. The Farm once again possessed its own poultry, with a commercial free-range unit of 500 birds, introduced in 1932.[31]

Above all else, the years from 1922 to the outbreak of war were difficult ones for Harper Adams in matters of finance, despite increases in fees and grant allocation from the County Councils in 1923.[37] As Crowther said on his retirement in 1944: *'Looking back over my twenty-two years I see them mainly as a continuous struggle for the greater part against difficult circumstances so far as the economic welfare of the College is concerned.'* In 1932 total revenue stood at £33,500. As regards expenditure, the wages bill alone ran to £17,000.[33] By that summer, despite economies, Crowther spoke of all reserves being exhausted and the accumulated deficit balance running to nearly £3,000, with the prospect of it rising. Most serious was the matter of the bank loan borrowed to finance the building and modernization programme of the early 1920s. By 1936 the loan amounted to over £19,000.[34] That year, with no sign of undue haste, the Ministry of Agriculture convened a meeting of the contributing County Councils and suggested the 'superior credit of the Counties might be utilised to secure some reduction in the Interest charges to which the College is subject in respect of Bank Loans'.[34] As a result of this, in October 1937, a loan of £17,000 was advanced by Shropshire County Council for a period of twenty-five years.[35]

The picture so far sketched of the College during this period is one of struggle, but, on his retirement, Crowther described the difficulties 'as but trivial specks upon a background of great good fortune and of congenial labours not ill rewarded in their final outcome'.[1] Indeed, student memories paint a far from austere picture. In part this can be attributed as much to student determination to enjoy life, as to the College's small size, coupled with its spirit of comradeship. Harper Adams had a quite distinct character at this time.

The College day began at 6.15 a.m. with a wake-up call for those students on farm duty.[36] Those on duty proceeded to various tasks, such as milking, mucking out the animals and feeding and watering poultry. All agricultural students took a turn at farm work and the poultry students similarly worked a rota at the NIPH.[21] Martin Cross, student 1937-40, remembered early duty having its advantages: *'The great draw was that the pantryman, George, was in there brewing some jolly good tea so we all drank tea and ate biscuits. Then we went and milked our cows.'* Those on farm duty resumed work in the late afternoon, finishing by the late 1930s at around 5.30 p.m.[11] The student rota for farm work did not stop at weekends.[9] Breakfast for all began at 8.00 a.m.[36] and in the late 1930s consisted of porridge or cornflakes, with bacon and egg and toast and marmalade.[11] Morning lectures ran from 9.00 to 1.00.[36] There were lectures on Saturday mornings. The whole of the routine was punctuated by bells.[36]

Gongs summoned all to meals.[11] Dinner was taken at 1.00, with high tea of bread and butter and cake at 4.30 and supper at 8.00.[37] Numbers were such that everybody could be accommodated in the dining room on long benches on either side of two long tables. The dining room is now used as the board room. A third table was provided for the resident staff, presided over in the 1930s by Archie Campbell, Warden and Lecturer in Agriculture.[11] At dinner, however, Crowther would preside. The tables were waited upon, with the Butler in charge of operations.[11] Food was plain but substantial, 'with lashings of home-produced butter and cheese and vegetables were plentiful'.[38] Bob Tupper, student 1943-45, could recall the superb taste of the cheese to this day. Women students dined separately at Ancellor House, which was run as a separate institution with its own warden and staff.[39] From 1922 [40] there was a student canteen to supplement the diet, which in the mid-1920s was stocked with cigarettes, pipe tobacco, chocolate, marshmallows and candles, the latter to assist mental sustenance.[38] Willard Moore, student 1939-40, managed the canteen for a time in early 1940 and made sure 'there was "Three Nuns" for Dr Crowther's pipe'.

Bottled beer could be bought to drink with meals, but otherwise alcohol was not permitted.11 Cross recalled an unofficial bar was set up by the students in Old Hostel in the late 1930s for the End of Session Dance. In the days of Hedworth Foulkes no alcohol

7.5 Dining room, c. 1935: A. J. Guy.

7.6 Archie Campbell explaining the points of a Dairy Shorthorn heifer to students, 1936: Sport & General.

whatsoever had been allowed in College.[41] Despite relaxation of the regulations, students continued to bend them. Warden Archie Campbell was in the front line. He had been appointed Lecturer in Animal Husbandry and Farm Accountancy in 1926 and was considered 'a somewhat reserved and reticent man, [...] with a rare sense of humour on occasion'.[42] Paula Large (nee Johnson), student 1943-46, recalled that if anything went wrong he would say 'You'll walk in the canal until your head is under water'. Arthur George remembered attempting to get the better of Campbell one night: *'[At roll-call] we had four other students in the room when Mr Campbell noticed that our water tumblers were in various degrees of emptiness, and containing beer. On being warned that it was "against the rules to drink intoxicants on the premises", poor Mr Campbell was told that we were only drinking our tea without milk as he himself always practised. Nothing more was heard of the incident.'* Public houses were out of bounds, but in reality were used, although those frequented by staff were naturally avoided.[11]

Afternoons were filled with laboratory work, farm walks or demonstrations. Wednesday and Saturday afternoons were given over to sport. Saturday work, sport commitments and the difficulties of transport meant that students tended to stay at weekends.[11] A feature of the weekend was the Sunday walk around the farm. Cross declared such walks were fascinating and excellent appetite providers. Students were encouraged to attend Edgmond Church and some male students were part of the Choir.[43] As for transport, in 1925 Frank Ward stated there was only one car and one motorcycle owned by students.

In the late 1930s students still tended to use bicycles or walk to Newport.[11] Once a month, Cross would cycle home to Shrewley, Warwickshire, for a weekend. It was a journey of fifty miles and took four hours. Arthur George cycled back to Pembrokeshire at the end of each term to save the rail fare. Most students would come to the College by rail, arriving at Newport station and then being fetched by the College's pony and trap.[44]

The day came to an end with roll-call, prior to lights out at 10.30 p.m.[36] Roll-call could be a precarious affair for staff, as Rowley wrote: *'One was never sure over whose door the bucket of water was delicately balanced, as you peeped inside the room to make sure it was a body in bed and not a set of pillows artistically arranged.'* The final event of the day was the locking of the iron gates 'that had wicked rotating spikes on top'.[45] The gates barred the entrance to the quadrangle between Old and New Hostels.[46] Those entering the hostels after hours and therefore 'by other means than the gates', usually by means of ground-floor windows, were fined.[11] In the late 1930s the fine was one pound.

The quality of the accommodation at Harper Adams was not high, although the exuberance of youth probably made this failing less noticeable. Martin Cross raised no objections to hostel life in the late 1930s, even though the wash-basins had only cold water [47] and the bath water ran at near boiling point. Old Hostel was antiquated and seriously lacking in washing and toilet facilities.[48] Since its opening in 1921, rooms in the notorious New Hostel had been improved with the conversion of as many pairs of rooms as possible into potential double rooms, thus achieving space without losing accommodation. Crowther appreciated,

7.7 Pony and trap belonging to the College: Ann Warner.

however, that the real need was for a new modern hostel. He was concerned that the overall bad impression given by the accommodation was losing the College students to its rivals. It was Crowther's suggestion that Old Hostel should be extended towards the dairy, which led to the opening of the Jubilee Hostel in 1951. Ancellor, the women's hostel, was similarly in need of improvement. As late as 1944 the only washing facilities at Ancellor were located on the ground floor and the only heating was by means of coal fires. Inspectors from the Ministry of Agriculture frequently criticized the conditions.

Some improvement in living standards arrived with the decision to change to public electricity.[49] As a new student, Cross had come well equipped with candles: *'Until 1937 the College generated its own electricity and so the electric lights were supposed to be put out at 10.30. Joining instructions advised students to provide themselves with candles to help with late studies.'* It was an end of an era when connection with the public supply was made in 1937. William Neate, as a boy in the 1930s, recalled visiting the old battery house with his father Billy Neate: *'You had the big acid battery house, with all the lead and acid batteries, rows and rows and rows of them. They were storage batteries which kept the supply going.'* His memories clearly predate the age of Health and Safety Requirements:

> *That was frightening when I went. Father took me in. When he first took me in there and the switchover [for a time the College continued to use its own supply for the daytime] was about half past nine at night and it involved pulling this big lever out and straight down into the connection at the bottom, and as you pulled the lever out so it flashed. Of course, that used to frighten me. It was a long time before I would go in there again.*

The Neate Family had a long association with the College. William Neate Snr arrived at Harper Adams in 1902, coming as Engineer at the invitation of the Principal. The two had been acquainted in Reading. In 1913, at the age of fifteen, his eldest son Billy began work and in 1916 they were joined by Aubrey, then fourteen years of age. The whole family ensured the engineering side of both the College and the NIPH ran smoothly for a total of sixty-five years. The brothers retired in 1967 on the introduction of the modernization programme.[50]

These were the days of copious domestic staff. They made beds, cleaned the student rooms, served at table, did the washing-up and undertook the laundry. Students provided their own sheets, pillowcases and towels, but all linen and clothes were laundered on the premises, as is still the case.[11] Even student socks were darned.[4] Domestic staff had to call students 'Mr'. They were never allowed to be alone in a room with a student and were only allowed in the hostels, for example to clean the rooms, before noon.[46] The young unmarried female servants were under the Matron and were offered accommodation in a dormitory in the Main Building.[4] As a pantry boy in the late 1920s Rowley well remembered the rigid hierarchy 'below stairs' and the hard work expected of the domestic staff. His very first day at the College began at 6.15 a.m. and ended at 9.00 p.m. He was paid a weekly wage of six shillings which included meals.[45] The Butler Maydew, nicknamed 'Pop' or 'Mildew', terrified him: *'He was a typical butler. If you think back on what a butler would have looked like, his paunch came out to about here, like this and he was exact, he had to be ... no flies on him. All of a sudden there would be such a commotion, you wouldn't believe. Something would be wrong, according to his approach and everything was upside down.'* Maydew had a reputation for eccentric motorcycling and was killed in a road accident in 1935.[51]

7.8 Library, c. 1935: A. J. Guy.

As regards study, the majority of the work was simply dictated notes. Wilson was especially famous for the amount of material he dictated. Twenty years later Ian Gedye as a student well remembered Wilson *'with his pipe and his gruff voice and old yellowed notes'.* Cross declared you could *'take his notes whilst completely detached and thinking of other things'.* In veterinary science alone, Cross took 400 pages of notes. There was no essay writing and no tutorials.[11] An important element of the work was the compilation of a farm diary, but even here the information was provided. Bill Thompson, the Farm Director, issued weekly information on 'what had been happening on the farm and what had happened at the markets'.[21] There was heavy reliance on the set textbooks. Students were expected to purchase either Watson and Moore's 'Agriculture' or Fream's 'Elements of Agriculture'. As James Scott Watson was an examiner at Leeds, the examination centre for the NDA, the choice was simple. Another recommended book was Lord Ernle's 'English Farming Past and Present', which Cross thought excellent for pressing flowers. For agricultural botany there was 'Common British Grasses and Legumes' by Thomas and Davies and the wise became acquainted with Robinson's 'Leguminous Forage Plants', the work of the College's own Dr Robinson. Textbooks were also recommended for book-keeping and surveying. Cross used the library on the first floor of the Main Building: *'Books, such as they were, carefully housed within glass cupboards and formally issued out on a weekly basis.'*
The library gained some importance late in the Second World War as conditions made books increasingly difficult to obtain and Malthouse took on the role of Librarian and

began the task of cataloguing. By 1945 he was calling for more space, although student use of the facilities does not seem to have increased.[9] The library is now a committee room, but its ancestry is still recognizable.

Barbara Blackburn, student 1927-30, was later to achieve national fame as the television personality and dog trainer Barbara Woodhouse. In her autobiography, she declared that the syllabus was very comprehensive and that she loved every minute spent at the College, learning about everything from the harvesting of crops and buttermaking to 'the secrets of the four-cylinder engine'. She appreciated the manner in which 'the subjects were put over in an interesting way and [she] was not treated as a sub-standard object.'[52] She clearly enjoyed the practical character of the course. The practical side of agriculture had always been an important part of instruction at Harper Adams and this continued under Crowther.[53] As Crowther proudly stated: *'All the courses of instruction at this College combine, to a degree hardly reached elsewhere, practical instruction with the more academic work of the class-room and the laboratory.'*[36]

The College offered a full range of courses. Students could study for the intermediate and final examinations of an external degree in agriculture taken in association with the University of London. Student files show that a handful of students from the very early days, before the First World War, had sat for external university degrees. There were courses for the National and College Diplomas which took three years. Harper Adams in the Crowther years also offered the opportunity to study for the NDD (National Diploma in Dairying). The NDA was directed towards those students not intending to go farming. As Sir Daniel Hall had explained in 1922: *'The National Diploma nowadays is perhaps not so much a guide to the education of the future farmer as a useful commercial asset to the student who wants to become a teacher or an official.'*[54] The examinations for the two parts of the NDA were taken in Leeds at Easter in the second and third years.[11] All papers were accompanied by vivas, the experience of which Cross well remembered:

Veterinary Science, with the inevitable question on the structure of the horse's hoof. The dear old man who examined us and conducted our 'vivas' had never been known to fail anyone. He gently chided me on not having attempted a question on the Notifiable Diseases, [...] with the words 'I don't know whether I will be examining here next year. I may be dead, but if I am I will look down from heaven and will know that there will be a question on the Notifiable Diseases. Run along young man, run along!'

Bob Tupper never forgot the experience of visiting Leeds for the first time: *'I've never been in such a filthy place in all my life. [...] You could put a clean, white collar on when you got up. By midday it would be filthy again.'* It should be noted that those taking the external examinations were still very few in number. Results for the NDA in 1934 were considered good and the figure amounted to just six students.[55] As regards internal courses, apart from the College Diploma, there was the College Certificate of two years duration. The College was also prepared to offer tailor-made courses of a year or even less.[56] Cross, for one, did not have a high opinion of the students in this latter category: *'The College was probably happy to get the money from them. In each case they were the sons of wealthy parents ... had motor cars. Time wasters really.'*

There were various requirements for the courses. It was recommended that students had some practical experience before entering the College. In the case of the NDA a year's practical work was a necessity. However, this could occur prior, during or after the end of the period of studies.[57] The children of farmers were exempt.[9] The College also sought a good general schooling, although student shortages meant that educational requirements were not always rigidly adhered to.[58] Crowther relied on personal judgement rather than evidence of formal education, preferring 'not to be fettered by "statutory" regulations'. However, for the NDA there was the requirement of the School Certificate [59] and for the degree the Higher School Certificate.[9] The minimum age on enrolment was sixteen for men and eighteen for women.[60] Women were eligible to join all courses at Harper Adams. Most female students, however, were enrolled at the NIPH. Barbara Blackburn was the sole woman on an agricultural course at the College when she joined in 1927. She wrote in her autobiography: *'A girl at an agricultural college taking subjects like building construction and surveying, veterinary science and engineering, was almost unheard of.'*

The lighter side of College life was not neglected, although low student numbers did reduce the diversity of the leisure pursuits on offer. The poultry students were very much a part of College activities, although the Poultry Institute Club, started in 1928, did organize various specialist talks, debates, trips and social events.[61] The famous cock-crowing competitions were a regular feature of its calendar.[51] All students of the Crowther years could play a variety of sports and in early June an annual sports day was held on the cricket pitch in front of the Main Building. Many of the villagers were invited and tea was served.[62] The College sports included hockey, rugby (from around 1926 the official college sport [63]), soccer, cricket, cross country running, boxing, table tennis and tennis. For a time in the 1920s there was even a club for golf at the College.[64] In 1935 it was

7.9 Lecture room, c. 1929.

7.10 1928-29 College mixed hockey team, Barbara Woodhouse (nee Blackburn), front row fourth from left.

said that the Shooting Club 'probably catered for the entertainment of more students than any other SUC activity'.[65] That year a female student once again won the annual shooting competition. Shooting practice took place in the large lecture room. This room was the venue for all manner of activities from boxing to plays and dances,[45] with the sliding doors opened for additional space.[11] On 10 October 1938 it was the scene of a fatal accident, when a Canadian student, Larry McLaughlin, was killed.[63] Another student 'not understanding the use of the sights on the rifle [...] fired his first round and shot and killed the student who was spotting near to the target (where he should not have been)'.[11] This proved to be the end of the indoor rifle range. Billiards was an extremely popular game. Jack Large, student 1944-46, recalled that T. C. Ward, Chairman of the Governors 1942-56, would invite students to play at his farm. The billiards table was situated in the student common room. By the late 1930s the common room ran to a radiogram 'with a limited supply of records', as well as the customary newspapers and periodicals.

As Cross concluded: *'We must have been easily satisfied in those days.'* Women possessed their own facilities at Ancellor and did not go to the men's common room.[9]

7.11 Rag procession, 1939.

For those less sport minded there was a bridge club and a debating society. Dances were a prominent feature of College life. There were three organized dances over the session, with the most important being that at the end of session.[11] The shortage of lady partners meant that outside guests were permitted.[61] However, Cecil Warner, student 1932-33, clearly remembered that fraternization between students and local girls was usually banned in his day. There were also several impromptu dances on Saturday evenings,[66] often with the radiogram or piano [67] and sometimes with an amateur jazz band.[68] In the session 1933-34 the College boasted its own band.[55] An annual variety show was performed in February or March for the benefit of all connected with the College and their friends. Afterwards the students waited on their guests.[67]

The origins of amateur dramatics at the College lay in the early 1920s, when staff members Robinson, Jary and Preston would perform a concert, featuring a 'burlesque operetta' based on Gilbert and Sullivan. One year the story-line was loosely based on 'The Mikado' and revolved around the discovery of the foot and mouth germ in the branch of a willow tree.[67] From 1929 there was a full-length play performed by the students in December and produced for many years by Robinson. Receipts went to the local hospitals.[67] These plays, although a result of just a few weeks of rehearsal, were always praised for their high standard. In 1930 the review stated: *'Miss Blackburn [...] did everything, including rolling off the settee, with her customary enthusiasm.'*[68] Cross performed in 'The Amazing Mr Clitterhouse' in 1938. That year's performance was especially memorable as rats had eaten through a lead pipe causing a cascade of water to pour down mid-performance.

The Farm Diary of Martin Cross 1937-39

As a student at Harper Adams, Martin Cross was required to keep a Farm Diary. Fortunately, the Diary, covering the period between October 1937 and June 1939, has survived in the careful keeping of its author. The following extracts from the Diary include later comments from Martin Cross, viewed from the perspective of a lifetime spent in education as a lecturer in agriculture and extra-mural organizer at Hertfordshire and Warwickshire Colleges of Agriculture.

7.12 Reaper-binder in use at Ancellor, early 1940s: G. S. McCann.

Week ending 9th October 1937. [Lifting potatoes] was done by piece-work which was measured by a peculiar measurement known as the Shropshire rood. Now [...] a rood is really 1/4 acre but the one used for potato lifting in Salop was completely artificial and was approximately 72 sq. yds i.e.12 rows of potatoes 8 yards long. It was done as a family undertaking, father exposing the potatoes with a fork and his wife and family picking them into buckets and baskets and throwing them into a series of carts passing by to carry them to the 'clamp'. This worked out at a lifting cost of about £4 4s 0d an acre. A hard working family might lift nearly an acre in a week and the chance to do this type of work was valued.

Week ending 7th November 1937. [The Dairy Herd was] at that time housed in a spacious shippon adjoining the Dairy, to which the milk

7.13 Martin Cross, student 1937-40: Bertram Sinkinson.

was carried before being made into butter or cheese. The cows c.30 were milked by hand. It was one of the early morning and afternoon tasks of the First Year Agric. students. The winter ration on offer was a mixture of one part decorticated groundnut cake, 1 part palm kernel cake and two parts of maize meal fed at the rate of 3 1/2 lbs per gallon of production after the first gallon for which about 21 lbs hay was fed. The cows were yielding about 700 gals per lactation.

Week ending March 6th 1938.
Manuring in preparation for sugar beet:
10 tons per acre farmyard manure plus about 1/2 ton poultry manure from the laying trials and 2 cwts Muriate of Potash,
3 cwts 35% Superphosphate,
2 cwts Sulphate of Ammonia.

Week ending 8th May 1938. W. B. Thompson's invariable method of weed control and tilth preparation: 'Potatoes' he said 'should be turned round in the bed three times before they come up.' By which he meant that as soon as the potatoes had been hand planted and ridges formed down the row, these ridges should be harrowed down - the first turn. Some days later the ridges would be re-formed - the second turn. Just before potatoes begin to emerge, the ridges would be harrowed down a third time - the third turn. Each cultivation having the effect of killing weed seedlings. Indeed, he was not averse to giving the crop another good harrowing when the shoots were well up, before the final ridging-up as leaves began to meet in the row.

Week ending 20th June 1938. Charlock spraying (in cereals) [...] was done with a hefty pump mounted in a 36 gallon beer barrel mounted in one of the farm carts with a spray bar behind. Pump worked by student power - me.

Week ending 19th February 1939.
A man engaged laying hedges at 5s per chain (22 yds) complete with staking and heatherings.

Week ending 14th May 1939. This week we have sheared the sheep, starting with the tegs which carry a heavier fleece and will be better without it in the warmer weather. It was a little early to shear the ewes. These are generally sheared about the end of the month. "We are going to give him 8s a score. [...] This is a generous payment, 6s 8d is a good price."

Chapter Eight —

Beginnings of the Advisory Services and the NIPH 1922-39

The 1920s and 1930s were extremely difficult years for Harper Adams in its role as a provider of education. Yet, in other significant areas of work, the College was assuming a higher profile on the regional and national stage. Bright spots there were in the College's history at this time and one of these was the Advisory Services. Despite falling student numbers, by the 1930s the rapid growth in advisory staff was leading to a serious problem of a lack of space. In a memo to the Governors in 1933 Crowther wrote: *'Every inch of available space has long been adapted for purposes of work, much of which is being carried out under conditions of serious disadvantage.'* The academic staff, including Advisory Officers and their assistants, now numbered thirty and, with the inclusion of clerks and outdoor staff, the total number of employees at the College exceeded one hundred.[1]

Crowther considered education to be the priority of the College, but education in the widest form. From the beginning, perhaps due to his long background in research and his recent experience in the commercial world, he emphasized the importance of the College's advisory role both within the Province, still consisting of Shropshire, Staffordshire and Warwickshire, and nationally. Within months of Crowther's appointment, a newspaper report on his welcoming address to a group of visiting farmers stated: *'He would put forward the best effort he could to make the College really useful, not only to the district surrounding it, or to those counties that contributed to its support, but to the whole country. [...] He wanted the farmers to feel that the College was there to help them, and that the farmers on their part were ready to help the College staff.'* [2] Crowther was astute enough to realize that the harsh climate for farming had created a golden opportunity for the College. At a time when farmers were keen to find a way out of the mire of the Agricultural Depression and Harper Adams was beating its own path to popularity, the College's Advisory Officers out and about in the community with their scientific knowledge packaged in practical advice epitomized the hopes of both parties. It is true that smaller farmers were often caught in the trap between high costs of production and low product prices, but larger farmers were increasingly turning to the Advisory Services.[3] Frank Ward, student 1925-27, was convinced that it was the College's advisory role which won it a sound reputation among local farmers and which won it support, even encouraging recruitment. As the post-war drive to promote the Advisory Services throughout the country gained momentum in the 1920s and 1930s, Crowther ensured the College seized every opportunity.

When Crowther arrived in 1922, the situation as regards the Advisory Services at the College was not auspicious. The Governors had just asked for the transfer of the College's one and only advisory officer, Roebuck the Entomologist, as a result of the crisis of 1922. The Advisory Services at the College were in effect non-existent. Two years later Crowther could declare that farmers either had no knowledge of the College's advisory role, or if they did, they had no idea that the advice was free and the costs of analyses reasonable.[4] There was little doubt in Crowther's mind that things had to be turned around. By the time of his retirement in 1944 the situation was very different. Advisory officers were squeezed into every available space. The Advisory Economist was housed in one of the army huts, previously used for student accommodation

8.1 Dr Clegg and staff undertaking bacteriological work with milk, early 1940s: G. S. McCann.

and lectures,[5] situated between Old Hostel and the dairy, on the site of the later Jubilee Hostel.[6] The Advisory Chemist worked in the Chemistry Department, despite the fact that it was space originally intended for instruction only. Advisory Mycology and Entomology were in the attics of Old Hostel. The Advisory Dairy Bacteriologist occupied two of the former dormitories in the Main Building.[7] As regards advisory veterinary work, for the majority of the period, the original veterinary laboratory in the east wing on the first floor had to cope with a growing amount of work. Not surprisingly, Crowther underlined the real need to erect a separate building for the Advisory Services and to return the space used by them to teaching purposes. This was a national problem, for although the Ministry of Agriculture paid the salaries and expenses of advisory officers, it made no provision for their accommodation.[3] The Luxmoore Report of 1943 ensured, however, that the future of the Advisory Services ran a very different course, quite divorced from educational institutions such as Harper Adams.

In 1923 the Advisory Services at the College were reborn with the employment of the College's first Advisory Mycologist, Norman Carl Preston,[8] who was to earn the nickname 'Pongo'.[9] He received national recognition for his research into the control of club root in brassicas.[10] He remained at the College until the Advisory Services moved to Tettenhall, Wolverhampton, in 1946. He was a skilled pianist and his departure meant a great loss to entertainment at the College.[11] In October 1923 the vacant post of Advisory Entomologist was filled by S. G. Jary.[12] Jary remained until 1935, when he was succeeded by H. C. F. Newton, who stayed until the department moved to Tettenhall. 'Newt' was described 'as a quiet individual with a great sense of humour and a mind so active and mercurial that it was at times impossible to keep up with it'.[13] He and Preston formed a partnership in the Province famous for its talent and eccentricity. Both posts had been agreed as early as 1920.[14] It was not until 1924 that a new post was created, that of Advisory Chemist. As no suitable person could be found, the Head of the Chemistry Department, Morley Davies, assumed the role.[15] By the session 1929-30 the prospectus could list advisory officers in mycology, entomology, chemistry, economics and dairy bacteriology. With these officers went a whole range of research assistants.

The situation as regards the Advisory Veterinary Officer was somewhat different. The College's veterinary lecturer had always given help and lectures to the farming community. It was not until the retirement of Wilson in 1928 that the College sought financial recognition from the Ministry for this continuing role.[16] William Rowan Kerr was appointed Veterinary Lecturer in 1929 with a dual job description and shared funding.[17] In 1931 K. D. Downham was appointed Veterinary Lecturer and Adviser.[18] He was a very tall man, who 'sort of walked and moved as though there were no bones in his body. Always seemed to be made of rubber'. It was soon apparent that the funding arrangement left the Province far short of its veterinary needs. In 1937 the Ministry took over the complete funding of the post, on the understanding that the officer's teaching role would be minimal.[20]

The range of work covered by the College's advisory staff was vast. The majority of the work was to do with handling enquiries, making visits, giving talks and undertaking surveys and investigational activities, the latter invariably linked with national work carried out at other research institutes but based on local farming problems. There were recurrent problems with lack of staff and financial resources, but this situation was not peculiar to Harper Adams. It was a nationwide problem. The Ministry of Agriculture exercised little overall direction and control.[3] In hindsight there is a tendency to draw a clean-cut line between the teaching and the advisory members of staff, but this was rarely the case at the time. Students such as Tony Gale made no such differentiation. A small proportion of the time of advisory officers was given over to teaching students. Tony Gale, for example, remembered going to the bacteriology laboratory to find out how clean his hands were. Leonora Grice talked of lectures with Newton: *'Here [in the Zoology Department] Mr. Newton, a trim, slim, white-haired man with glasses and a waspish temper on occasion, dealt with the life cycles of various pests.'* Similarly, those on the teaching staff 'undertook experiments and advice in such times as could be spared from teaching'. [21] Robinson especially was involved in advisory work, notably the Ministry's scheme for the botanical examination of grass fields managed for grass drying and he also acted as an adviser on the laying down of grassland for the National Laying Test at Milford, Surrey. However, a prophet is not without honour, save in his own country. In the College his was the task of improving the sports field and tennis courts.

Perhaps, the most important work undertaken by the Advisory Services at Harper Adams was in the area of alerting farmers to the benefits of analysing soil fertility. This was the demesne of the Advisory Chemist, Morley Davies. Members of the Ward Family were pioneer farmers and Frank Ward remembered his father, Stephen, taking Morley Davies's advice on soil structure. He would give Morley Davies Sunday dinner and then they would walk around the fields. Rowley similarly emphasized that advisory visits could often take on a very sociable nature, enjoyed by all. Robinson,

8.2 Morley Davies carrying out plant nutrition experiments, early 1940s: G. S. McCann.

while praising Morley Davies's skills as a scientist, noted he was *'very practical, with the ability to get on well with all sorts of people'*.[22]

In contrast, many advisory officers had little local knowledge and found it very difficult to win the respect of local people.[3] Of particular interest to Morley Davies was the use of lime to improve soils. This was of considerable relevance to the local area with the predominance of light soils and an interest in growing sugar beet. From the late 1920s a whole series of liming experiments was undertaken at Harper Adams [23] and on neighbouring farms.[24] Pioneering work such as this led to the Government's Land Fertility Scheme of 1937, which encouraged greater use of lime and basic slag by giving subsidies on their purchase.[3]

Morley Davies was also a pioneer of soil surveying. In the 1920s there was a growing demand for more exact information on the nature and distribution of soils and from 1926 Morley Davies received finance from the Ministry of Agriculture to develop soil survey methods in association with specially appointed surveyors.[25] It was at this time that a soil survey of North Shropshire was started.[26] In the early 1930s 'the eventual stabilization of the methods used in the mapping and classification of the area took place' under the surveyor Gwilym Owen and 'the identities of most of the [soil] series were established'.[25] The results of detailed mapping were reported by Morley Davies and Owen in 1934.[26] The survey of the Wem District was published in 1954 as part of the Memoirs of the Soil Survey of Great Britain.[27]

From the late 1920s sugar beet provided a valuable opportunity for the College to work for the local community. Morley Davies made a considerable contribution to contemporary knowledge of the growing of sugar beet. In 1928, for example, investigations took place into matters such as manuring, ridge and flat cultivation and suitable spacing.[23]

Concurrently, regular conferences were held at Harper Adams to distribute the information stemming from research. Frank Ward attended such conferences and testified to their benefits to farmers. The development of the sugar beet industry was one of the few bright spots in British farming at this time and of noted importance to Shropshire. It was a cash crop which could also provide fodder for livestock in the form of beet tops and pulp returned from the factories and it could act as an essential cleaning crop in the normal rotation. In 1925 the Sugar Beet Act made finance available for a chain of beet factories and offered subsidies to farmers for the production of the crop over ten years while they learnt the techniques of growing it. With the opening of the beet factory at Kidderminster in 1925, progressive farmers in Shropshire were becoming interested in growing sugar beet.[28] Their determination led to the opening of the beet factory at Allscott in 1927.[29] An old student of Harper Adams, G. K. Parkes, was the factory's first manager.[23] The Allscott Factory gave further impetus to the growing of the crop in the area. Significantly, Shropshire was the only county in the western half of the country to develop a sugar beet industry. In 1934 the acreage under beet in Shropshire was 16,017.[29]

The Advisory Economist also undertook a growing workload, as the idea gained ground that efficient management was not simply a matter of good practice and sound science. As Sir Daniel Hall wrote in the 'Journal of the Royal Agricultural Society' in 1922: *'The teaching of farming should be not only descriptive of the customs of existing farmers, but should pass on to discussion and criticism, checked step by step by determined costs and results. The consideration of costs, whether of materials and labour, must everywhere dominate the treatment of the subject.'* In 1913 the Agricultural Economics Research Institute had been established to compile farm records and to calculate costs of production. Economics, as opposed to simple book-keeping, was becoming part of the syllabus for agricultural students, for example as part of the syllabus for the University of London's B.Sc. The subject was first taught at Harper Adams in 1923.[8] Probably, many students felt as Leonora Grice did during the Second World War. She considered economics to be 'a lot of waffle [...] about what was only common sense'. The work of the Advisory Economist at the College was very much a matter of costings, whether studying poultry production or contributing to the national milk costings survey. Cecil Warner, a local poultry farmer and former student, remembered in the 1930s providing monthly returns in order for his costs of egg production to be calculated. He spoke of the costings service as being highly thought of and very useful. Similarly, Harper Adams was involved with the profit and loss sheets for the farm poultry unit at the Lancashire Institute of Agriculture at Hutton. These proved of considerable use to the staff there.[30] In 1938 the College saw both the appointment of a new Advisory Economist, J. J. MacGregor, who took over from F. S. Dennis, and the start of a national farm management survey to investigate the financial position of farmers over a five-year period. Throughout the West Midland Province 180 farms were studied.

The Dairy Bacteriologist, Dr A. L. Provan, began work at Harper Adams in the session 1928-29.[5] Provan was mainly involved with routine sampling and advisory work in association with clean milk. His workload increased dramatically with the development of a national market for milk in the 1930s. Such a development required the uniform administration of regulations concerning the condition of cowsheds, the health of the cows and the hygiene of the milk. In 1935 there was a massive seventy-five percent increase in the number of samples being analysed, rising from 2,655 to 4,630, due to the inauguration of Accredited Registration by the Milk Marketing Board. The Bacteriology Department undertook the whole of the examination of Shropshire samples taken under the scheme and also gave help to Shropshire County Council in the inspection of farms applying for a licence. That same year 250 farms were visited as a result of problems arising from the designated standards for dairies and parlours. Warwickshire similarly used Harper Adams for much of its sample analysis, but Staffordshire County Council possessed its own public health laboratories and did not call upon the College for assistance. The amount of work which this involved meant that there was little time available for research. Ironically, it was the fees from sampling which were intended to finance any research, but the time taken up with the sampling allowed no time for such an activity.[3]

The Advisory Veterinary Officer was similarly stretched to capacity. In the 1930s as well as routine post mortems, an important part of his work concentrated on the use of preventative medicine to control mastitis, contagious abortion, tuberculosis and various poultry diseases. The difficulties of meeting the volume of work were exacerbated by the inappropriate location of the laboratory on the first floor of the Main Building. In the late 1930s Cross recalled the distinctive noise each morning as the vets tapped down chickens for post mortems in the old laboratory as he sat in the room below. In 1934 a total of 2,087 post mortems were performed on turkeys as part of the research into blackhead. Jack Wickstead remembered the carcasses of calves being carried upstairs to the old premises 'and back again' in order for post mortems to be performed. In 1938 an extension to the Advisory Veterinary Department was built at a cost of £460, with half the building costs and all the equipment being financed by the Ministry of Agriculture.[31] The building is now used by the National Institute of Agricultural Botany. The extension consisted of a diagnostic laboratory, a post mortem room, an office and a small animal house.[32]

Thousands of guinea pigs were bred and used for milk testing.[33] The old laboratory was retained for practical demonstrations and lectures, blood testing, sectional work and the preparation of culture media. The year 1938 was significant for that year the Ministry of Agriculture created a national veterinary service, centralizing the campaign to improve animal health. With the continuing drive against tuberculosis, veterinary duties under the Tuberculosis Order were transferred from Local Authorities to the Ministry and Harper Adams was made the diagnostic centre for the biological examination of milk, with samples for diagnostic purposes gathered from Shropshire, Worcestershire and Herefordshire.

As has been seen, the College had sound links with the farming community, but there was the wider need to forge an interrelated network for agricultural education. Sir Daniel Hall envisaged a college as 'the intelligence centre of its area' [34] and Crowther adhered to this view. Thus Harper Adams promoted regular conferences within the Province, gathering together teaching staff, advisers, county agricultural organisers and instructors and representatives from the Ministry of Agriculture.[12] By the mid-1930s such conferences were held four times a year. Similarly the College sought to maintain

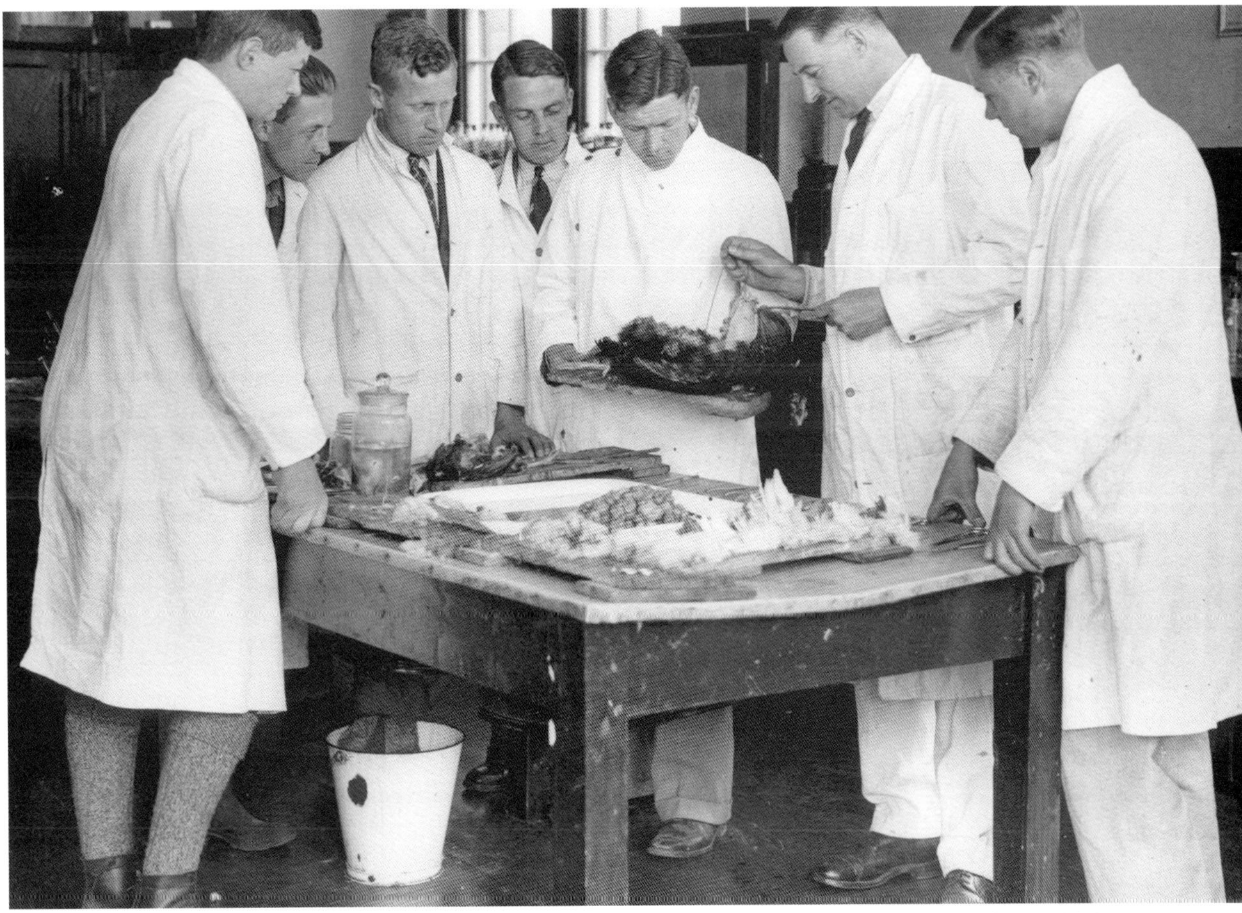

8.3 K. D. Downham, Advisory Veterinary Officer, explaining diseases of chicken to students, 1936: Sport & General.

links with County Councils and the Farm Institute at Rodbaston and to promote joint exhibitions at various venues.[12] However, despite such efforts, co-operation within the Province was not of the highest order. In matters of grant aid the contributing County Councils were not generous to the College, giving a total of £1,850. In contrast Seale Hayne received £4,000 from the Councils of Cornwall and Devon and the Midland Agricultural College £5,100 from the surrounding counties.[35] In a report of 1934 Crowther expressed concern at the low entries for students from Harper Adams' contributing counties. His figures for 1934-5 showed that for the College the proportion of these local students out of the total student body was twenty-three percent, while for Seale Hayne it was seventy and for the Midland Agricultural College, now centred at Sutton Bonington, it was eighty-three. In 1935 the Block Grants Committee of the Ministry of Agriculture emphasized the lack of council support for several colleges, including Harper Adams. It recommended changes in the provision of agricultural education in the West Midland Province, whether in the form of increases in the maintenance grants and scholarships financed by the County Councils, or in co-operation as regards advisory facilities. In July a conference of representatives of the Three Counties and the Governors of Harper Adams was held at the Ministry. Agreement was reached on improving local recruitment and promoting better use of the facilities at the College. Yet, there was not a great improvement. During the Second World War Crowther complained that the County War Agricultural Executive Committees did little to ensure 'the fullest possible benefit for farmers' from the College's services. He continued to put his faith in the strong links forged by the College with the farmers themselves. Crowther's diligent support for and encouragement of the Advisory Services was bringing in sound returns.

Within the category of advisory work there is the significant contribution of Bobby Boutflour, a figure of national importance on the farming scene. Boutflour arrived at the College in October 1926 as Director of Dairy Husbandry to the Ministry of Agriculture.[5] It was a post created specially for him and possessed the considerable annual salary of £1,000.[36] At this time the Director of the NIPH was being paid a salary of £750 a year, plus an accommodation allowance.[15] Prior to his arrival Boutflour had been Chief Agricultural Adviser in Wiltshire. There Boutflour had won a national reputation for achieving high milk yields. After chairing a meeting with Boutflour at Shrewsbury, Crowther made the suggestion to the Ministry of Agriculture that a post should be specially created for him at Harper Adams, enabling him to travel and to spread his expertise.[37] Within five years Boutflour had made Harper Adams 'the Mecca of dairying enthusiasts from all parts of the country'.[5]

Boutflour's research centred on maximizing profits from a smallholding by means of high milk yields. He took over Ancellor Farm as a model smallholding of forty acres, complete with cowshed, silo, dairy, Dutch barn and outbuildings.[38] The cowshed still survives and externally, traces of Boutflour's feed mechanism, once on rails around the walls, can still be seen. A herd of twenty tuberculin tested pedigree Friesians was purchased in September 1927,[39] with the aid of a grant of £1,000 obtained by Crowther from the British Friesian Cattle Society.[37] Most of the cows came from the famous Terling Estate Herd belonging to Lord Rayleigh.[36] In the past Ancellor Farm had been devoted to soiling experimentation. Now the land was devoted to commercial cropping on a three-course rotation, with crops such as potatoes, sugar beet and silage and a saleable catch crop such as cabbage.[23] Profits, rather than the crops, were utilized to run the dairy enterprise. Cows were pastured in summer and the grassland was managed on the new intensive system involving nitrogen

8.4 Bobby Boutflour's Friesian herd at the Experimental Dairy Farm, c. 1929.

fertiliser and close grazing.[21] Grass had been the least considered crop of the time, but from the 1920s valuable research was being undertaken at Jealott's Hill by ICI, producers of the new synthetic nitrogenous fertilisers. Here demonstrations took place illustrating how well grass could grow with the use of nitrogen, phosphates and lime and with the encouragement of grazing in rotation to maintain an even production of young foliage. Over winter the Ancellor cows were rationed according to individual milk yield and fed on concentrates, hay and silage according to Boutflour's own theories on feeding.[40] His ideas revolved around control of the dry matter fed and reduction in roughage as concentrates were increased and precluded any feeding of roots when roughage was reduced. One student turned the Boutflour system into verse: [40]

Who'd get from cows a larger flow
Of milk in Boutflour fashion,
By giving lots of H_2O
And cutting down the ration?

Then ploughing up the fields of roots
And having grass seed sown,
And feeding with the tender shoots
When just four inches grown.

Who thrice per day would then extract
The lactic juice which flows,
If to this system they're exact?
Well. Goodness only knows!

At a time when roots were liberally fed, Boutflour's ideas initially caused considerable suspicion and earned him the nickname

'No-Roots Boutflour'.[37] However, his great knowledge and enthusiasm, coupled with an informal lecturing style and humour, won him much admiration. Often he would start a talk by asking: *'Why have you all come here tonight?'* He would then proceed to answer: *'You don't know? - well I'll tell you, it's because you're a lot of b..... fools who cannot do your own job and I am here to tell you how to do it.'* [37] On another occasion, when asked what he would put in wet land, he replied: *'Drain pipes, you silly B...!'* [37]

As early as 1927-28 the Ancellor Herd achieved the highest average milk yield for the Shropshire Milk Recording Society with an average yield of 1,231 gallons for fifteen cows. As a consequence the herd was awarded a Silver Cup by the British Friesian Cattle Society.[41] The following two years saw the herd maintain their supremacy, with four cows having individual records of 2,000 gallons. 'Terling Eclipse 10th' was the highest credited, with a record 2,498 gallons.[41] One home-bred heifer, 'Harperadams Cinderella', established a new British record for first calf heifers with 2,356 gallons in 365 days.[41] In contrast, the College Shorthorns were giving on average 600 gallons per cow.[37] By the early 1930s the Friesian herd numbered almost sixty and included a prize-winning young bull, 'Tarvin Undaunted'.[42]

Boutflour travelled thousands of miles to inform farmers of his research. He gave talks and made advisory visits throughout every county.[37] The profits from his booklet, sold for one shilling, were used to finance this extensive travelling. Boutflour personally described the book as 'a successful venture, and very good propaganda for the College'. Frank Ward remembered his father taking his herdsmen to listen to Boutflour. He also had fond student memories of playing cricket with him at the

8.5 Harperadams Kernel at the Staffordshire Show, 1932.

College. Boutflour had a reputation for batting the ball through the windows of the Main Building.

By 1930 the Boutflour system had spread throughout the country and was a decided success. That year disaster struck. Boutflour had accepted a three-month travelling fellowship of the United States.[37] In his absence two cows were poisoned by lead through 'licking the newly-painted cricket screens and goal posts on the sports field'.[36] He returned to find the remaining herd in a very poor state. Amid the disappointment and with the belief that his work in spreading the word about balanced rations was more or less complete, he accepted the invitation to apply for the post of Principal at the Royal Agricultural College.[36] On 1 February 1931 he took up the appointment.[37] The research went with him. In November 1932 there was a partial dispersal of the Friesian herd and the remaining cows were merged with the College's shorthorns. The sale was a great financial success, with £2,000 being paid for forty-five animals. At approximately £45 per animal, it was the highest average price per head that year.[5] However, a great blow had been dealt to the prestige of the College and Boutflour's departure was sorely felt by Crowther, who had done much to support him.[43]

In discussing research at Harper Adams the work of national importance personally undertaken by Crowther in indoor pig nutrition and management must be mentioned. The Pig Husbandry Experimental Station was established at the College in February 1926, as part of a national scheme linked with the Animal Nutrition Research Institutes at Cambridge and Aberdeen. Experiments in pig nutrition to demonstrate the economic merits of various home-grown feeding stuffs had, however, started earlier on Crowther's arrival.[8] The unit consisted of a new brick piggery in the form of a quadrangle open to one side, built to the

8.6 Pig Husbandry Experimental Station, c. 1929.

west of the stackyard. Financial assistance for building came from the Development Fund and annual Ministry grants supported the research.[21] There were eight breeding pens, eighteen feeding pens, two reserve pens, food stores, weighing room, exercise yards and paddock. In 1939 a larger food store was built.[44] A Large White breeding herd of ten to twelve sows and gilts and a boar was maintained.[45]

The rearing of pigs was traditionally associated with dairy farming due to the feeding of animals with dairy by-products. Feeding with cereals alone produced poor results arising from protein and mineral deficiency and thus any break in the link with dairy farming was considered unprofitable. The feeding experiments carried out by Crowther effectively sought to experiment with the various economic values of different types of milk products and with the use of cheap protein-rich vegetable foods, such as soya meal, in place of the expensive animal foods, such as fish meal. Similarly, there were experiments to provide improved substitutes for traditional cereal, such as dried sugar beet pulp. Experiments were also carried out with mineral supplements and cod liver oil. There was a limited number of management experiments, covering everything from water supply to restriction of food and the benefits of day pasturage as against housing for post-weaners.[45]

Undoubtedly one of the brightest spots in the history of Harper Adams during the period of Crowther's principalship must be allocated to the National Institute of Poultry Husbandry. Indeed, throughout the years of the Agricultural Depression poultry keeping made a significant contribution to farming as a whole. Strictly speaking the NIPH was independent of the College during this time, but its origins lay in the College and from 1932 Crowther held the post of Director. He declared the title was honorary, but the fact remains that the NIPH was officially under the Principal of Harper Adams for most of the Crowther period.

8.7 Interior of the Pig Husbandry Experimental Station, probably early 1940s: Farmer & Stockbreeder.

Furthermore, the relationship between the two institutions in matters such as the utilization of staff and facilities of all types was extremely close. Agricultural and poultry students were, however, always educated separately.[46]

It was mild and dry weather on 3 November 1926, the day chosen to mark the official opening of the NIPH.[5] Panic and frustration had preceded the day and hard work and precise planning dominated it. There had been much scrubbing, cleaning and polishing. All this could mean but one thing; there was to be a very important visitor. The Duke of York, the future King George VI, had agreed to honour the ceremony. Altogether there were 1,500 guests, including the Rt Hon. Walter Guinness, Minister of Agriculture,[47] and 'everywhere was stiff with detectives and police'.[48] Stories were to be recounted of the minutiae of the day long after the event had been consigned to history. There was the story of the local reporter, known as 'Castagloom',

8.8 The opening of the NIPH in November 1926. W. H. Lander, Chairman of the Governors, reading the address, with the Duke of York sitting to his right: Graphic Photo Union.

flagging down the Lord Lieutenant's car and jumping in so that he could get his copy back to the newspaper office in record time.[24] Maydew the Butler, always inclined to panic, served the Duke from a homely cracked jug when thrown by the royal request for milk as opposed to cream.[49]

The opening ceremony took place outside the NIPH Administrative Building, now part of the School of Management. Three large covered stands had been erected in the event of rain. These, and the loyal greeting suspended from the roof, unfortunately obscured the building and gave the place the impression of being 'an aerodrome'.[50] Rowley, as a pantry boy, had memories of making tea for the guests. He made 'umpteen gallons of the stuff', using 'two thumping great urns', with all the hot water and the milk having to be transported by horse and cart to the table from the Main Building. Portable boilers had to be hired to supplement the kitchen's supply of hot water. The Principal's Room, now the Accounts Office, was turned into a dining room for the day, with 'silver worth a fortune' gracing the table. And, at the end of the day Rowley well remembered 'the state of floors after thousands of feet had imprinted them with mud' and 'the mountain of soiled crockery'.[48]

It was a momentous occasion. One national newspaper wrote that the event 'is a tribute to the general efficiency of the College, and

8.9 The NIPH, 13th May 1927: Aerofilms Ltd.

to the attention it has paid, in the past, to the Cinderella of agriculture. [...] Cinderella is no longer a person to be despised. At last the poultry industry is recognised as something more than "a pin-money industry for the women-folk" '.[51] In 1921 a scheme for the NIPH had been agreed.[52] It was to consist of six sections, with the majority associated with research and experimentation and the sixth being that section eventually established at Harper Adams. The latter was to concentrate on higher teaching in poultry husbandry and practical investigations of a direct commercial character. At this time probably at least half of egg and poultry production took place on general farms and it was thus felt important that poultry husbandry should be regarded as a branch of agriculture and that such work should be carried out in close association with an existing agricultural college.[53] In 1923 Harper Adams had been chosen for the site of the NIPH and the Governors agreed to pass twenty-six acres of land, over the Shrewsbury Road from the College, to the new Institute, along with the existing poultry plant and livestock.[47]

The NIPH was built at a capital expenditure of nearly £15,000.[54] This sum was raised in part by a contribution from the Poultry Industry, organized through the National Poultry Council. The remainder of the cost and maintenance requirements were met by the Ministry of Agriculture.[47] The architect of the new buildings erected on the site was Major H. P. G. Maule.[50] The central feature of the plant was the Administrative Building with a lecture room, offices and an incubator room. Close by was the Service Building with food stores, a carpenter's shop, an egg room, a hospital and an engine room.[47] There was a large brooding house with a capacity for 2,500 chicks, two long laying houses, a house for male birds and special experiments, a multiple unit commercial laying house, a farm poultry house for 150 fowl and a large number of breeding pens. There were also about twelve acres of free range poultry. The stock was to comprise 1,800 fowl, 400 ducks, 40 geese, 75 pairs breeding pigeons, 750 rabbits and 12 goats.[38] The dawn chorus and the volume of noise at feeding time were distinctly remembered by William Neate.

The former Poultry Department, adjoining the Shrewsbury Road, was given over to waterfowl, pigeons, goats and rabbits. There was a duck house with a capacity for 350 ducks and breeding pens.[47] In the 1930s there were extensive and innovative investigations into the production of table ducks. The work involved was considerable, as Rowley, by then employed at the NIPH, recounted:

> *Enough records to paper a block of flats were required to detail all the information regarding feed consumption, individual body weights and mortality. Between March and September some two thousand table ducks were produced. The feed had first to be mixed on the floor with a shovel and then made into wet mash by hand. The allowance for each pen was weighed, three times a day for the first four weeks and twice thereafter, with anything up to fifty buckets at feed time. I have often wondered why my arms don't reach the floor, from the weight of countless thousands of buckets of water carried in a vain attempt to quench the thirst of these creatures. [...] You either pushed, pulled or carried everything, all sacks of feed were in 112lb., 140lb. or 240lb. and the only help available was a horse and cart and a mule and cart each Friday.*[48]

In the same area there were two houses for pigeons.[38] The first Director of the NIPH, Willard Thompson, had plans to investigate the breeding of the birds for table production. The idea derived from Thompson's homeland, America, but the consumption of pigeon in this country never caught on and the birds were eventually dispensed with.[9] Goats were kept on grass ranges to 'illustrate the usefulness of goats as an adjunct to farming especially poultry'.[38] Experiments were undertaken on milk production and some of the goat milk went to the nursing does.[48] The goats were far from popular with the staff and were eventually sold.

The most important section of these subsidiaries was the Rabbit Department. The former Poultry Department Office was given over to 220 hutches.[47] The Head of the Department was William King Wilson, who had joined the staff in 1925.[55] He remained in post until his retirement forty years later. He was a very tall and distinguished looking man and possessed of some wealth: *'The tale was he had a big Sunbeam car. At the time you had the [petrol] pumps with the big glass. He used to pump petrol in. It was rationed. He went to fill up with petrol and the chap said "Would you mind switching your engine off because it's gaining on me!".'* [30] He was devoted to his work and insisted on all who visited his rabbits being 'nice and quiet' so as to avoid disturbing them.[56] Rowley often acted as a guide:

> *Conducting these parties round was quite pleasant, until we entered the stud buck house, then it could be very embarrassing for many had the most ungentlemanly habits. As the ladies bent*

8.10 1925-26 first poultry course at the NIPH.
Staff in front row from left: Prof. Willard Thompson, Dr Charles Crowther, Herbert Howes, Mr Dixon.
Staff in second row from left: F.C. Bobby (fifth), W. King Wilson (eighth).

down to have a closer look at these splendid creatures, their large hats, decorated with artificial fruit, which were fashionable at the time, proved favoured targets for the little 'fiends'.[48]

The Department experimented with every aspect of breeding, rearing and feeding through to meat, fur and wool production. There was growing interest in the consumption of rabbit meat, although prior to the war the main direction of the research was fur production.[30] Not surprisingly, rabbit was eaten in the College dining room.[48] By the outbreak of war King Wilson had researched the carcass qualities of over forty varieties of rabbit. The Department was also involved with teaching, rabbit husbandry being an element in the various poultry courses.[57]

In the 1920s expertise in the poultry world was very much dominated by the United States of America. Not surprisingly, the first Director of the NIPH was an American, Willard Thompson, the Head of the Poultry Division of the New Jersey Agricultural Experimental Station and Professor of Poultry Husbandry at Rutgers University. His services were given on loan for two years, commencing in November 1924.[58] Thompson left late in 1926 and, after a gap of a year, he was replaced in October 1927 by another American, Professor Raymond Parkhurst of the Idaho Agricultural Experiment Station.[59] Parkhurst's distinctive American ways caused some unease amongst his colleagues: *'He was a bit frightening. [...] We didn't quite know what to make of him.'* [60] His staff was small. The 1927

Prospectus listed Herbert Howes as Assistant Director, Chris Bobby as Research Assistant, George Dixon as Senior Practical Assistant and Instructor in Carpentry, William King Wilson as Assistant in Charge of the Rabbit, Goat and Pigeon Sections and Statistician Frank Dudley. Whatever the difficulty of getting to know the ways of an American, on Parkhurst's departure in the autumn of 1932 the loss of 'a great man, a good "sport", and a first-class director' was keenly felt.[61]

Parkhurst's research experience gave excellent encouragement to the NIPH. The Institute's earliest research concentrated on nutrition and management problems associated with egg production.[61] Supplementing lighting to encourage egg production was one of the very early experiments at the NIPH and indeed, it was considered worthy to tempt the interest of the royal visitor in 1926.[9] The Duke of York was said to be intrigued by the use of an alarm clock to wake the birds one hour earlier than daybreak by turning on the light.[62]

Funding was dependent upon the Ministry of Agriculture, but on 1 March 1929 the Empire Marketing Board awarded the NIPH a capital grant of £5,000 and an annual maintenance grant, beginning with £1,581 and rising to £1,816 in the fifth and final year. The funding was to investigate the production and marketing of table poultry and eggs and to undertake experiments in duck, turkey and goose production.[61] Among the research topics of the early 1930s were the feeding of pellets, which sought to encourage a rapid feed and prevent excessive waste among shy feeders,[63] the use of home-grown foods, the importance of vitamin D to sex-linked crosses and the benefits of a slatted floor in poultry houses.[61]

As a result of the injection of money, between 1929-31 much improvement and reorganization took place. More land was purchased. By the close of the 1930s the plant covered forty acres.[64] More roads were built. It was in this period that the Administrative Building was revamped with

8.11 New incubator room, c. 1935: A. J. Guy.

offices, a research laboratory and a store room on the ground floor and a lecture room and a library on the first floor. Additions to the plant included a new incubator room, egg marketing rooms, a table poultry fattening building, a battery brooder section, an experimental brooder house and several laying houses, including a long intensive house. The waterfowl section also underwent change to handle the new research.[61]

On the departure of Parkhurst in 1932 Crowther assumed the responsibility of Director. This was to be a temporary arrangement due to the lack of suitable candidates, but the situation was not to change. A. J. MacDonald, previously Head of the Poultry Section at Rowett Research Institute, Aberdeen, arrived in October 1932 to take up the post of Deputy Director.[65] In 1934 the end of the Empire Marketing Board grant resulted in serious tightening of the financial belt. In 1935 the need for modernization led to an application for a grant from the Ministry of Agriculture. Some assistance did come from the Ministry and by 1936 poultry houses and pens were being refitted and a new laying house and two smaller breeding units were erected for the ducks.[66] A new battery house for 300 hens was a source of great interest to visitors. However, the shortage of money persisted. Margaret Brewer, student 1937-40, remembered some of the buildings in 1937 as being 'very tatty'. She had been advised by Tom Whittle of Sparsholt Farm Institute to study for the NDP at the NIPH with the words: *'The Midlands College had much better buildings than Harper Adams but Harper Adams has the best education. It may be grotty but the education's better.'* There even 'came the time when one leading member of the Poultry Press declared that he could not see the buildings behind a mass of uncut grass and weeds'.[67]

The situation confronting the NIPH was not helped by the severe problems facing the Poultry Industry at large. High feed prices and low poultry prices were worrying realities. Furthermore, the 1930s was a time before the use of antibiotic drugs and disease, such as Fowl Paralysis or Marek's Disease, was decimating poultry stocks in the face of intensification. The intensive rearing of turkeys at the NIPH, for example, was especially hampered by the incidence of blackhead.[30] Nationally, laying trials were especially badly hit by rising rates of mortality, peaking in 1937-38 at eighteen percent.[68] The College Laying Trials were no exception. The Principal's Report for 1935 stated that the death rate had reached twenty percent. Entries declined and in 1938 there was a real fear that the trials might have to be closed due to insolvency. Crowther sat on the Poultry Technical Committee, established by the Ministry of Agriculture to look into the reasons for rising mortality on poultry farms. The conclusion reached at the NIPH was that mortality was not linked to management and that attempts to introduce new blood lines had few benefits. It was considered better to effect a strict culling and breed from home stock with a proven ability to live on the plant.[66] The NIPH acted on its own advice and in 1939 the breeding stock was drastically culled and revised methods of rearing were introduced in a bid to restore the Institute's stock to a sound condition.

In 1938 MacDonald left.[69] Again there was difficulty in finding a replacement. Crowther thus continued in his capacity as Director. The decision was made not to appoint a new Deputy Director. Herbert Howes was to continue as Assistant Director and there would be a reorganization of the staff with the appointment of two new posts. Neil Jacobs was appointed Senior Lecturer in Poultry Husbandry, with responsibility for practical management, and Harold Temperton, a name to illuminate the future of the NIPH, was employed as Assistant Lecturer and Research Assistant.[70] Temperton was in fact returning to the NIPH, having obtained the Certificate in Poultry Husbandry there in 1933.

He obtained his NDP in 1934 and a degree in agriculture at Durham University in 1936.[71] That same year he obtained the NDA. He then undertook a travelling scholarship in America and lecturing work at the Lancashire Farm Institute, Hutton and as a County Instructor.[72]

The range and quality of research at the NIPH should not be allowed to overshadow the Institute's role as a provider of education. The NIPH opened for instruction in 1925.[5] By 1929 students numbered thirty-one, with the majority studying for the one-year Poultry Certificate.[73] Advanced and Elementary Certificates were awarded by the Institute and took two years and one year respectively.[74] The former had started in 1928 as a specialized course for those going into advisory work as county poultry instructors and was considered an ideal background for the NDP.[75] The Elementary Certificate consisted of one year of study and was associated more with practical poultry keeping.[76] In addition students could study for the NDP. In September 1931 the National Poultry Diploma Examination Board passed regulations extending the NDP course from one year to two years, making it with the year's practical requirement a three-year course.[77] At the same time a new one-year Junior Certificate course was started by the NIPH in line with the requirements for the new National Certificate in Poultry Practice, introduced by the National Poultry Diploma Examination Board in 1931.[77] Henceforth students at the NIPH could study for one year in order to obtain the Junior Certificate and the National Certificate, or for two years to obtain the Senior Certificate and the NDP. These changes and problems in the Poultry Industry led to a drop in poultry recruitment in the session 1934-35, especially on the Junior Certificate course intended for practical

8.12 Poultry students in the NIPH lecture room, c. 1929.

8.13 1932-33 NIPH staff and students: G.S. McCann.

R.D. Scott, R.F. Hall, W.J. Lintin, H. Bolton, L.C. Cathles, G.T. Williams, F.J. Brooker.

H.G. Hughes, G. Robbins, P.B. Wright, C.A.H. Warner, J. Bennett, H. Temperton, K.M. Bryan, C.W. Bartlett, R.J. Darneley, P. Mottet.

Misses G.R. Marshall-Barnes, M.C. Bennett, E.M. Roberts, E.P. Warlow, H.M. Stumbles, M.A. Green, S. McCowen, M.V. Marshall-Barnes, K. Bachari, E.B. Smith.

K.D. Downham, Miss H.M. Molyneux, I.W. Rhys, H. Howes, Dr C. Crowther, A.J. Macdonald, F.J. Dudley, Miss M.R. McMurray, Miss V.K. Tallent.

Misses M. Montgomery, E.M. Hosford, B.E. Andrews, D. Taylor, W. Allison, C. Smith, E.M. Lidstons, E. M. Rix, M.O. Westley.

poultry keepers. It was this course which faced the most competition from farm institutes with their lower fees. However, demand for places on the NDP course was growing as a result of good employment prospects for these students.

By the late 1930s Harper Adams was weathering the storm and, despite persistent financial restraints, there was cause for optimism. With the improving farming scene the future of Harper Adams looked far better than at any time since the end of the First World War. However, the days ahead were not destined to be tranquil. In the summer of 1939 the war clouds were gathering. On 24 July Crowther was notified by the Ministry of Agriculture that in the event of war, the College, along with the majority of institutions teaching agriculture, would close for the education of students in agriculture and would concentrate on the training of the Women's Land Army.[78] Poultry students were to continue to be taught if demand dictated, with the NIPH becoming the only centre for such instruction in the country.[74] With the arrival of war Harper Adams was plunged into an uncertain future.

Harper Adams
Power Farming Club 1933-35

An important development at Harper Adams which was to have national repercussions was the instigation of the first power farming course. This opened on 2 January 1933 and ran for six days.[a] Machinery demonstrations had always been a routine feature of the College. In 1925, for example, visiting farmers were impressed by a demonstration of a mole plough and trench digger.[b] In the early 1930s William Neate, son of Billy Neate the Engineer, was often invited by dealers to drive their tractors. As a young boy, he was considered a good illustration of the simplicity of the process.[c] From events such as these developed the annual power farming courses, a mix of talks and demonstrations concerned with the latest developments in farm machinery. Crowther saw such events as an opportunity to involve local farmers and gain their support. However, he also recognized the wider implications as a growing interest in machinery drew attendance from around the country.

The second power farming course in January 1934 was accompanied by the establishment of the Harper Adams Power Farming Club for 'the interchange of information and ideas'. There is little information on the precise origins of the Club, but Crowther was very quick to declare that Harper Adams would willingly assume 'the responsibilities of paternal interest implied in the title given to the

8.14 William Neate illustrating the ease with which a tractor could be driven during a demonstration at the College, early 1930s: Private Collection.

Club'. The Secretary of the Club was the Lecturer D. N. McHardy and the Treasurer was Cyril Block, the College Secretary. The Journal was produced at the College. In the foreword of Volume One of the Club Journal Crowther talked of the Club as 'fixing and extending the mutual contacts established by those who have come to us for shorter periods'. He went on: *'To me this development presents the germ of a movement that may ultimately be very potent in its reactions upon the effectiveness of agricultural educational work throughout the country.'* As a good business man he set about forging lasting links with his customers, with the product on offer being the educational work of the College itself.

Within three months membership ran to over one hundred and in July 1934 membership was expanded to include manufacturers. In January 1935 the third course was held. Demonstrations were bigger and wider in scope and all bode well for the future. However, the presence of many experts from the research institutes at Oxford foretold of trouble to come.

A paragraph in the October issue of the Journal said it all: *'In view of the fact that a Power Farming Conference is to be held at Oxford, under the auspices of the School of Rural Economy, the Institute for Research in Agricultural Engineering and the Agricultural Economics Research Institute, the Authorities of the HAAC have decided not to hold a Power Farming Course this January.'* Crowther must have found little consolation in the knowledge that imitation is the greatest form of flattery. The start of the Oxford Conference on Mechanization in Mixed Farming, to be renamed the Oxford Farming Conference in 1938, wrecked all aspirations that Crowther might have had. Even the conference programme was a blatant copy of the 1935 course at Harper Adams.

8.15 Roye Adams of John Bromley & Co. demonstrating an International tractor at the college, 1930s: Private Collection.

Chapter Nine —

World War Two 1939-45

The Second World War broke out on 3 September 1939. Fortunately for Harper Adams, despite chequered fortunes in the early months of the war, it did not have to submit to the Ministry's scheme to curtail agricultural instruction. Indeed, as the war progressed, students came to the College in numbers far greater than in the 1930s. For those young men and women who came to Harper Adams, the College must surely have been among the best places to spend the war. Its location in rural Shropshire meant that there was little danger from bombing or attack. There are no memories of air raid shelters at the College. Damage to the College amounted to little more than broken windows.[1] For the most part, the war seemed to have passed the College by. Willard Moore and his future wife, Margaret Brewer, stood 'out in the road and watched the bombs at Coventry'.[2] Those planes that did fly over were destined for other places: *'One particular night there were seventeen flares round about [...] and of course, they got Tern Hill [the airfield] and it certainly rattled the place up a bit.'*[3]

Many staff and students were members of the Home Guard, but their closest encounter with the war was the dowsing of the fire at Tibberton Manor Farm in August 1940 as a result of incendiary bombs.[4] When asked what he would have done if the Nazis had invaded Edgmond, ex-student Bob Tupper replied with both sound sense and good humour: *'Got under the bed I suppose.'*

Life was safe, but was it comfortable? That essential item food was rationed. All the ex-students of the period have memories of keeping their sugar rations in individual jars on the ridge of the panelling on the lower part of the walls around the dining room, although 'a trip around Allscott Sugar Beet Factory soon rectified' any dearth.[5] With an adjacent farm and garden and the NIPH, students 'were better off than most people'.[6] As Tony Chart, student 1943-45, explained: 'Everything that had grass on it had to be growing food.' Food was not in short supply, but meals could be predictable.[7] A daily egg became a substantial part of the student diet: *'If you didn't have it for breakfast, you knew you were going to have it for tea.'*[5]

Leonora Grice supplemented her diet with gourmet flair, eating 'pears growing on a tree against the College garden wall' and 'quantities of watercress from a nearby stream'. Above all, she recalled the ice cream served at the End of Session Dance:

> *The food was lavish and delicious, though simple and straightforward by today's standards. The triumphant end-result of reserving and combining all those left-over sugar and butter rations from the preceding months perhaps? And, to crown the occasion, the most sumptuous treat of all - home-made ice cream! [...] The ice cream at Harper Adams was made right there on the premises with super-fresh eggs from the Poultry Department and rich cream from the dairy among the other ingredients. I don't honestly think I have tasted better.*

Student social life was not unduly spoilt by the war, although the range of activities on offer was limited. Paula Johnson, student 1943-46, had fond memories of the camaraderie and Bob Tupper emphasized the enjoyment to be had from simple conversation. Leonora Grice stressed her contemporaries were accustomed to making their own entertainment:

> *Don't imagine for one moment that we didn't have a lot of fun! Before the days of television we had a goodly share of practical jokers, of potential comedians and comediennes who could reduce us to helpless laughter with their witty comments. Before ghetto-blasters and walkmans, out of studying and working hours, the corridors of the hostels would often echo with the heart-felt and very tuneful singing of current popular melodies by certain male students - and the hits of those days heard again still reduce me to nostalgia.*

9.1 Harper Adams Revue 1945. Back row from left: L. T. Chase, F. B. Cottrell, F. Dickens, J. C. Harrison, J. Long, L. Watney, M. H. Stewart, A. N. Gale, P. Bailey, R. White, M. Walker, J. Poulter, J. R. Young, R. Thompson. Front row from left: N. Broadhurst, Sheila Boddy, G. E. Pester, E. J. B. Greenwood, P. Smyth, A. P. Rodway, A. J. Wynne, N. Elcock, J. R. N. Vaughan.

Brenda Willy, student 1943-45, put the matter succinctly: *'No doubt the idea of playing rounders in front of the College followed by "sardines" in the farm buildings after supper on Sundays (in the summer) would seem rather strange to present-day students.'*

Dances or 'hops' were held nearly every week throughout the war in the large lecture room. The End of Term Dance on 30 March 1944 involved the College and around 130 guests dancing to the music of local artist Fred Lawton. There was a raffle for a chicken and the proceeds of the evening, ten pounds, went to the local hospitals.[8] Students would travel on their bicycles to Edgmond, Newport or Lilleshall for dances.[7] Newport was also the popular destination for those seeking the cinema, although in the session 1943-44 the Film Society was formed at the College.[8] For bicyclists there was the thrill of dodging the local policeman, who was ever alert in his search for unlit bicycles: *'We used to ride with one light at the front and another with a light at the rear and the rest of us in the middle. We didn't meet any cars you see.'* [7] Grice revealed these dances were anything but dull: *'Most of the dances involved closer physical contact between boys and girls than many of us had experienced so far in our lives, so there was a heady excitement in the air, flushed faces and giggling or whispering.'* And, for those male students determined to cut a dash, there was always the opportunity of sitting in a chair in the quad and having one's hair cut by the ever versatile Albert Rowley.[7] There were skits and plays performed by the students and a number of concerts relying on 'the considerable talent available at Harper Adams'.[9] Members of the thriving Music Society could listen to records.[8] During the 1942 Rag an unpleasant prank directed against a student led Crowther to give the male students 'the raw edge of [his] tongue in a fashion that [he] had never used before to any body of students'. He considered the incident to have been 'disgusting and disgraceful in the extreme'.[10] The Rag was discontinued for the remaining years of the war. The 1942 Rag had shown due respect for wartime conditions, with the parade being composed of bicycles rather than cars and the customary free for all, when flour bombs and the like would be thrown, taking a different form due to the paper shortage and an associated lack of paper bags.[11]

Sport and sports days continued throughout the war. Hockey, tennis, rugby, football and cricket were all popular. Wartime conditions could create their own complications. Tennis in 1944, for example, was hampered by the poor quality of the grass courts and a scarcity of balls.[8] The students often played teams from the Forces. The latter games were eagerly accepted, not least for 'a jolly good feed with them in the canteen'.[7] Sport was essentially for enjoyment and the size of the College meant that team standards were variable. The Forces, in contrast, could have players of international standard.[7] Tony Chart remarked: *'We were always thrashed comprehensively everywhere we went, but that was alright, we enjoyed it.'* Dora Gale (nee Saunders), student 1942-45, remembered, however, that the College hockey team did defeat the RAF.

By the middle years of the war, College life had settled into a routine and was none too harsh. At the beginning of the war, however, such comfortable predictability did not seem a possibility. In the summer of 1939 there were widespread rumours that the College would be closing down.[12] In contrast to the First World War the Government had prepared plans for the intensification of food production in the event of hostilities.[11] Substitute and supplementary farm labour would be provided by the Women's Land Army (WLA).
All agricultural education institutions, with the exception of Reading University and the women's colleges of Studley and Swanley, were to close for the instruction of students

9.2 1942-43 College rugby team: Bertram Sinkinson.

9.3 Sports Day 1945: Private Collection.

College Ayrshires in the 1950s.

The former Poultry Husbandry Teaching Block, 2000.

Harper Adams Day 1959: Private Collection.

Mrs Price's ornamental pond, used by students for ducking, 1950s.

Student play, c. 1955.

The College, c. 1966: Aerofilms Ltd.

The Harper Adams victors in the Seale Hayne Run, October 1966.
From left: Dick Beal second, Gordon Tweddle third, Ian Hollows fourth, Trevor Parnham fifth, Dick Hookes sixth, Bill Wilder seventh, John Hartland eighth: Private Collection.

Bulgarian students on a TEMPUS programme visit, the Soil Hall, 1992-93.
From left: Richard Bruce, Head of the Enterprise Unit, Dr Nickolay Naidenov, Senior Lecturer in the Department of Agricultural Engineering, the Technical University of Rousse, Bulgaria, Dr Wynne Jones, Vice Principal and Ian Gedye, Head of the Department of Agricultural Engineering (extreme right).

in agriculture and were to concentrate 'on the provision of short courses of instruction, essentially practical in outlook, intended to qualify inexperienced "recruits" for agricultural labour', namely the WLA. On the 'belated discovery that Reading could not provide the needs of the poultry diploma', the NIPH was made the national provider for poultry education.[11] Crowther argued that staff and facilities would be under used and such an arrangement could only be wasteful of resources. The Government prevaricated and in the meantime the WLA descended on Harper Adams.

Unlike the First World War when women trainees for agricultural work had offered a life-line to the College, the Land Girls were viewed by Crowther as a nuisance.[11] Quite simply, they were taking up places which he considered belonged to bonafide students. Twenty-one poultry students [11] were under instruction, after a delayed start to the term, [2] but no new recruits to the agricultural courses had been allowed. Crowther was none too pleased as enquiries had reached their highest level for many years.[11] Furthermore, thirty-five existing Harper Adams agricultural students had been sent to Reading, where they were unhappy with the teaching arrangements.[13] Martin Cross remembered a trip back to Harper Adams to request the Principal to pull strings to get him and his colleagues back. The first forty-six Land Girls from London and Lancashire started a four-week course at the College on 11 September 1939. A second course with forty-four women began on 23 October.[11] However, even before the first intake had left, it was clear that the demand for labour had not materialized, a situation partly caused by the Phoney War. Of the first intake only twenty-five women found employment on completion of their training. The figure for the second course was little better with twenty-seven gaining work.[11] Bill Thompson believed these figures were due to the unsuitability of the recruits.[12] As a consequence of the poor employment figures, by early December the WLA had given notification that no further recruits would be sent.

The Land Girls were housed in New Hostel [11] and Ancellor House, with the large lecture room used as their common room.[3] According to Crowther's jaundiced view they soon settled in: *'Whatever may have been the varying degrees of skill acquired in farm work, there were none in the whole-hearted way with which they seized upon the rare opportunity of having a "summer holiday" at Harper Adams.'* [11] The poultry students were warned by Crowther 'to be very careful with [their] dealings with them'.[3] The Land Girls, described as possessing 'charm', did indeed fully involve themselves in the social life of the College, enjoying dances, social evenings and concerts.[12] Willard Moore went to outside dances with them: *'There was one girl there. She was a mannequin. I can see Reggie Wrench [a fellow student] now as she came down in this topless [strapless] dress. He said "Good God, what's holding that dress up Gwen" [...] She said "Willpower".'*

Doubts at the Ministry of Agriculture as to the wisdom of closing down the majority of agricultural courses had set in by the autumn of 1939. A factor in favour of resuming a wider range of agricultural education was the fact that the call up age had been agreed at twenty, as opposed to eighteen. This meant that demand for education was greater than had been originally estimated. By October degrees and diplomas were reinstated at most universities. Harper Adams was next in line to benefit from the change of heart, as normal courses were allowed to be reinstated at agricultural colleges subject to the allocation of places to the WLA. However, the extent of the demand could not be estimated and the Royal Agricultural College, for example, was closed throughout the war. Old students of Harper Adams did not arrive back until January and it was too late to recruit afresh

9.4 Members of the Women's Land Army training at the College, September 1939. The third lady from the left, Nancy Dawson, was a professional dancer from New Zealand: Wellington Journal & Shrewsbury News.

for the session. Therefore, for two terms the College ran with thirty-six agricultural and twenty-two poultry students.[11] The temporary suspension of agricultural courses and the ensuing reduction in numbers 'caused a serious and irrevocable loss of fees'. Furthermore, for the agricultural students the work of three terms had to be undertaken in two. With the added complication of an outbreak of measles, tutorials were instigated as an emergency measure. They were to prove very popular.[12]

The summer of 1940 brought with it its own set of difficulties. With the fall of France it was clear conditions would worsen and a greater emphasis was placed on food production. In a broadcast on 3 June the

Minister of Agriculture gave the impression that all agricultural colleges would close. There followed a circular from the Ministry which encouraged all students to go farming and instigated a system whereby staff were to be seconded to advisory duties in association with the County War Agricultural Executive Committees. There were, however, to be no compulsory closures. As a consequence the term was shortened at Harper Adams and there was the real possibility that the College would not reopen. Enquiries and recruitment suffered in the confusion. It was not until September that the Governors were able to make the decision that it was viable to remain open. The session 1940-41 opened with thirty-four agricultural and thirteen poultry students.[11] With the gloomy prospects for large-scale poultry husbandry under wartime conditions poultry student numbers were low. Similarly, of the twelve Agriculture Diploma students who should have returned for their third year, only three did so. However, as Crowther stated: *'With the goodwill and co-operation of all we managed to keep all our various flags bravely flying.'* [11]

By the session of 1941-42 the College had 'weathered the storm remarkably well'. Every bed was occupied and there was also a small waiting list.[11] Crowther attributed this to wartime conditions stimulating an interest in agricultural careers. Students totalled seventy-one, but there had been a further decline in the number of poultry students. In the session 1942-43 student numbers reached eighty-three, the largest entry for many years. Poultry numbers had stabilized at around eleven.[14] For the session 1943-44 the College ran into problems of over demand. Places were available for approximately forty-five new students, but the list of applicants numbered 112, with two names already down for enrolment in the session 1945-46. By the close of the war the total number of students had stabilized at around ninety a session.

The majority of students during wartime were around sixteen to eighteen years of age and straight from school, with a number of older students, who had been classified as unfit for service in the Forces.[5] There was some anxiety when the age of call up was reached, but in the early years the authorities were reasonably generous in allowing deferment.[11] However, in the session 1942-43 thirty-four students were obliged by the demands of military service to interrupt their education and there was some tightening of entry standards. Henceforth, all new students were required to possess a School Certificate. A further demand that periods of deferment should be as short as possible led to the shortening of courses in the session 1942-43.[14] The Agricultural Diploma course was shortened to two years and the Agricultural Certificate course to one year. There were no changes as regards the poultry courses. In the case of the Certificate the effect was not serious, with simply less weight being placed on science. The Certificate course had always had a more practical emphasis. As Bob Tupper explained: *'Anybody doing practical farming, I would have thought that one-year course [the Certificate] would have been every bit as good as the two-year course [the Diploma].'* In December 1943 the decision was made to withdraw it temporarily due to increased teaching pressure on staff and within months the Certificate was officially dropped. The main reason given was that the planned reopening of farm institutes in October 1944 had introduced considerable competition for any short course offered at the College.

The reduction in the length of the Diploma course to two years was not so easily accomplished as the national syllabus was already dictated and examination standards remained unchanged.[15] Consequently, students were required to increase their hours of private study and teaching became more intensive.[7] Tupper and Chart remembered the emphasis being very much on lectures in the years

9.5 1942-43 staff and students: Bertram Sinkinson.

1943-45, with the practical side of things focussing on observation, farm walks and the farm diary. Paula Johnson commented upon the 'awful lot of subjects'. The second year students of 1942-43 faced the most daunting prospect of all. They were presented with the choice of either taking both parts of the NDA examination in 1943 or leaving without a qualification. Despite this difficult situation Crowther was proud to recall that twelve out of the thirty-seven students awarded the NDA at Easter 1943 were from Harper Adams.[14] No other institution had secured more than seven NDAs. The situation eased slightly in 1944 when nationally NDA examinations were moved to a summer slot, allowing greater study time. The war was characterized by the increased popularity of the higher courses. The requirement that all seeking deferment must possess a School Certificate raised the standard of entrants and of their choice of course. In the session 1943-44, the last year when the Certificate was awarded, fifty-eight of the eighty-three students were studying on either the Diploma or the Degree courses. Numbers of recruits for the Degree course, run in association with London University, were noticeably above the pre-war level.

Very few students came from farming families due to labour shortages on the land, but there was often some link to agriculture in their background. Jack Large, 1944-46, was not untypical. His father worked in a steel mill, but he helped out at his grandfather's small-holding and took part in work on local farms:

> *I used to go out on holidays, weekends and this sort of thing, onto the local farms. There wasn't the health and safety regulations then. Children were about. We used to go in the cowsheds, help them with carrying the milk or feeding the calves, loading the hay in the fields and stooking the corn, this sort of thing. At the same time, in the latter years at school, we went on agricultural camps doing work for the farmers when we were allowed a fortnight off school to go. [...] I was at farming camps [as part of the war effort] helping out with the harvest or with potato picking.*

Increasingly, students were willing to contemplate a managerial or advisory career instead of farming.[6] The years of the Agricultural Depression had discouraged many farming ambitions and the more favourable climate for agriculture in wartime had its part to play in creating new opportunities. As Bob Tupper explained: *'When I went to Harper, Mother [he was an orphan] had no idea what I should do in life. She didn't think there was much chance of me getting a farm and if I had to be a manager, well, I'd got a bit of paper to wave.'*

Larger numbers of students were women, perhaps influenced by the fact that agriculture was a reserved occupation.[5] More women were also choosing agriculture in preference to poultry, although Dora Saunders found there was still considerable prejudice 'that girls couldn't cope with agriculture' because of the heavy work and thus she went in for poultry. Paula Johnson, who studied agriculture, found no such problems and received encouragement from her headmistress. She believed the Women's Land Army had done much to overcome the prejudice. For the session 1943-44 there were twelve women students in comparison with seventy-one men and in 1944-45 twenty women compared with seventy-one men.

The financial situation of the College was improving throughout the war. From the beginning of the war the Ministry of Agriculture and the Three Counties maintained their grants at pre-war levels. Things continued to get better. On Crowther's retirement in August 1944, he wrote that 'the financial burdens of the past have been removed and the position of the College made stronger in all respects than ever before in its history'. In March of that year the decision had been made to repay

9.6 Measuring phosphates in soil. From left: H. E. Griffiths, Assistant Soil Analyst and P. W. Arnold, Land Fertility, early 1940s: G. S. McCann.

the whole of the balance of the Shropshire County Council loan of 1937. In July Coventry City Council had agreed to pay the College an annual grant of £100 on condition that students from the city received reduced fees and that free advisory services were made available. In spring 1945 Shropshire and Staffordshire County Councils increased their respective grants from £750 to £1,000, in recognition of the increased general expenses facing the College. In April Worcestershire County Council agreed to an annual grant of £450.

Up to the summer of 1940 the academic staff had remained intact, although there was some part-time work for the County War Agricultural Executive Committees (War Ags), set up by the Ministry of Agriculture to increase food production.[11] At the beginning of the war seventy percent of the country's food was imported. Throughout the war, Crowther was involved with national work on animal nutrition and had a far less visible profile amongst the students at the College.[6]
From the summer of 1940 members of staff

were regularly seconded by the Ministry of Agriculture to work with the War Ags. That summer eleven members of staff, including some in the Advisory Services, were seconded to the War Ags of Worcestershire, Shropshire, Staffordshire and Warwickshire, although Morley Davies and Thompson were part-time appointments only. Pressure of educational and advisory work led to the recall of three personnel by the autumn of 1940, but apart from the part-time employment of Malthouse as a substitute for Robinson there was little new recruitment. For much of the war the College juggled staff, borrowing them back from the War Ags on temporary and part-time arrangements when the need arose. It was not uncommon for ex-students to be given junior appointments. In contrast the impetus to advisory work as a result of the war did involve a conscious policy of new recruitment and a large number of semi-skilled and unskilled assistants were employed at the College.

As part of the considerable intervention by the Government to stimulate home food production, the Advisory Services were

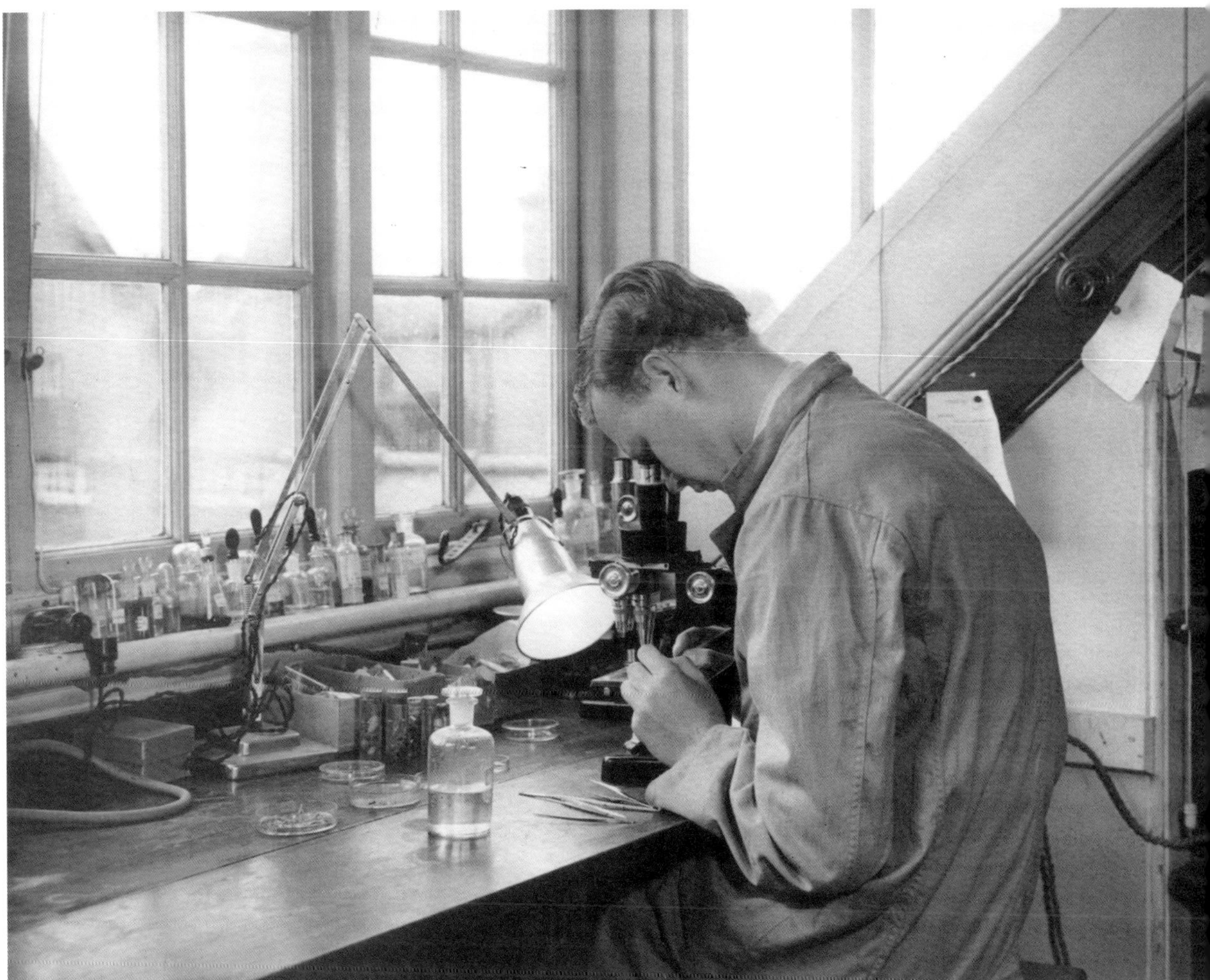

9.7 J. E. Satchell identifying wireworms for the wireworm survey, early 1940s: G. S. McCann.

extended to a nationwide scale. At Harper Adams there was pressure on the Advisory Chemist with growing demand for soil analyses and the continuing work on the survey of soils. Advisory Entomology received a special grant from the Ministry of Agriculture for a survey on wireworms in selected areas of grassland scheduled for ploughing. Some 500 fields were investigated. This was to be of great assistance to those selecting fields for the ploughing up campaign. The first wireworm survey at Harper Adams had taken place on the College's own land in 1914, initiated by the ploughing up campaign of the First World War. The figures obtained by Roebuck, Lecturer in Botany, were the first of their kind in Britain, with clear implications that the breaking up of permanent pasture and the cultivation of arable crops altered the soil environment and jeopardized the survival of insects normally found there. Roebuck considered five years of cultivation would reduce the number of wireworms to a level not harmful to crops. The rapid field method of sampling wireworms used in the Second World War originated with Roebuck.[16]

The Dairy Bacteriology Department was also undertaking work of national importance. Shropshire had been one of four counties chosen by the Ministry of Agriculture to undertake trials of a new milk testing and improvement scheme and in June 1942 this was absorbed into the National Milk Testing and Advisory Scheme. The Department became responsible for the supervision and systematic testing of milk deliveries to all depots in Shropshire, Staffordshire and Warwickshire. By 1943 the milk from 11,000 producers in the Province was being tested regularly. This work was undertaken in the old veterinary laboratory on the first floor of the Main Building.[17] In 1944 the Ministry even contemplated erecting a building to the east of the Caynton Road, opposite Old Hostel, to accommodate the Dairy Bacteriology Department, now inundated with work.

The Advisory Veterinary Officer continued the work on the prevention and control of common animal diseases. Interestingly, in 1942 investigations began into diseases of rabbits on behalf of the British Rabbit Council. The subject had hitherto been neglected and in the first year of research alone 674 post mortems took place. Advisory Economics was seen as the Intelligence Department of the Ministry. From the summer of 1940 a farm survey was undertaken across the country. It was vital to assess the ability of each farm to play its part in the food production campaign. In 1941 this developed into the National Farm Survey, a 'Domesday Book' of British agriculture, with a longer term policy of providing data for post-war planning. The Advisory Economics Department at Harper Adams undertook the checking, filing and statistical analysis of the records collected by the War Ags of the Province. The summary report was published in 1946.[18]

The rapid expansion in the work of the Advisory Services put great strain on the resources of the College. A serious accommodation problem was contributing to the situation. In the session 1940-41 Ancellor House had ceased to be the Women's Hostel due to the small number of students. The Army wanted the building, but the Ministry of Agriculture opposed the move. Meanwhile, the College sought to rent the house, even offering it as a Dr Barnado's Home, but to no avail. In the spring of 1941 it was requisitioned by the Shropshire War Ag. for the use of foreign refugee labour.[11]
In 1943 the Shropshire War Ag. discussed the possibility of turning Ancellor House into an Approved School, but the College would have none of it. With the Ministry's support, in July 1943 Ancellor House passed back to the College. There was a clear need to restore it to student accommodation, but the shortage of domestic staff (Crowther declared the College was 'continually on the verge of crisis' because of their inability to find

domestic staff) and the pressure on space felt by the Advisory Services resulted in the Advisory Economics Department taking over the house. It was not until the autumn of 1945 that Ancellor House returned to its former use as a hostel. With Ancellor no longer a hostel, wartime women students were accommodated on the top floor of New Hostel.[11] They had a separate entrance from the men, still in residence beneath, and their own common room.[19] Leonora Grice remarked upon the communal bathroom with 'an ample Victorian style bath-tub' and the fact 'it was not done to lock the bathroom door in spite of the affronted modesty of some of us when we realised that fact'. Yet, even with these arrangements and careful planning more and more students had to be accommodated. Lodgings in Edgmond, the rooms of resident staff and even the Principal's House all came into use.[14]

The history of the NIPH during the war was very much that of a curate's egg. The work of the NIPH was seriously affected by the war, both in matters of education and research. The root problem was the difficulty in

9.8 Poultry students with Herbert Howes, early 1940s: G. S. McCann.

obtaining poultry feeds with the introduction of rationing by the Government.[20] Commercial firms were threatened and consequently were not recruiting and this in turn dampened any enthusiasm for enrolment. Thus, despite the fact that the NIPH was the national training centre, the session 1942-43 saw just ten poultry students and for 1943-44 there were eleven. The serious dearth in poultry feed made an impact everywhere. The shortage of poultry feed led to difficulties running the Laying Trials and in 1942 the College Authorities reluctantly brought them to a close, although there was the hope that they would resume under more favourable conditions. They never did resume. The Utility Poultry Journal was taken over as a means to publish experimental results undertaken by the NIPH. This ceased in 1948.[20]

During the war, research at the NIPH suffered. Stock numbers had been reduced early in the war and by 1943 adult stock represented a mere fifty percent of capacity. The experimentation that did take place concentrated on matters of wartime importance, most notably the use of alternative feeds. Everything from tapioca meal, dried town waste, horse chestnuts and artichokes was tried. Tottenham Pudding, consisting of house scraps boiled up and made into blocks, was another experimental food, but one that had to be treated with respect.[3] Rowley described the danger inherent in preparation: *'Wolverhampton had its own ovens and we received fresh consignments every week. Mixing the stuff by hand, it was prudent to wear gloves, I lost my finger nails through not doing so - it was a lucky dip in many ways containing sink stoppers, pins and needles, knives, forks, spoons to mention just a few "contraries".'* [21] Temperton also managed to continue experimentation in hybridization.[5] In 1940 the war situation made it impractical to hold the Poultry Conference. This was the first time in its history that it did not run. It was not revived until 1953.[22] The whole appearance of the NIPH reflected the stagnation as lack of labour and building materials meant that essential repairs were neglected. By 1944 sections of the plant were in a very bad condition. Rowley concluded: *'The immense amount of trial work done at this time was conducted in dilapidated houses and worn-out equipment. The war years were lost years, it was all work, the plant staff was a skeleton force and mainly female labour and every spare moment taken up with Home Guard or A.R.P. duties.'* [21]

In contrast the Rabbit Department grew in strength as a result of the renewed interest in rabbit meat due to wartime rationing. Interestingly, many of the NIPH's rabbit carcases were sent to the London hotels at this time.[3] A considerable number of poultry breeders transferred their allegiance to rabbit production.[11] At the close of the 1930s plans were formulated to revamp the Rabbit Department and, with the war, such was the interest in rabbits that in the summer of 1940 the Ministry proposed to expand the work of the Department and to develop a separate National Rabbit Institute. Dr J. N. Pickard from the University of Edinburgh was to become Head of the Department.[23] The scheme was to come into force on 1 January 1941, despite Crowther's concerns that the additional expense could not easily be met in wartime. However, Pickard's inability to find a house for himself and King Wilson's inability to find housing for the man's rabbits led to the plans being shelved, never to be revived. As with poultry, much of the wartime experimentation was to do with alternative feeds. The rabbits were indulged in everything from coffee grounds to tea leaves. King Wilson's advisory role also expanded as enquiries and personal callers multiplied greatly.[11] Amongst those to whom he gave advice were army officers and prisoners of war. Many of the most common questions were answered by King Wilson in his broadcast talks.

9.9 Students picking potatoes, c. 1944.

The changes on the Farm imposed by war proceeded smoothly. The main problem encountered, not unexpectedly, was a lack of general labour. Jack Large and Paula Johnson remembered the students being expected to help when labour was short and time was of an essence. Such work was quite distinct from instruction and involved activities such as topping and tailing sugar beet, digging carrots and picking potatoes. As Jack Large said: *'If you can put thirty students into a potato field, good luck to the lecturers controlling them.'* The overall appearance of the Farm was significantly changed by war conditions. In 1941, for example, about 200 trees were felled in order to improve the arable land. In 1939 the Government had introduced subsidies to encourage the ploughing up of old grassland and the ploughing up of College land for arable use began early in the war. By 1941 fifty-five acres of permanent pasture had been broken up. Nearly thirty-eight acres followed the next season, increasing the area given over to wheat by forty percent, to oats by 250 percent, to barley by fifty percent and to potatoes by two hundred and twenty percent as compared with pre-war acreages at the College. By the summer of 1943 the ratio of grass to arable had been completely reversed at the College.[14] The pre-war balance in favour of grassland was not unique to the College. In Britain, at the beginning of the war, around nine million acres were under crops, eighteen million acres were old pasture and a further sixteen and a half million acres were rough grazing land.

As regards mechanization, there had been tractors on the College Farm since the First World War. During the 1930s tractors and implements were borrowed from the major companies.[13] However, in 1938 Ford ceased to lend tractors to agricultural institutions and for a time the College was without a tractor.[24] This was not remedied until the autumn of 1939 when a Fordson tractor was issued to the College by the Ministry of Agriculture for the training of WLA recruits. Thompson, the Farm Director, did not encourage the use of tractors due to his belief that they impacted the soil.[25] He was seen by many students at the time as being old-fashioned in his emphasis on the horse.[6] In defence of Thompson, it should be remembered that even on the outbreak of the Second World War there was little mechanization on most farms. The majority of work was done by horses and even those larger farms possessing a tractor tended to use it in close conjunction with horses. The mechanization of root crops was virtually non-existent. Corn was cut and tied by a binder. The steam threshing box was still in use. The first trailed combine had appeared in this country in 1928, but it was to be the Second World War before the first self-propelled combines appeared in any number. Early tractors had much to their disadvantage. They were unreliable, lacked suitable implements and were of limited use and thus uneconomic for the smaller farm. Although, by the mid-1930s, developments such as pneumatic tyres, the hydraulic lift and three-point linkage ensured that tractors could be used for most jobs on the farm, availability was limited and in 1937-38 there were still only about 50,000 tractors in this country. In contrast there were some 700,000 horses at work on farms. It was the ploughing campaigns of the war which led to the rapid adoption of tractors. By the close of the war there were 140,000 tractors at work. By this date the College Farm had two tractors.

Thompson declared with pride in 1943: *'At no previous period has the farm contributed more to the purposes and repute of the College than in its present condition.'* The Farm was fully fulfilling its role as an 'efficient food production unit'. That year livestock on the College farm amounted to 101 cattle, including 32 milk cows, 40 ewes, 156 fattening sheep, 1 sow, 4 gilts in pig, 9 feeder pigs and 150 poultry. A high state of fertility was evidently being achieved with little reduction in livestock. The herd of pigs had, however, been reduced by half.[11] At the start of the war nine million tons of animal feed had been imported each year. Rationing was inevitable and priority went to dairy cows. The allocation of feed to pigs was restricted. Home-grown feed rocketed in price. The four million pigs in Britain in 1938 were reduced to just one and a half million by 1945. Thompson maintained that production had only been enhanced with the assistance of the scientific staff and advisers, highlighting matters such as soil problems and the preservation of grass. Similarly, he attributed improvements in the dairy herd to veterinary investigations into cattle fertility. He declared: 'Truly this is a scientists' war.' [14] This was not a situation unique to the College. By the close of the war, increased appliance of science, better management and greater mechanization was advancing farming beyond all recognition. At Harper Adams growing confidence was reflected in the decision of 1943 to purchase approximately forty-eight acres previously rented from Colonel Sykes.[26] The sum of £2,400 was paid. Colonel Sykes, Chairman of the Governors since 1927, had died in January 1942.[11] The argument in favour of buying had no doubt been strengthened by the wide-ranging system of guaranteed prices instigated by the Government as a wartime measure.

The 'purposes and repute' of the College were intricately associated with conveying advice and information to farmers and the role of the Farm in this was intensified by the war. In 1943 Thompson wrote of extra cultivation and manurial demonstrations and of an open day the previous August held in association with the Shropshire War Ag. The Open Day highlighted everything from silage making and potato clamping to thatching and spraying and placed special emphasis on bovine health and clean milk production. Alongside practical events were talks and 'academic displays'. Between 2,500 and 3,000 people attended and it was described by Crowther as the 'greatest and most successful event of its kind in our history'. Petrol rationing limited the number of visitors beyond the organized open days, but groups from the Shropshire Chamber of Agriculture and the Young Farmers' Clubs still made their appearances. Much interest was shown in soil problems resulting from ploughing up and Morley Davies was usually on hand to give advice.

Experimentation at the College was restricted during the war, mostly due to land utilization for the war effort.[14] The experiments that took place were mostly to do with the need for alternative animal feeds. There were valuable experiments at the Pig Experimental Station into feeding grass cuttings. However, on the departure of Crowther in 1944, the Ministry closed the Station and the buildings passed to the commercial herd. In 1924 the Ministry of Agriculture had requested that the College, along with five other centres,[27] should supervise special crop variety trials which would be carried out on behalf of the National Institute of Agricultural Botany, established in Cambridge in 1920.[28] A special assistant, financed by the Ministry, was duly appointed at the College. Initially the trials took place at Leegomery, but from January 1928 they were established on twenty-five acres of College land.[29] Despite the war the NIAB trials continued, but in 1941 the cereal variety section stopped and in 1943 a crop recorder was no longer appointed for the limited amount of work undertaken. Thompson took over responsibility until normal working resumed in 1945.

Crowther retired as Principal in August 1944 [30] and settled to life in Surrey, writing for publication, serving on committees and sitting on the Council. He died in 1964 and his ashes were interred at Edgmond Church.[31]

On Crowther's retirement, Archibald McVicar was appointed Principal. Britain was still fighting the war and the search for a replacement to Crowther had been a difficult one. McVicar had spent his career in county advisory work.[9] He was greatly respected for 'his shrewd judgement and sound knowledge of farming'.[32] As the Chief Agricultural Officer for Shropshire County Council since 1937, he was well known and well liked, but from the start it was clear he was not happy in the post.[9] In his obituary it was said: *'It is doubtful if McVicar was ever really at home in this office with its many administrative duties, and relatively few opportunities for making direct contact with the problems of the practising farmer.'* Neither was his character suitable for the task in hand. A retiring man, he preferred informal occasions, where he was 'delightful, humorous and charming' and always 'perfectly natural and genuine'.[32] In 1946 he resigned, despite attempts by the Governors to get him to stay, and went to work as Director of the Northern Province of the National Agricultural Advisory Service.[33] He died at the early age of fifty in 1952.

9.10 Archibald McVicar, Principal 1944-46.

McVicar's main achievement during his short time at the College was the introduction of Ayrshire cattle *(see photo page 143)*, popular at the time as a breed free of tuberculosis.[34] As early as 1943 the decision had been made to raise the College Herd to Attested Status, a move considered 'essential to the reputation of the College'. Late in 1945 twenty in-calf pedigree Ayrshire heifers, licensed for the production of tuberculosis tested milk, were bought. The dispersal sale of the existing Friesian herd took place in September 1945. The forty-one cows and served heifers, nineteen empty heifers and three bulls found buyers from all over the country and fetched the sum of £8,061.

It was during McVicar's principalship that peace returned. On Monday, 7 May 1945 the whole College gathered around the radio in the common room to hear the news that the next day would bring peace to Europe.[9] Hurried arrangements led to an impromptu dance that evening, which 'to say it was a glorious success was a gross understatement'. The next day, VE Day, was a holiday and many students left for home.[5] That afternoon the staff and the remaining students gathered around the flagpole to hoist the Union Jack and to hold a two minutes silence. At night the College was floodlit as 'a true symbol of peace'. Dora Saunders, Tony Gale and friends celebrated by bicycling to Wellington and riding up the Wrekin. On 15 August VJ Day was celebrated as victory was won over Japan. A short year later the College found itself with a new leadership. Both Britain and the College were entering a very different world.

The Home Guard at Harper Adams 1940-44[a]

Those at the College of call up age and on deferment from the Forces, whether students or staff, were required to give some time to National Service.[b] Originally the Local Defence Volunteers, first formed on 14 May 1940 as a result of a call to arms by Anthony Eden,[c] were divided into Zones, which in turn were divided into Groups under a Group Organizer or Commander. Patrick Davies, Local Auctioneer and then a teenager, remembered his father, the local Company Commander, buying about twenty broomsticks and training the men on his tennis court at The Grove, Edgmond. Martin Cross, as a student and then as Crop Recorder at the College, was a member of the early Home Guard. He vividly recalled the days of 1940: *'We were under the command of Lt. Col. Traill, the Bursar (who seemed to shed years as he got the scent of powder in his nostrils). Clad in denim overalls to start with, but later changed to regular khaki uniforms and armed to the teeth with American Springfield rifles and bayonets and with pouches filled with ball ammunition, we stood guard at Tibberton cross roads where we stopped and interrogated sundry innocent motorists. [...] The spells of sentry duty always being accompanied by the dismal coughing of Mr. Belcher's sheep in the neighbouring field.'*

In July 1940 the Local Defence Volunteers were renamed the Home Guard[d] and the Groups were renamed Companies and Battalions. Harper Adams became Battalion and Battle Headquarters for the 11th Battalion of the Shropshire Home Guard, under the command of Lieutenant Colonel H. L. B. Lovatt, owner of Meeson Hall. 'B' Company was based at the NIPH, under the command of Major R. J. (Dick) Davies, with No. 1 Platoon at the Institute and No. 4 Platoon at the College. The other Companies were at centres distributed around the area.[e] Willard Moore, then a student, remembered his platoon practising rifle drill in the NIPH's lecture room and loosening the blocks on the floor.

He also remembered one particular night in August 1940, when Morris Belcher's farm at Tibberton was bombed. The Home Guard was using the cottage at Bobaston Bank as a look-out point when incendiary bombs fell on the farm nearby. Probably, this was to light a flare path towards Liverpool. They jumped into action and quelled the fire. As a reward for their sterling work, the Home Guard was treated to 'a slap-up supper at the Lamb' at Edgmond.

No. 4 Platoon, the College Platoon, was formed in December 1941, meeting above the carpenter's shop in the workshop building.[f] The College Platoon became a 'crack fighting

9.11 Eleventh Battalion of the Shropshire Home Guard. Front row from left: A. W. C. Johnson (Transport Officer), Lt Colonel H. L. B. Lovatt (Commander), Major J. Sowerbutts (Medical Officer), Cpt H. J. Kendall (Liaison Officer), C. G. W. Block. Back row from left: Lamb first, 'Porky' Blakeman fourth, Arthur Baker fifth, Owen Jones (Intelligence Officer) sixth, Jack Hicks seventh: Private Collection.

unit',[g] one of the best equipped and efficient platoons in the Battalion due to the efforts of the original Platoon Commander Lieutenant Don Salmon and Sergeant Turner.[f] This good leadership and the youth of its members ensured its place as one of the chosen elite platoons. Salmon left the College and Sergeant Turner was commissioned in September 1942 and became the replacement Platoon Commander. With staff moves during wartime, rapid changes of command were inevitable and in January 1943 command of the Platoon passed to Captain Mark Carpenter and Sergeant Frank Peregrine of the Advisory Economics Department.[f]

Carpenter 'should really have been a General, or at least a Major, with his regulation moustache, a regal bearing, and utter disdain for any poor blighter who could not throw a hand grenade far enough away to escape the blast. He drilled, bullied, shouted and swore at us so effectively that had the enemy been silly enough to have landed within a hundred miles of Harper Adams they would surely have slaughtered us all, for most of the time either we were not trusted with arms, or they were all being shipped to Russia!'.[h]

By 1944 the Home Guard did not involve a large number of students.[i] Nationally, the Home Guard was stood down in December 1944.

Chapter Ten —

Confidence and Charisma 1946-62

Bill (W. T.) Price was appointed Principal in the summer of 1946. During the principalship of Price, Harper Adams had very good reasons for optimism. The Government was smiling benignly on agricultural education, encouraging and supporting progress. The College had a secure place in the national hierarchy and was, moreover, highly respected in its role, as the growing queue of prospective students testified. After years of restrictions, expansion was taking place. Principal Price was a man of presence and panache, ideally suited to take the College into the future. His background and character well qualified him for the task in hand. As well as his experience of lecturing in agriculture and estate management at Harper Adams in the years 1922-24, he brought to the post a wide range of experience and contacts. He possessed a degree and both the NDA and the NDD. From Harper Adams he went to Wiltshire, where for twenty years he worked for the County Council as Agricultural Organizer. During the Second World War he became the Chief Executive of the Wiltshire War Ag.

Price's great talents lay in his charm and enthusiasm. He was an instantly popular figurehead for the College and with effortless ease and presence he was able to raise its profile. Maurice Lancaster, student 1951-53 and a lecturer at the NIPH from 1955, remarked that 'he was a showman'. Thus all visitors to the College Farm received a copy of the Farm Guide, which, although distributed free of charge, was printed with the price of one shilling. Price's masterstroke ensured that all left feeling that they had been given something for nothing.[1] Bob Bradley, farm pupil and student 1957-60 and appointed to the Farm Mechanization Department in 1965, recalled Price's great ability to conjure up stories about the College, 'all these plausible stories, all sounded so wonderful'. Principal Price got on well with everyone, both students and staff.[2] He could be seen at all times around the College, accompanied by his golden labradors, giving praise and encouragement.[3] Bob Bradley remembered a visit made to the College by his parents and grandparents when he was a farm pupil. Although not a planned visit, Price invited the party into his sitting

10.1 The College, 1950: Aerofilms Ltd.

room and gave them tea: *'My grandparents were so impressed by this outgoing person and such a charming man and I am sure if he had said to my father "Would you mind paying for the tea", he would have done because he was just that sort of bloke.'* Price was a great believer in a friendly conversation over a drink in his study solving many a problem and pacifying many an irate member of staff.[4] It was in his day that the College had its own-label sherry, specially bottled by Addison's and, typically of Price, chosen by consensus of opinion once all had shared the delights of a blind tasting.[1] The College and the man became indistinguishable in many people's eyes. Moreover, 'the energetic drive, initiative and spirit of adventure which has always characterized Mr Price's work' became very much a feature of the College.[5] As Irene Doody, Office Junior and later Principal's Secretary 1959-86, recalled: 'Mr Price would immediately jump onto anything new.' And, above all 'there was more fun around'.

With the appointment of Principal Price the College gained a bonus in the form of his wife, Fanny Louise Price.[6] Her keen interest in all manner of things, above all people, coupled with her conscientious approach to her role as the Principal's wife was to filter throughout the College. Thus, she was instrumental in designing the furnishing of the renovated library, opened in 1950.[7] A lover of flowers, she did much to enhance the Principal's garden and to bring about the prettification of the grounds. Fresh flowers were always placed around the College and were very much in evidence at important events. She was a big supporter of the Harper Adams Club *(see photo page 145)* and did a great deal to keep it going.[1] She oversaw the arrangements for their meetings and it was her idea to encourage members to bring their families, tempted by all manner of things to keep the youngsters happy.[8] It was due to Mrs Price that the friendly atmosphere long evident at the College continued and blossomed. She took an interest in all the families of employees at the College, greeting the wives arriving as newcomers, awaiting news of their children's births and marriages and helping them in sickness and trouble.[9] The Prices introduced a variety of social gatherings to the College, which did much to create a sense of community among the personnel as well as providing opportunity for much enjoyment. These included Christmas parties for the children of the College staff and pantomime trips for all, when chocolates and cigarettes were liberally handed out free of charge.[1]

Mrs Price had an all-powerful presence. Although always frail in health as an asthma sufferer, she had 'a rod of iron for her back'.[10] She was unflappable, even on the occasion when a visiting girlfriend left a student literally holding the baby: *'[She] acquired a big wicker laundry basket and there was the baby safely tucked up in the house there. Mrs Price rushing round with bottles and the nappies were hanging outside in this little yard.'* [1] On occasion her concern for the College's well-being could be regarded as interference. For example, she opposed the clearing of the old orchard at the NIPH and had all work stopped.[11]
Mrs Price especially liked the College to look picturesquely rural, a preference which at times could clash with commercial interests.[2]
Her influence over her husband was clear. However, she never involved herself with College policy nor lingered in the limelight.[4]
Her intentions were always for the best.
With the formidable charms and talents of Mr and Mrs Price at the College the stage was set.

The mid-1940s saw agriculture and agricultural education very much in the forefront of the Government's thinking. The war had found many farmers unprepared for the demands made by the food campaigns. As a consequence the war years had seen massive intervention by the Government into farming. With the end to hostilities, productivity continued to be important in a world where an acute shortage of food necessitated the continuation of rationing. The Labour Government was determined to produce an efficient farming industry capable of producing Britain's needs. No longer would agricultural interests be sacrificed to desires for cheap imported food. The Agriculture Act of 1947 thus continued the wartime policy of intervention and gave assurance of long-term support for farming. The main thrust of the Act was an annual price review to determine the levels of minimum prices to be guaranteed by the Government the following season. If prices obtained by farmers were less than the guarantee announced, the deficiency was to be paid by the Government. Floor prices were also set for the year after that so that farmers could plan for the future with confidence. In this policy, the importance of advancing skills and knowledge was not overlooked and the Government was determined that training and education would rise to the task in hand. Coupled with this, there was the

10.2 Biology class with Drs Janet MacLagan and Norman Gill, probably 1950s: Shrewsbury Chronicle.

desire to learn, encouraged by the security engendered by the Government's support for farming. As Peter Luscombe, Lecturer in Animal Husbandry, declared: *'The quest for knowledge was great'*.

The future of agricultural education was shaped by the Luxmoore Committee on Post-War Agricultural Education in England and Wales of 1943. The Committee's Report found a confusing picture of farm institutes, agricultural colleges and universities offering a range of courses which very often overlapped and resulted in unnecessary competition. The agricultural colleges, including Harper Adams, were severely disadvantaged, threatened by both farm institutes and universities. Shropshire's own farm institute was to open at Walford in 1949.[12] The Report recommended that a clear three-tier system should be imposed, with colleges concentrating on Diploma courses of two years' duration and the option of one- or two-year specialist courses for those already possessed of a qualification. At the close of the war, in 1946, the Loveday Committee on Higher Agricultural Education in England and Wales took on board the recommendations of the Luxmoore

Report, supporting the three independent agricultural colleges, namely Seale Hayne, the Royal Agricultural College and Harper Adams, in their national role as providers of Diploma courses. The Loveday Report envisaged that the future of the Diploma lay in its practical nature and suitability, therefore, for the working farmer seeking a broader and deeper education than the kind on offer at farm institutes.

The response to the Loveday Report from Harper Adams was generally favourable. The College welcomed a clear definition of its role and could look to the future with confidence. With a clearer stratification being given to agricultural education the Report recommended that there was an end to the teaching of one-year certificates and degrees at college level. Indeed, by 1945 circumstances had dictated that the one-year course had stopped at Harper Adams. As part of a natural progression, any work better suited to a farm institute had been dropped because of fears of competition. However, the recommendation that degree teaching should end was not so willingly taken on board. The session 1947-48 did, however, see the end to enrolment for degrees. The number of students involved had never been large. In the last year to possess both first- and second-year degree students, 1946-47, eight students were enrolled. Yet, the decision was made with reluctance and the blow was softened by making it a temporary move.[13] Significantly, as late as 1957 the prospectus was still listing the degree as being temporarily suspended.

The educational contents of the Luxmoore and Loveday Reports were in line with the situation which was evolving at Harper Adams, but the Government also looked to the future role of the Advisory Services and here the recommendations were not welcome. The Luxmoore Report's recommendation was that all advisory services should come under unified national control with the creation of the National Agricultural Advisory Service (NAAS). The thirteen provinces would be reorganized and each county would become a unit within a province under the direction of a county organizer. The new West Midland Province would be enlarged to include Worcestershire, Cheshire and Herefordshire. The new Provincial Director would be W. B. Mercer, former Principal of Cheshire School of Agriculture at Reaseheath and an ex-student of Harper Adams and author of the diary reproduced on earlier pages. Most importantly, the headquarters of the new province would no longer be the College.

As the war drew to a close it seemed more and more likely that the recommendations concerning the Advisory Services would be adopted. The Governors issued complaints, concerned that such a move would be detrimental to the College's prestige and would deprive the College of essential teaching staff and equipment. It was to no avail. NAAS was born in October 1946 and the College's Advisers gradually left. Many went to the new provincial headquarters at Tettenhall, Wolverhampton, but others left for other appointments. For a considerable time the College still relied upon advisory staff for specialist lectures. Veterinary studies and bacteriology were still being taught by NAAS officers at least as late as 1957, if not into the early 1960s.[14]

Back in 1943 Crowther had anticipated that the end to overlapping courses within agricultural education as first advocated by the Luxmoore Report would result in a significant rise in student numbers at the College and, indeed, time proved him to be right. Student numbers at agricultural colleges rose by thirty percent between the pre-war period and 1957-58.[15] For the session 1945-46 student numbers at Harper Adams reached 103, but three years later there had been an increase of fifty percent with a figure of 151. Initially enrolment figures were influenced by large numbers of ex-

10.3 1956-57 staff and students: West Midland Photo Services.

servicemen, but numbers just kept on rising. This was despite higher entry requirements introduced after the war to comply with the demands of a diploma course. Students now needed to be eighteen years of age, possessed of higher qualifications than those required for the farm institutes and to have had at least one year's practical experience. By the session 1956-57 student numbers were again up by fifty percent to reach 222. From then until the early 1960s numbers stabilized around that mark. The number of female students doubled in the Price Era, although all were now studying poultry husbandry as men only were accepted for agriculture.[16] For the session 1945-46 there were seventeen women and in 1956-57 the number reached thirty-two, the highest figure in the decade.

Initially ex-servicemen formed the overwhelming majority of students. The Government insisted that from October 1946 men from the Armed Forces or those who had been engaged in reserved occupations were to receive priority places. For the session 1947-48 only ten students out of 123 were not ex-service personnel. As a result the average age of students was high at around twenty-four [8] and the new students possessed an attitude to work which was praised for its enthusiasm and sense of responsibility. By the summer of 1950 the ex-service personnel were dwindling in number and there was genuine regret at their departure.[17] Willard Moore as a lecturer remembered them as 'grand lads' who 'drank hard and played hard because they'd had five years of war and action and they were raring to go'. His fellow member of staff, Peter Luscombe praised their attitude to study:

Generally speaking students were there because they wanted to be there and they wanted to know and learn, mainly because immediately after the war [...] there was a fair number of ex-war servicemen who were followed by National Servicemen. All of which meant they were well disciplined and they wanted to learn and get out of college and back into earning their own living. Therefore attitude and interest and applying themselves to their courses was very good.

In the face of these older, more worldly students, many of the College's regulations had to be abandoned, especially those pertaining to alcohol.[9] There was still no bar in the College, except at the major dances,[1] and beer could normally only be brought in for Sunday dinner.[18] However, pubs were no longer out of bounds and were a popular destination.[19] As a script in the 1948 Cat-A-Mountain put it:

There is a cult at Harcol which is called Booze, and many follow it. And they do go down to certain houses in Edgmond, and thus is the word of the prophet fulfilled: 'They shall lie down at the Lion and the Lamb, and they shall be lifted up, for they shall not be able to lift themselves'.

However, discipline was self-imposed and problems were rare.[3] Derek Turner, student 1947-49, was one of the few scholarship students at this time. He remembered the ex-servicemen having 'some wonderful things like Alvises and big Vauxhalls' and he also remarked upon their highly polished shoes. During the 1950s many of the students came to the College having completed National Service, which did not end until 1960. It was possible to obtain deferment.[1] Again military discipline and experience characterized many of these students. Ian Gedye, student 1956-58, noted:

[Discipline] wasn't needed so much by the staff, because the older students tended not to be very tolerant of these youngsters who were being stupid. A prank was a prank and one expected that there was a certain amount of high spirits and tom-foolery went on, but I think it was modified by the fact that you had a proportion of older students there anyway. They wanted to learn. They had a responsible attitude towards learning. They wanted to get a good qualification. They wanted Harper Adams' name to be good from that point of view.

The rise in student numbers at agricultural colleges was intricately linked with the Government's encouragement of education. It is clear the colleges were rising to the expectations required of them. In 1955 the College received extracts from the confidential Keith Murray Report on agricultural colleges.[20] The Ministry of Agriculture felt it was not proper to publish the Report nor to issue it in its entirety to the colleges.[21] Dr Murray, Chairman of the University Grants Committee, had been requested by the Ministry to examine the work of the four grant-aided agricultural colleges, namely the Royal Agricultural College, Seale Hayne, Studley and Harper Adams.[22] Due time had elapsed since the implementation of the Loveday Report and the moment was right to reassess. Considerable praise was heaped by the Murray Report on the work of the agricultural colleges, which were seen to be most capably supplying the growing demand for trained students: *'We have been most favourably impressed by the general tone of the establishments and the energy and devotion of their heads. They are undoubtedly contributing to agricultural leadership.'* With reference to Harper Adams the Committee was 'well satisfied' with the standard of teaching and learning.[21] The Committee had no doubt that the colleges and both college diplomas and the NDA had essential roles to play in agricultural education, with no prospect of falling demand for courses from students or for students from Industry.

The quality of the teaching at Harper Adams was highly regarded. Tony Harris, Lecturer in Crop Production in the mid-1950s, remembered the degree of freedom allowed the academic staff in those days:

> *I remember arriving and the Principal saying to me, this is the syllabus, which was the national syllabus for the NDA and which was two sides of paper on crop and grassland production. That was it. What level one had to attain, or what content the teaching was up to the individual. That was the peculiar freedom that one had, but it was important that you got the students through the National Diploma examination which was external and set and marked externally.*

A few years earlier, in 1949, the newest teaching aids for biology consisted of an epidiascope and a series of film strips. Students were taught in their year groups and of an evening would read and supplement the given notes. The library was used, but prior to 1955 it opened just two evenings a week. Thenceforth, it was opened daily.[23]
Books were often bought. There were no written assignments.[24] It is interesting to note the high record of achievement for Harper Adams students on the national stage in the NDA examination. All students sat the same examination regardless of their college and thus the NDA was regarded as a significant test of standards.[25] In 1957 fifty-five Harper Adams students sat the NDA and fifty-four passed, giving an impressive ninety-eight percent success rate in comparison with sixty-six percent for other colleges. In 1960 the pass rate had risen to one hundred percent and among the successful students were two with first-class honours.[5] As Ian Gedye said: *'The students went out with their heads held high.'* Interestingly, many students and staff considered the College Diploma to have been of a higher standard than the NDA. Thus, 'the feeling was, that if you could get the College Diploma, you'd certainly have no trouble getting a National Diploma.' [25]

During the Price Era the agricultural courses at Harper Adams were the College Diploma and the NDA. Most students sat both examinations by this date, although there was a choice as to whether the NDA was taken.[25] The College did accept a small number of students on a special one-year agricultural course. These students, for whatever reason, often lacked the necessary academic qualifications for the Diploma.[1]

10.4 Library, probably 1950s.

Arrangements would be made on an informal and individual basis, often through personal acquaintance with Principal Price, and subjects and levels varied according to the student's requirements.[26] Steven Colman, for example, had a background of education abroad. As the son of an agricultural engineer intending to follow his father, he attended Harper Adams for a year in 1950 to obtain a knowledge of farming. Those on special courses were never many. In the entry for 1955-56 they amounted to just eight.

The poultry courses, namely the Senior Certificate and the NDP, were the equivalent of the two agricultural diplomas.[27] The Senior Certificate was to assume the title of the Diploma in Poultry (HAC).[8] The one-year Certificate course in poultry husbandry had been dropped at the end of the war due to increased demand for higher courses.[28]

The College Diploma, to be renamed the Diploma in Agriculture (HAC),[8] was marketed as a farm management course with special emphasis on agricultural engineering and farm machinery. The bias towards farm management and machinery was highlighted by Price in 1961 as a very significant move, 'well ahead of rivals'.[29] Clearly, the College was showing due respect for the recommendations of the Loveday Report and in a time of great technical advances was responding to the needs of the working farmer. In return those wanting to go farming saw the College as fulfilling their needs and enrolled.

Throughout the Price Era most students intended to go farming. In 1957 Price put the figure of students gaining 'practical posts on farms' as high as sixty-five to seventy percent. Two years earlier the figure had been the same, with around twenty-five percent taking careers in 'official appointments, teaching and administration'.[30] A student survey of 1959 put the figure for those intending to go into farming at fifty-four percent, with a career in advisory work being the second most popular choice with twenty-five percent.[31]

Harper Adams had always concentrated on the practical character of its teaching and with the Loveday Report this policy received endorsement for the future. There is little doubt that the College had wide appeal, attracting students of high academic calibre who could have enrolled on a degree course.[25] Peter Luscombe's 'unique style of lecturing which combined practical experience with a thorough understanding of the scientific principles [...] endeared him to students and farmers alike'.[32] However, around 1950, the actual requirement for students to undertake work on the College Farm ceased.[33] A year's practical experience on a farm had changed from being desirable to becoming the rule. Thus, the importance of working on the College Farm was diminished. Growing student numbers had also meant that farm classes had grown to an impractical size.[2] The change was welcomed by the Bailiff, Tom Fordham, who found the lack of expertise among students could decimate a crop. Agricultural students were now encouraged to observe and discuss and a great deal of time continued to be devoted to the farm walk.[33] Ian Gedye was especially appreciative of the opportunities to go outside the lecture room:

What I do remember [...] was the botany side of it, because Janet MacLagan particularly used to take us round the fields and we'd look at weeds, identify weeds, grasses and identify grasses. She was really keen on that and did a very good job in

10.5 Peter Luscombe and students, 1950s: Private Collection.

teaching us our botany, which is often the underpinning of our crop production work and of course, Tony Harris' [Lecturer in Crop Production 1956-58] farm walks were always very interesting, looking at the management of the crops. Bill Lightwing [...] used to take us round, talk about different machines that had been used.

The Farm thus continued to give invaluable support to the College's educational role.[34] 'The amalgamation of practice and the classroom' was all important at Harper Adams.[33] The continuing small-scale character of poultry farming for much of the 1950s meant that poultry students continued to be involved in far more 'hands on' activities than their agricultural colleagues. These activities included everything from carpentry (the students undertook considerable work helping in the renovation of the NIPH)[3] to plucking and vaccination,[35] but, even with the poultry students, involvement with the routine running of the plant had stopped by around 1950.[3]

From 1948 [36] a number of would-be students, men and women, worked for one year on the College Farm or at the NIPH as farm pupils.[2] They paid their own board and lodging and received no payment, beyond an honorarium of twenty pounds at the end of the year.[37] Their time was entirely taken up with farm work, but they benefited from a teaching environment where the instructors fully explained the practicalities. Almost all progressed to the Diploma course.[4] The Keith Murray Committee of 1954 was none too happy with the existence of farm pupils at Harper Adams.[38] Such a scheme the Committee felt was more suitable for a farm institute. After a sound defence from Price the farm pupils were allowed to continue. Price argued that the scheme was 'one of the best things' the College did, as places on farms providing the necessary variety and completeness of instruction were in short supply and invariably the farm pupils, with their sound practical knowledge, became some of the best students.[37] As Luscombe wrote: *'The training was excellent in every way as a very useful blend of practice and technology was obtained under expert supervision in both fields.'* [32] Bob Bradley, farm pupil 1957-58, recalled the range of work:

You had a fortnight milking cows [...] and putting the milking machines together and washing the yards down and included in that would be the young stock, feeding calves and so forth. We had a fortnight with pigs and the pigs were farrowed outside and then we had the inside fattening side of things. So, there was a fortnight where we learnt about rations and castrating pigs and ear marking them and we learnt how to identify the pigs by notches taken out of their ears [...]. Then we worked on the arable side. [..] It could be working with Bill Burrell in the NIAB trial plots. It could be singling sugar beet by hand with a hoe. [...] We had a complete overview of farming, whereas a lot of the students that came in had milked cows for twelve months or they had driven tractors, but there were very few students coming in who had had a complete practical education through all the farming enterprises.

The scheme, along with the one-year specialist course, ended in 1959.[31]

The Luxmoore Report had advocated that colleges should offer optional one- or two-year specialist courses for those already possessed of a qualification. Harper Adams acted upon this recommendation. Early hopes of a post-diploma course in poultry husbandry did not materialize,[39] but from the late 1940s Harper Adams offered the College Diploma with Honours.[40] Those students possessed of a Diploma could proceed to one year's post-diploma work based on a special investigation linked to one of the departments. From 1957-61, for example, there was a one-year post-diploma course in livestock husbandry with special reference to pigs.[41] Numbers enrolled were never large and already by the third year there were difficulties recruiting.[42] Of all the post-diploma courses, perhaps the most interesting development was the National

(NDAgE) and College Diplomas in Agricultural Engineering, which started in 1952.[43]

The birth of an agricultural engineering course at Harper Adams was an important if short-lived development. The immediate origins of the specialist course go back to the late 1940s when the College sought to revitalize the College Diploma with greater emphasis on machinery. The study of farm implements had been part of the syllabus since the beginning of the College and the subject still went by the name of Farm Implements in the late 1930s. Wartime students benefited from the growing popularity of mechanization as food production took centre stage. Their practical knowledge continued to be limited. No student, for example, was allowed to drive the College tractors. However, the theoretical grounding they were given in what was a rapidly developing area was sound.[44] In the last two years of the war the College was clearly not ignoring the growing importance of machinery to the farmer. Lecturer John Robertson, for example, arranged visits to mechanized farms and approached forty firms in America and England for equipment.

In 1951 thoughts turned to establishing a special course in agricultural engineering in line with a new National Diploma 'in the science and practice of agricultural engineering', a course which would be the first of its kind in the country.[45] The course would be open to those with a diploma or degree and would run for one year. The Governors, concerned at the cost of additional buildings, must have been heartened by a letter, written in August 1951 by Alexander Hay, Chairman of the Examining Board in Agricultural Engineering:

> *The facilities you have at the College in workshops, storage accommodation for implements and demonstration halls are much more than is usually found in many university agricultural departments and agricultural colleges; further, you have an excellent mechanic [Reg Hollis] who is responsible for the maintenance and repair of agricultural machinery and who would be a favourable addition to the staff for practical instruction.*

The workshops and machinery sheds were scattered about the farm and in 1947 Boutflour's cowshed at Ancellor had been converted into a modern machinery hall for larger equipment.[24] Students were to have the benefit of a new drawing office, now the Farm Records Office, although this was not completed in time for use by the initial intake during their first term.[46] The course opened in the autumn of 1952 with eleven students.[43] New member of staff Arnold Lackie and Bill Lightwing, Robertson's successor, undertook the lecturing. Unfortunately, continuous low enrolment numbers led to the abandonment of the course in 1956.[47]

The growing popularity of Harper Adams brought with it problems, namely the need to renovate, modernize and expand after the strict economies of the 1930s and the special circumstances of war.[8] It was not so much that the College was in a neglected state: *'It wasn't the fact that it was not tidy. [...] You could see it was just stopping. It hadn't moved on.'* [11] The rapid growth in student numbers especially highlighted the urgent need for accommodation.[48] In 1946, such was the need for rooms that even the Principal's bathroom was earmarked for six students, presumably after some modification.[49] Much minor work was started straight away, but, before any major action could be taken, the money had to be found.

In February 1948 the Ministry approved the schedule of renovations and alterations submitted by the College. The list was long and covered everything from 150 Windsor chairs, cutlery and repairs to gutters through to renovations at Ancellor House and an extension to Old Hostel.[50] The Ministry undertook to grant around £25,000, representing seventy-five percent of the overall costs involved, which amounted to around £33,500.

The College for its part was to meet the sum of approximately £8,000, which was to come from the Endowment Fund and was to be paid back to that fund. By 1951 costs of the building programme had risen greatly, one figure putting it as high as £40,237.[51] The Ministry's contribution now ran to around £27,000 and subsequently, the College had to increase its commitment substantially.[52]

By the early 1950s the College site had been greatly improved in many ways. The library, for example, was refurbished in 1950, with the assistance of the Harper Adams Club, in memory of the students who had died in the Second World War.[7] Unfortunately, a list of those students who died was never compiled. Since the war the library stock had been greatly extended to around a thousand volumes under the guidance of Dr Norman Gill, Head of Biology, Warden and Librarian. At a time when supplies of wood were scarce, oak for the bookcases, tables, chairs and memorial tablet was donated by the Earl of Dartmouth. He had been a student in 1919 and, as Lord Lewisham, had served as President of the Harper Adams Club 1920-27.[53] The work was carried out by the College's carpenters, Frederick Shakeshaft and Jack Wickstead.[7] 'Shaky' had been employed at the College since 1919 and before that he had been involved with the original building work. He was an extremely talented carpenter and joiner who could turn his hand to anything from a simple repair to a table.[24] However, somewhat temperamental, 'he did what he wanted, when he wanted and how he wanted it'.[11] Shakeshaft and Wickstead were also responsible for the panelling of the entrance hall in 1954. The window on the main staircase with the stained glass crest of the Adams Family was also installed at this time. An immediate improvement was made throughout the Main Building with a change of decorative scheme, from chocolate to cream and grey paint with natural woodwork.[46]

Accommodation needs had been the top priority. Old Hostel, now renamed Adams Hostel, had been extended. The new wing ran parallel to the Caynton Road and received the name Jubilee Hostel. The new hostel could accommodate twenty-five students [54] and also possessed bathrooms, staff rooms and a large common room.[46] The architect was Mr Byram of Wellington [55] and the original cost was estimated to be £8,662.[56] All was of the most modern design with built-in cupboards, divan beds and even 'wall plugs for table lamps'.[57] With his considerable charm Price had managed to obtain a contribution of £500 towards the furnishing from British Oil and Cake Mills Ltd (BOCM). Ancellor House had been extended and renovated to accommodate fifteen[16] women.[58] Ancellor improvements included cream as opposed to chocolate colour paintwork, a new bathroom on the top floor and most importantly, central heating.[46]

The finances of the College were finely balanced. As an autonomous college, income was reliant on the Endowment Fund, fees and an annual grant from the Ministry of Agriculture, which had stood at £4,000 since 1930. In 1947 the grant was raised to £5,000 and by 1950 the year's allocation was £7,000, but the grant for 1952 was cut to £5,000.[59] The system of quinquennial grants had ended in 1939 and grants were now on an annual basis. An unforeseen cut in the grant could impair solvency or overturn long-term planning.[38] Furthermore, it was a time when the cost of living was greatly increasing and in particular rising catering costs and staff salaries were a considerable problem for the College.[59] Fees also increased, steadily rising over the next few years. For the session 1946-47 fees for residence and tuition stood at £125. In 1951 fees rose to £250 a session, but such action made little impact on the worsening situation and ran the risk of discouraging students.[60] The next year the College's investments and reserves were declared to be

10.6 Jubilee Hostel, 1950s.

almost exhausted.[59] The bank overdraft was mounting. During the period from June 1953 to February 1954, for example, it reached £8,981.[61] The spiralling financial crisis facing the College was stemmed, at least temporarily, by the Ministry's decision late in 1953 to award an ad hoc grant of £4,000. This was on the understanding that the College would contribute a similar sum to reduce the bank overdraft. In future, all capital schemes were to obtain the approval of the Ministry whether a grant was involved or not and an upper limit would be fixed on any overdraft.[62]

Harper Adams' relationship with the Contributing Counties was undergoing change at this time. Increasingly, the College saw its role as a national one, recruiting students from around the country. In 1949 the Contributing Counties agreed to end the privileged position of lower fees for students resident within their boundaries.[63] This had been a contentious point, as financial benefits for the College from the arrangement seemed few. That same year, 1949, the financial considerations allowed by the College to twenty-nine students on scholarships awarded by the Contributing Counties were greater than the grants contributed by the Counties by £400.[64] However, in return the Counties recommended the awarding of only token grants on the assumption that increased fees and the end to differentiation would alleviate the need for greater payment.[63] The importance of grant aid from the Contributing Counties was already decreasing. In 1950 Shropshire and Staffordshire contributed £250 each, Warwickshire £150, Worcestershire £300, Herefordshire £175, and Birmingham and Coventry £100 each.[65] In contrast, in the closing year of the war, awards had been far higher, with Shropshire and Staffordshire, for example, making grants of £1,000 each. After the Department of Education took over Harper Adams in 1964 the College was regarded as an independent institution financed by direct grant.[66] Thenceforth, the Counties were no longer tied to financial contribution and in 1967 the last county to make a grant, Shropshire, stopped doing so.

The Keith Murray Report of 1959 acknowledged the decline of financial support from the Contributing Counties. However, it emphasized the importance of the Contributing Councils in continuing to give indirect support to the College through student scholarships and it recommended that the Counties should continue to be represented on the Governing Body.[42] Indeed, local representation was strong. In 1951, in recognition of the increase in the Contributing Counties to five,[16] the Harper Adams Foundation Scheme had been revised. The number of Governors stood at thirty-one.[67] There would be six Co-optative Governors and twenty-five Representative Governors. The latter would be: four from each of the County Councils of Shropshire, Staffordshire, Warwickshire and Worcestershire, three from Herefordshire County Council, one from the Council of the University of Birmingham, one from Birmingham County Borough Council, one from the Council of the RASE, one from the West Midlands Advisory Council for Technical, Commercial and Art Education and two from the Ministry of Agriculture.

In the mid-1950s Government thinking on funding was revised. The Keith Murray Committee on Agricultural Colleges of 1954 advocated a return to quinquennial grants as a means to promote consistent policy and these were duly reintroduced in 1954.[38] On an annual basis over the next five years, as soon after the beginning of the financial year as possible, an agreed sum as a proportion of the quinquennial grant would be paid to the College. During the five-year period the College would have full control over its accounts, an all-important point. Despite the reforms in the system of grant allocation, the College's financial plight did not ease. It was not helped by the continuation of the renovation programme. It is clear that the College was not always acting with ministerial sanction let alone with ministerial funding. Inevitably such action led to debts.[20] By the end of 1958 the College had been obliged to arrange a bank loan of £4,760. The Ministry pointed out that the deficit was 'a source of embarrassment to both the College and the grant-aiding authorities'.[42] Not all was the fault of the College, the reintroduction of quinquennial grants itself resulted in problems. It was virtually impossible to assess grants over a period of five years in a time of inflation. Original figures had been reached on the assumption that prices would be stable, whereas between 1954 and 1959 prices rose by fourteen percent. Only in the case of academic salaries were supplementary grants made by the Ministry.[42] At Harper Adams the rising costs of everything from food to fuel meant that for the 1956-57 session it was estimated that the deficiency in income would amount to £46 50s per student.[68] Neither did quinquennial grants take into account the dramatic rises in the number of students and the need for additional finances to cope with such increases. Actual numbers of diploma students throughout the country were eighteen percent higher in 1958-59 than actually estimated.[42]

In April 1959 the Ministry of Agriculture again appointed Keith Murray to chair a committee to look into agricultural colleges.[42] The Keith Murray Report of 1959 recognized the difficulties implicit in the funding system and praised the colleges for the 'prudent management of their affairs during the quinquennium', a period which was described as 'very testing'. The Report recommended the continuation of the system but with the allocation of higher funding. This would take into account rising costs, most notably salaries, and the growing numbers of students. As a consequence of the recommendations the grant allocation over the next quinquennium was to be generous. The recurrent grants were fixed at a total of £72,600, varying in annual allocations from £11,000 to £17,000.[20]

Of great importance to the future of the College, the Keith Murray Report recommended financial support for further expansion. Over a two-year period from 1959 a total capital expenditure of £209,760 was approved. Towards this sum the College was expected

10.7 Construction of the Queen Mother Hall, c. 1962.

to contribute half of the estimated trading profits for the quinquennium, namely £7,500. Central to the building programme was the construction of a new assembly hall and dining room and kitchen. As early as 1953 Price had stressed the need for one assembly hall and one dining room to house all.[69] There was also the need to provide more accommodation, with agreement on the building of two hostels. Throughout the 1950s many students had been obliged to live out. Houses in the village, including Edgmond House and Hill House, had been leased as hostels. As regards the teaching facilities the plans looked to more space being gained in the Main Building and to improvements to both the biology and chemistry laboratories.[5] The future growth and development that now lay ahead was naturally welcomed, but few at Harper Adams could have foreseen or truly have comprehended the vast scale of the changes that would take place.

Amid all the significant advances of the 1950s and the ever worrying financial situation, it is heartening to note the human element was not missing. Reading the Minutes of the Finance and General Purposes Committee of March 1956, the needs of Bursar Block come to the fore. After much debate, it was agreed that the College should purchase a car for his use. It was specified that it must be second-hand and intriguingly, it was stated that it was to be used 'on somewhat similar conditions to those applying to the County Council Dentists'. Block could drive happily into the rosy future which beckoned the College, safe in the knowledge that man's love of sweet things would inevitably result in tooth decay and secure a future for dentists and their cars.

Principal Price's Animals

Bill Price is always remembered as the man with his labradors at his side, but his dogs were not the only animals to have a place in his life. There was Sally, a very fecund and popular sow. Together Sally and the Principal made numerous television appearances in the 1950s spreading the latest word on pig management. Visitors to the College were always taken to meet her in the orchard and to receive news of her latest litter 'and how many tons of meat she had produced'.[a] In many ways Sally epitomized the College in this period. She was a friendly and vivid image, which kept the College alive in popular memory long after the more erudite points of research had faded.

10.8 Sally: Private Collection.

Gert and Daisy were two Highland cows, bought for Mrs Price as pets.[b] It was a time when coach loads of Women's Institute members would descend upon the College and invariably, they would be photographed with the College's Highland cattle.[c]
As Ian Reekie, farm pupil and student 1953-56, related: *'It was commonly held that their sole purpose in life was to impress non-agricultural visitors. This no doubt scurrilous rumour was given some credibility as the "Highlands" seemed to spend most of their time in the fields adjacent to the front of the college.'*

10.9 Highland cattle and visitors with Principal Price, probably 1950s: Wellington Journal & Shrewsbury News.

Chapter Eleven —

The Eggs and Bacon College 1946-62

At the dinner in July 1961 to celebrate the Diamond Jubilee of Harper Adams, Principal Price looked back over his time at the College.[1] He drew attention to the first year of his principalship, 1946, as being a significant watershed in the history of the College. This has been seen to be very much the case. However, one point which Price emphasized was not so much concerned with new directions, but centred on the defence of an existing state of affairs. After the war there were plans by the Ministry of Agriculture to move the NIPH to Reading. Such an idea seems incongruous today, so closely interwoven is the story of the NIPH with that of the College.

It is true the war had left the NIPH seriously dilapidated and action of some kind was vital. Financial pressures and a lack of labour and materials had led to serious problems. Everything from dangerous wiring to leaking poultry houses and inadequate fences had to be fixed. All this required expenditure and the continuing need for deficiency payments from the Ministry put the whole financial viability of the NIPH under the spotlight. Student numbers were low at the close of war and prospects appeared little brighter, as poultry food continued to be rationed until 1953.[2]

It was against this background and with the encouragement of the NAAS [3] that the decision was made that the NIPH should go to Reading.[4] However, in 1946 renovation was already underway at the NIPH under the expert guidance of the newly appointed Lecturer in Building Construction, Willard Moore, and with prudent use of materials and money, improvement was soon evident.[5] Moore acted as both draughtsman and craftsman and was highly respected for his ingenuity, skill and common sense approach to poultry keeping.[6] Much of the work itself was undertaken by

11.1 Poultry students undertaking reconstruction work at the NIPH after the Second World War.

the students as part of their experience: *'When they [some of the girl students] went home their mothers were complaining about their bulging muscles from wheeling barrows of hard-core and stuff like that to make roads.'* [5] Such activity weakened the case for removing the NIPH. By June 1947 Price had been able to convince the Ministry of Agriculture that the NIPH should stay at Harper Adams.

In 1948 a further step was taken to secure the future of the NIPH. That year the Ministry of Agriculture made the proposal that the NIPH and College should merge. All capital assets would pass to the College and the Ministry would pay a maximum capital grant of £25,000, with some fine-tuning of financial detail, to cover the modernization programme.[7] For a period of three years it was agreed that the College would not be required to meet any expenditure regarding the maintenance of the NIPH. In future one of the two nominees of the Ministry on the Governing Body would be a representative of the Poultry Industry. Thus, from 1 October 1948 the NIPH became a Department of the College, keeping its original title.[8] The Principal became the Director of the NIPH and the Head of the Department acted as Assistant Director.

Herbert Howes became the Head of the newly formed Poultry Department. On his retirement in 1951 he was succeeded by Harold Temperton.[9] From 1955 Temperton acted as Deputy Principal and in 1957 he received the title of Vice Principal in recognition of his role. In 1949 Temperton had obtained his doctorate from Durham University for research undertaken at the NIPH in association with Levers on carotene, vitamin A and xanthophyll in poultry diet.[10] Many attributed the success of the NIPH to Dr Temperton: *'He was very well known and respected, a real figurehead and this must have attracted a lot of people to Harper.'* [11] He was similarly referred to as *'a powerful man, [...] for whom there was nothing but the utmost of respect'* and who was especially praised for his ability to marry the academic with practical application.[12] At the same time he was always a charismatic man, whom both students and staff found very approachable.[13] The role of Temperton's personality in inspiring his small but dedicated band of workers should not be overlooked. One article in the Press commented on the *'ability, enthusiasm, and team-spirit of the staff. [...] Comprising barely enough persons in number for the multifarious duties of administration, teaching, and research study,*

11.2 Cartoon from Cat-A-Mountain 1951.

their accomplishments in the direction of creating an up-to-date centre of teaching and demonstration are in themselves remarkable.' [14]

The session 1947-48 saw the beginning of an upturn in the prospects of the NIPH. Students now numbered a healthy thirty-one,[15] rising to thirty-eight two years later.[16] That session the Ministry awarded an initial grant of £6,000 for renovation, modernization and building work.[15] Further grant aid, as agreed in the transfer scheme, was to follow. The scheme, running over several years, was all pervading, covering everything from increased storage in the Poultry Husbandry Library, which contained the important collection of books donated by Sir Edward Brown, to an overhaul of the incubator room.[6] It was the time which saw the first battery housing at the NIPH. The earliest use of battery cages on a commercial scale had appeared around 1937 at the Lancashire Farm Institute at Hutton. Significantly, Temperton, then working at Hutton, was involved in the design of the 1,000 layer building.[2] At the NIPH one of the most interesting early battery houses was a semi-outdoor unit, with mechanically driven non-stop 'food and drink pavilions' passing in front of the cages.[6] It soon received

11.3 Semi-outdoor laying battery unit at the NIPH, c. 1950: The Farmer & Stockbreeder.

the nickname 'Pneumonia Avenue'.[13] Among early developments was a new poultry and egg marketing building, completed in 1951. Here there were killing, plucking and trussing rooms, egg storage and feather storage rooms, a quick freeze apparatus [17] and the 'unusual feature' of a refrigeration room capable of holding 250 birds.[18]

The 1950s saw the NIPH in a very healthy state. In 1951 Cat-A-Mountain could declare with pride: *'There is not the slightest doubt that when all the proposed alterations and renovations have become an accomplished fact the NIPH will be the leading Institution of its kind in the world.'* All aspects of poultry husbandry were catered for and every recognized system of housing of young stock and adult birds, whether under extensive or intensive conditions, was represented. There was even a representative collection of rare breeds to create interest and to act as a gene store. White Leghorn, Buff Rock and Light Sussex were now the major breeds used in laying stock improvement.[19] The year 1953 saw the revival of poultry conferences at the NIPH.

11.4 Rearing range in front of of the NIPH, showing standard night arks, 1950s: Horace Hall.

As the Press said, something of the glamour for which the NIPH was famous prior to the war had returned.[20] The conferences, held biennially,[21] were very important events for the Poultry Industry.[3]

The renovation programme continued, spurred on by the end to rationing of poultry feed in 1953. Nationally, no longer were there restrictions on table poultry production on a commercial scale.[2] In 1954 a broiler house for 1,000 table birds was constructed at the NIPH, incorporating the very latest thinking on heating and on the colour and intensity of artificial light.[22] The year 1954 saw the introduction of broilers to Britain.[23] Broilers had been developed in the United States during the Second World War. Chickens were reared in high density housing under controlled environmental conditions and were fed high energy diets to maximize their growth.
As Rowley explained: *'A new name for which an alternative has not been found, the "Broiler" came on the scene and was very soon a billion dollar industry, making chickens cheap and readily available in vast numbers every day of the year, instead of the luxury item for festive occasions.'* [24]

Temperton sought to lead the field not to follow. He was especially skilled in obtaining financial support and winning Industry over to his side. Thus, on 1 April 1957 a new table

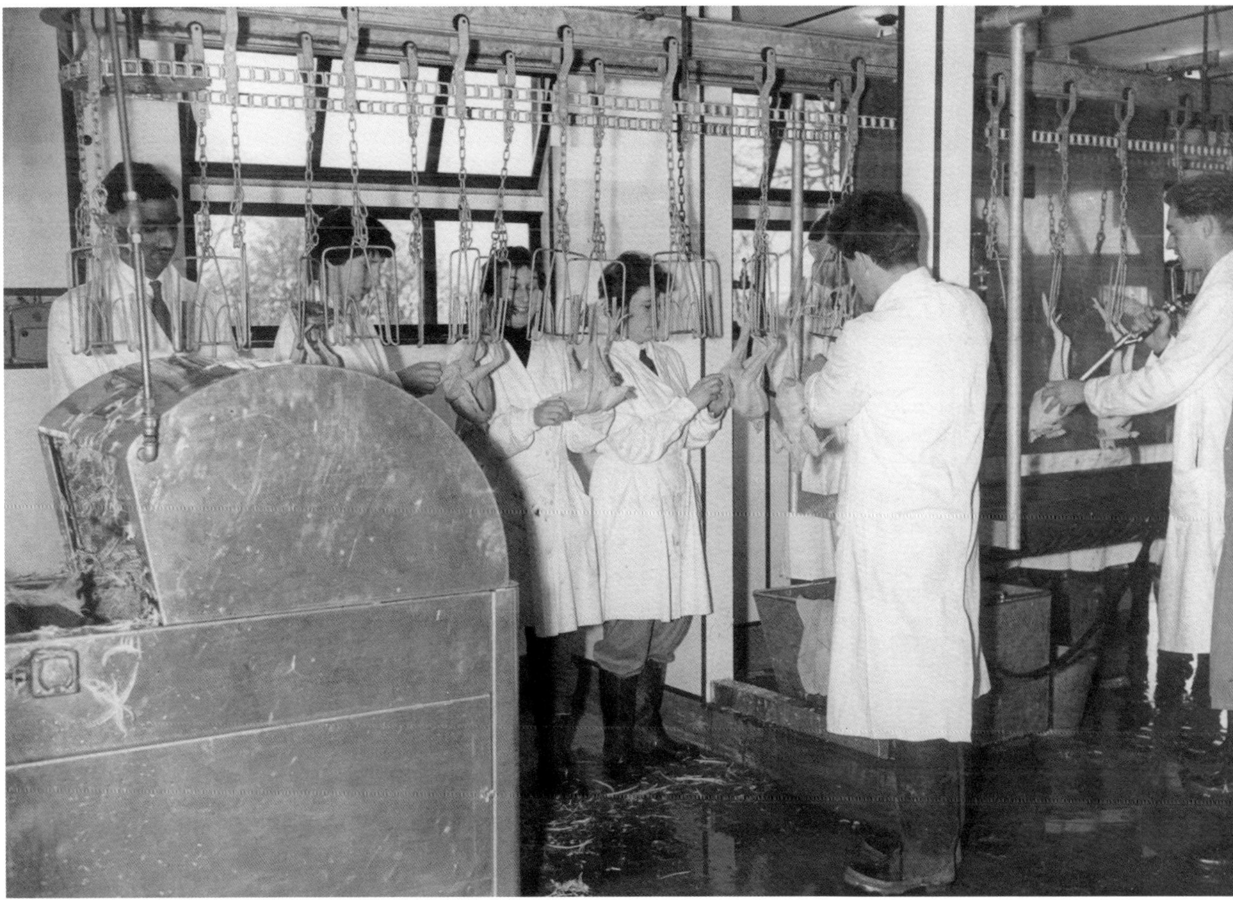

11.5 Table poultry processing line, 1957.

poultry processing unit was officially opened.[25] It had been given on permanent loan by Gordon Johnson Stephens & Co. Ltd of Gloucester, with equipment provided by several other firms. The total outlay amounted to £5,345, of which an impressive £4,395 came from donors.[26] The new processing building was utilized as a mini packing station with a continuous processing line, using overhead conveyors, in the American fashion. The processing line could handle 150 birds an hour.[25] A vital consideration, as by this time the NIPH was moving rapidly towards intensive indoor production.[27] In 1959, for example, an intensive breeding house was built.[28] It possessed 'unorthodox features', such as walls without windows to allow exploitation of lighting patterns and weld mesh floors which allowed greater density.[23]

By 1960 a two-storey intensive brooding and rearing house for 3,000 birds was under construction,[29] making it hard to contemplate that just over thirty years before the entire original stock of chickens at the NIPH was to be 1,800.[30]

Research was very much part and parcel of the developments at the NIPH. Perhaps the most important of early investigatory schemes was the housing of 1,500 layers in fold units around the College Farm, commencing in 1948.[15] Folding poultry was a method fast gaining popularity on general farms. The poultryman, when not engaged in collection of the 700 eggs or care of the birds, must have been comforted by the thought that the daily task of moving all sixty fold units could be accomplished in around sixty minutes.[31] During the session 1956-57 the fold unit came

11.6 Fold units, 1949: Modern Poultry Keeping.

to an end, having proved uneconomical.[25] Another early scheme attracted media attention but was altogether less practical. In August 1949 thirty-six very special eggs were brought by the Daily Express from the United States. The United States led the field in pioneering work in breeding.[3] These were the eggs of wingless chickens. Nine chicks were successfully hatched, of which three were absolutely wingless, the first of their type to be seen in Britain.[32] The theory was that being wingless, the birds would have more meat on other parts. This proved not to be the case. Furthermore, the chickens had one major disadvantage: *'If they go over on their backs when chicks they cannot get up by themselves'.* The wingless characteristic proved a recessive factor and experimentation stopped in 1953.[33]

Throughout the 1950s the investigative work undertaken by the NIPH was very wide ranging, everything from breeding pheasants to the development of the chicken sausage.[34] Around 1954, various methods for sexing day-old chicks were tried.[35] In this the NIPH led the field.[36] Albert Rowley undertook most of this work and declared requirements were 'exceedingly good' eyesight and the ability to make a quick decision. However, as Willard Moore recounted, not everything was initially a success:

> *They were never off our doorstep coming to see the methods we were using: the first comb [?] plucker, then the first wet plucker, then the deep freezing and wax plucking. Harold Temperton pioneered it in Lancashire with an old primus stove heating the wax. They did these waxed birds, let them dry, went home and had dinner, had a pint on the way, by the way, and when we came back, we started to take the feathers off. The wax had gone*

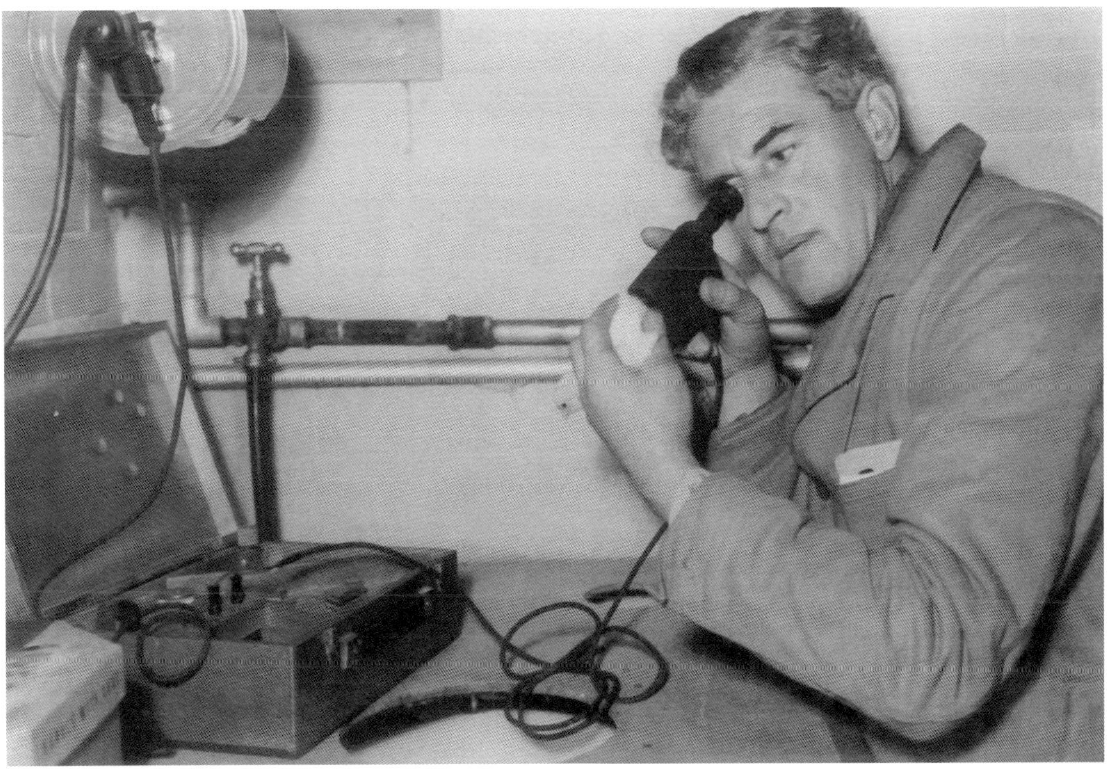

11.7 Albert Rowley sexing chicks, 1950s: Private Collection.

that hard that we pulled the skin off the ducks as well.

Persistence was rewarded with success in this instance and the widely adopted 'wax defeathering of ducks and geese can be claimed to have been originated by the NIPH'.[24] Ducks were now being reared for table production on a semi-intensive basis, although as a very small part of NIPH activities.[13]

Essentially the practical needs and questions of the poultryman were kept in mind. The Press declared: *'[Work at the NIPH] is restricted to plain honest-to-goodness husbandry of the type which the Industry needs and needs badly.'* [37] A great deal of the equipment was designed and built at the NIPH by the staff with a view to its ease of construction and use by the poultry farmer. Thus the Small Holder wrote:

The institute had designed a range shelter with many excellent qualities, a standard night ark for use for every purpose from brooding day-olds to housing layers, a turkey straw-yard, an outdoor battery cage unit, a manure storage shed, an automatic trough watering system and many other things too numerous to remember or record. These are produced in the ordinary course of poultry farming. A need arises, an appliance is thought out to meet it, a prototype is designed and tried out, then modified if necessary, before being finally put into operation. If it then has any likely value to the average farmer or smallholder - and most of these appliances have - detailed drawings and specifications are then prepared, which anyone interested can obtain for a small sum on application directly to the institute.[38]

The teaching side of the NIPH was always in evidence. Already by 1952 there were insufficient poultry students to meet the needs of Industry, whether as advisory officers or as employees of the growing number of feeding stuff and poultry firms.[39] With an end to the rationing of poultry feed in 1953, the expansion of the Industry occurred at a rapid pace and the need for trained staff became all the greater.[23] Numbers of students were, however, growing. For the session 1958-59 a record number of seventy had been reached.[28] It has been claimed that the NIPH can be held *'directly responsible for the education and training of almost all the leaders of our industry at that time'*.[40] With the start of the 1960s, plans for expansion within the College very clearly included poultry husbandry. As many as 150 poultry students were envisaged, similar to the number of agricultural students.[29] Furthermore, under the direction and encouragement of Temperton, a new advanced course was proposed. It had become evident that the rapid technical developments in the Industry had led to a demand for such training. The Diploma in Poultry Technology (DPT) was to be science based and aimed at advisory staff, teachers and senior technicians. It would run over three years and its intensive syllabus would make many regard it as comparable with a degree.[41] Students were to possess A-Levels in chemistry and either biology or zoology.[42] The first students, numbering four,[41] were admitted in the autumn of 1962. To provide facilities for the growing number of students the new poultry husbandry teaching block was opened for the session 1961-62 [43] *(see photo page 144)*. The students who would use that building would have little comprehension of how far the NIPH had progressed since the end of the war and talk of its removal to Reading.

In this period the Rabbitry faced a patchy time, albeit with some peaks of success, but throughout King Wilson was at its helm. By 1952 some of the renovation grant from the Ministry of Agriculture had been used to replace the dilapidated army hut, inherited years before from the original Poultry Department, with two modern concrete buildings.[5]
Of special pride, were the unique metal battery hutches with wire flooring and automatic cleaning, designed by King Wilson.[19] For a time rabbit meat became big business along the lines

11.8-11.9 The old Rabbitry and the new, built in 1952: Wellington Journal & Shrewsbury News.

of the broilers in the Poultry Industry and a lot of money was made and a lot lost.[44] Moreover the Rabbit Department gained independent status as an institute in its own right.[19] Its main task was teaching and this included preparation for a diploma awarded by the National Rabbit Council.[35] There was a growing demand for small livestock officers and their work included rabbits.[43] King Wilson's own advisory role did not waver, with a constant stream of enquiries and radio broadcasts.[35] As late as 1962 it was impossible to meet all the demands for advice, lectures and demonstrations.[4] Research continued into the diet, housing and breeding of rabbits for a time after the war. However, in the mid-1950s research stopped due to a lack of funding and henceforth the commercial side predominated, more especially the felt and fur trades in the face of myxomatosis.[45] In the early1960s interest in rabbit meat again emerged.[46] However, despite continuing interest in rabbits, the need for modernization and improvements [47] and the illness of King Wilson led to the running down and eventual closure of the unit following his retirement in 1965.[46]

During the 1950s Harper Adams was known as the Eggs and Bacon College due to the degree of activity involving both poultry and pigs.[48] At the close of the war the future did not seem bright for pig production either at the College or nationally. The piggery had been neglected since Crowther's departure and in late 1945 Fordham remembered just four sows and some 'porkers'. Nationally pig numbers had declined throughout the war, but the country's post-war need for meat meant that pigs with their high prolificacy and growth rate could provide the quickest answer. Gradually, indoor breeding, yards and fattening houses were to gain in popularity. The 1950s saw the appearance of large herds and high density housing with environmental controls. The Pig Husbandry Experimental Station reopened in 1947, but it ran to limited capacity with, as Luscombe remembered, 'a half-hearted experiment going on under Cambridge University Genetics Department investigating the effect of inbreeding in pigs and I had four sows which were four-fifths themselves by inbreeding'. On his arrival as Lecturer in Animal Husbandry in 1948, Peter Luscombe found the whole place old-fashioned, dirty and overrun with mice.[49] The time was opportune for pig husbandry to be regenerated at Harper Adams.

Price was especially interested in pigs and was affectionately known as 'Piggy Price'.[50] In 1949 Luscombe took over responsibility for the Pig Department. Structural alterations got underway and restocking began with eighty pigs.[49] The partnership was a success. Under Luscombe and Price the piggery became internationally known. The most famous achievement was the creation of the Harper Adams Pig Parlour, also jokingly called the *'Harper Adams Pig Palace'*, which was developed in June 1951 [35] to designs by Price.[17] This prototype fattening house was built on to the outside wall of the west wing of the old piggery. It provided accommodation for sixty extra pigs.[19] In essence the Parlour was 'really the old cottager's in-and-out sty with certain modifications'.[51] The Harper Adams Pig Parlour was cheap and simple to build, effective at keeping the pigs healthy and warm in winter and efficient to run, with labour-saving devices such as the outside run needing infrequent cleaning. Later developments included the abolition of windows and the addition of air extractors.[52] Harper Adams' name for pig husbandry was ensured. On one occasion Price appeared on television and offered to send out plans for various houses, including the Pig Parlour, on request:

There were hundreds [of requests]. We'd got about five hundred of these things printed off and then the letters started coming in. Miss Agates [Secretary] opened all the post - it took until lunchtime. The desk was piled up high in the morning with people asking for these plans. I found myself having to type these envelopes and stick these plans in and they just kept coming and coming and coming. I had so many of them. They had to have reprints to cope with the demand.[50]

Over the 1950s the piggery buildings slowly grew, so that the main types of housing were demonstrated. The piggery was mainly for fattening, with outdoor farrowing as was the custom of the period.[53] It was not until 1962 that the College seriously considered an indoor breeding unit of sows. Experimentation was never isolated from the piggery.[54] Every building was utilized for research and every pig was expected to make money.[55] Indeed, the pig herd became 'one of the most profitable enterprises on the College farm'.[49] Funding was limited, with money coming from the Agricultural Research Council (ARC) and commercial firms.[54] In 1958, for example, the Solari piggeries were under

11.10-11.11 Harper Adams Pig Parlour, early 1950s.

construction.[27] In order to obtain funding Price was not averse to gilding the lily. Research was often given a veneer of science in order to win support. Essentially, the research was of an inherently practical nature and of great value and interest to farmers at a time when few of them indulged in experimentation.[54] The pig conferences held at Harper Adams in the 1950s were vastly popular. Luscombe at the time was only too aware of the overwhelming need for research: *'Looking ahead into the problems of the pig industry almost gives me an inferiority complex due to the inadequacy of our knowledge. There are so many aspects requiring investigation that it is difficult to know just where to start.'* [27]

Experimental work looked in the main at nutrition and breeding. In 1951, for example, there was an investigation into fish solubles as these were not rationed and in 1952 the spotlight fell on antibiotics.[35] Later, controlled water intake was researched, to see if unlimited access to water damaged the conversion ratio of food to flesh.[21] As regards breeding there were experiments into dual mating, namely serving the sow at the same heat period with two boars of different breeds to produce

11.12 Outdoor pig rearing at Harper Adams showing the College designed 'crark', half crate, half ark, with Frank Rugman the pigman, mid-1950s: Wellington Journal & Shrewsbury News.

pure-bred and cross-bred pigs in one litter.[56] There was also research into hybrid vigour, with a variety of breeds being kept in the early 1950s, such as Essex, Wessex, Large White and Tamworth.[55] By the early 1960s the pigs were predominantly Large White with some Wessex Saddlebacks and Saddleback and Large White crosses.[42] Some of the most interesting work was concerned with the creation of a virus-free herd. Pneumonia was a great problem in pigs at this time and there were few such herds.[54] Farrowing took place in isolation and ways were developed of taking the piglets away from the sow as soon as possible. This led to the opening up of the new field of artificial rearing.[57] It was to be the late 1950s before the College obtained a virus-free herd.[42]

The successful pig unit was quite independent of the Farm. But, again, the success story here was dependent upon the excellent relationship Price maintained with his colleagues.
The running of the Farm benefited from the successful partnership of Price, acting as Farm Director, and the newly appointed Farm Bailiff Tom Fordham. For many years the Bailiff had been an ex-student keen to gain experience. He would stay a short time and then move on. In contrast Fordham was a man of considerable experience, who had spent the war on reclamation work. They were men with mutual respect for each other. Both had the well-being of the Farm very much at heart and both had boundless enthusiasm for farming. Together they could be seen of a Sunday morning, walking the fields and discussing the work to be done, their respective dogs tagging along beside.[58] Never a cross word passed between them, even over the issue of the use of herbicides. The first selective herbicides became available for general use in 1948. Price was not alone in considering their use to be a lazy way of farming, but Fordham was all in favour and around 1953-54 he forged ahead. The successful results did much to convert the Principal.[58]

The post-war period, with the passing of the Agriculture Act of 1947, was an exciting time for British farming. For the first time there was a sense of security as a future of higher and more stable prices than existed on the free market beckoned. Farmers felt safe to invest in the future, to borrow to begin new enterprises or to adapt existing ones and to exploit technological advances. Soon the Harper Adams Farm would be caught up in this spirit of progress, but initially there was the immediate problem confronting the College of the direction the Farm should take. In 1947 there was nothing less than open hostility between the Governors and the then Farm Director and Lecturer in Agriculture, Kenneth Campbell.[59]

In January 1947 Campbell set out his plans for the future development of the Farm. In the past, he accurately declared, policy had been to spend as little as possible on the Farm. Now the need was to spend in order to increase profits and to enhance the educational value of the Farm. He argued that the pedigree Ayrshires would be the main source of income and thus due emphasis should be placed on a self-supporting dairy herd of around eighty to one hundred cattle. With the acreage such as it was, in 1951 it stood at 290 acres, there would be no room for beef and sheep.[60] Crops should be directed towards home-grown food for the dairy stock and high value cash crops. The Farm must improve, modernize but above all specialize: *'An educational farm should not be a hotch potch of all possible enterprises in the belief that each enterprise can have an educational value completely and entirely divorced from the economies of production in respect of each enterprise.'*

The Governors saw things in a very different light. The issue of the role of the College Farm was not a new one. Their views more closely followed those later produced by the Keith Murray Report of 1954. The Report declared that the Farm was first and foremost a teaching facility and, although it should set a good example by making a profit, this should not be to the detriment of the education side of the College. Bad feelings resulting from 'a cleavage of views', as so aptly described by Campbell, rumbled on through the year. He sought improvements, but he was not allowed to follow his convictions and act. The ultimate grievance came when the Farm Accounts for the year ending March 31st 1947 showed a loss of £649 in contrast to a previous year's profit of £5,650. Unfairly to Campbell, he was taken to task for figures which reflected farming during a period when he had only just arrived at the College and the new Ayrshire dairy herd was not up to full milk production. Clearly, the Governors wanted his departure. He was dismissed as Farm Director and invited to resign as Lecturer. On Campbell's departure at Christmas 1947 the Principal became Farm Director.[15] Price's character would ensure that there would be no further ruffling of the Governors' feathers.

Meanwhile, as issues of policy were resolved, Fordham set to work on the practicalities of farming. Campbell had assessed the Farm as being in a dilapidated state, a view with which Fordham agreed:

> *The first time I walked round the Harper Adams College Farm was on a rather grey day in late November 1945. I had just taken up the position of Farm Bailiff. The result of six years of war was only too evident, lack of man power and the austerity forced by war had meant that all effort had gone to produce food and maintenance work like hedging, ditching and draining had been reduced to a minimum. The farm buildings had fared no better and an enormous programme of rebuilding and modernisation was needed, it all made a rather sad picture.*[61]

By 1946 labour and material shortages appear to have been biting deep and the Farm was no longer the 'efficient food production unit' described by Thompson in 1943. Fordham worked with a will. Land was cleared, hedges laid and ditches cleaned. New equipment was bought and buildings were improved. The renovation programme was extensive, with the inclusion of the farm buildings at Ancellor and Flatt Farm. By 1949 the newly refitted cowshed had the capacity for between thirty-four and fifty-five milking cows in line with expansion of the herd. At this time the old piggery opposite the dairy became a compact dairy for the handling of milk. The end of dairy teaching at the College meant that the original dairy building had become redundant.[19]

War had altered the traditional emphasis on grassland and with a return to peace Harper Adams maintained a balance between the arable and the pastoral. Fertilisers were now subsidized and their increased use allowed an increase in the numbers of animals despite a reduction in the area of grass.[62] The College maintained a small flock of breeding ewes.[63] For a time in the early 1950s the College had a flock of prize-winning Clun Forest sheep.[58] Such was the growing interest in sheep [62] that in the mid-1950s a practical study was made at the College of breeding from multiple births. Fordham related: *'We would only breed from a twin or a triplet to see whether there was a link up between multiple births and progeny of multiple births. We had to tag all the sheep and keep records of birth weights and breeding weights and so on.'*

As regards dairying, at this time there was a dramatic rise in the size of herds, with a move towards intensive systems, which entailed greater capital outlay but less labour.[62] The College followed the national trend. In August 1946 the College Herd amounted to twenty-nine Ayrshires and by 1949 the herd consisted of sixty animals, mostly Ayrshires.[63] These were attested animals and Fordham well remembered those 'early days of testing': *'One of the first things we did was putting double fences all round the farm because we had to have so much space between you and your nearest*

11.13 Official opening of the new cattle yards, with left T. C. Ward, Chairman of the Governors, 12 December 1955: Dan Arden.

neighbour so the cows couldn't have contact from other herds that weren't tested. In the same way we had to put water troughs round the field because you couldn't drink out of the brook, because that could be infected. [...] We must have been one of the first herds round here that became tested.'

By the mid-1950s the Friesian was proving the more popular dairy cow because of its higher yield and the College made the decision to introduce Friesians and to experiment with Ayrshire and Friesian crosses. Some very good animals were bred.[64] The first Friesians arrived at the College in February 1955.

Altogether fourteen Friesians were given to the College in response to an appeal from Price and with the aid of the former Chairman of the British Friesian Society, H. M. Astley-Bell, an ex-student.[65] By the end of that year the College was celebrating the opening of new covered cattle yards. These had been erected through the generosity of feedstuff firms J. Bibby & Sons Ltd, BOCM Ltd and R. Silcock & Sons Ltd, each donating £500.[66] With the new yards it was possible to extend the herd to seventy-one, while enjoying the same labour requirements and overheads and ensuring healthier, higher yielding cows. The cattle yards allowed the cowshed to be used simply as a milking parlour.[58] In 1959 the College was selected by the Milk Marketing Board to be part of a pilot scheme for bulk collection of milk. By 1966 twenty-seven percent of the county's milk was being handled in this way.[62] At the close of the Price Era the herd was sixty in number, with a quarter pedigree Friesian, a quarter pedigree Ayrshire and the rest crosses of the two breeds.[42]

Crops on the College Farm consisted of winter food for the livestock and cash crops such as wheat, barley, linseed, potatoes and sugar beet.[63] The Shropshire Magazine for September 1950 gave average crop yields for the College Farm: 12-14 tons washed sugar beet per acre, 10-12 tons of potatoes, 33 cwts of wheat, 27-28 cwts of mixed grain and 12-14 cwts of linseed. These figures compared well with national averages. However, the crops grown at the College were not all of the conventional kind. In 1958 Ray Mortimer, newly appointed Senior Lecturer in Farm Management, introduced the idea of growing maize for silage after a period of post-graduate research on forage harvesting undertaken in the United States.[28] The idea of growing maize as a forage crop stored in a silo, as opposed to using it as green fodder, was all very new.[67] Maize had previously been grown at the College for pig food for a short time in the 1930s.[58] Price was always a great supporter of any new idea and Ray Mortimer's name soon became synonymous in this country in these early years with the growing of maize and the use of tower silos.[50] Maurice Eddowes, arriving in 1958 as Senior Lecturer in Crop Husbandry, had a strong interest in crop development and his background in research made him an ideal partner. The crop proved an initial success and brought back into use the tower silo which had been empty for many years.[28] The silo was equipped with a mechanical auger feed.[67] In 1962 a replacement concrete stave silo was erected. Cattle could be fed in ten minutes by pushing a button.[68] Battery feeding of beef in a covered yard was another idea brought from the United States and experimented with at Harper Adams.[67] In the early 1960s the College had about forty acres of maize. Maize was loved by animals and proved a rich energy source,[67] but it ran into problems with the British weather.[41] At this time Maurice Eddowes began trials comparing maize silage with grass silage.[68] However, the temperamental nature of the crop and the problems in handling it made it difficult to win converts from grass with its easy nature. Maize fell from favour in the 1970s as grassland techniques improved, only for farming interest to be revived in the 1980s with the introduction of new varieties.[67]

Not unexpectedly, one of the biggest changes on the College Farm during the 1950s was the move towards greater mechanization. The College Farm at the close of the war was considered by Fordham to be backward in this matter: *'We had two tractors on the farm. One was an International Farmall, the other was an old Fordson. Neither of them were very efficient. When I came here the Farmall was in dock having a new gearbox which meant we only had one tractor on the place for quite a long time.'* Unlike Thompson, Fordham, with a background in land reclamation, had every faith in new machinery. However, 'there was no money to go to town on new machinery' and so, most of the work in the 1940s continued to be done by horses.[58] There were three teams of horses and 'one horse who was what we called a shackler or trace horse'. In 1947 the College confirmed its faith in horse power by the purchase of a pair of Suffolk Punches.[61] The use of horses continued on the Farm until the mid-1950s, although in a declining role. Ian Reekie, student 1953-56, remembered using the last horse on the Farm to roll in carrots in the spring of 1954. By this date George Brown the Waggoner was driving tractors, although he continued to 'talk to his little grey Fergie tractor as if it were a horse'.[69]

11.14 Principal Price and Ray Mortimer inspecting the maize crop, 1961: C. D. Millington.

With the powerful combination of Price, a man enamoured of the new, as Farm Director and Fordham as Farm Bailiff, mechanization steadily made greater and greater inroads at the College. As the Principal reported to the Governors in 1949, *'mechanization was carried out to the economical and practical limit'*. Price, however, was always a man to push the limits. In 1949 the Principal *'with his amazing persuasive ways managed to get Mr Ferguson to donate a tractor and a whole range of machinery, on free loan, which was an enormous step forward. He had hydraulically controlled machinery which was new on the farming scene then. [...] It really put us forward years. Ferguson was years before his time anyway. We went from literally antiquated equipment to first-class modern machinery.'* [58] The arrangement was for the loan of a tractor with equipment to the value of £1,000. The tractor, a TE20,[70] was one of the first to come off the production line and had been used for a promotion at the Dorchester Hotel: *'All the fittings, the nuts and bolts were all chromium plated'*.[58] About the same time the College obtained a Ferguson hydraulic tipping trailer on loan and 'loading equipment for

11.15 Presentation of a new Fordson Diesel tractor by the Ford Motor Co. Ltd, June 1956. Tom Fordham, Farm Bailiff, and Bill Lightwing, Head of Engineering, to Principal Price's left: Dan Arden.

the front of the tractor which was a fork, on arms so you could load manure and so forth with the tractor and deposit it in the trailer'.[58] Both were considerable advances. Loans and acquisitions continued throughout the 1950s. In 1957, as a result of Price's visit to the United States, the College acquired a John Deere rotary chopper for silage making, the first of its kind in Europe.[25] That same year the Farm obtained a Mellotte reversible plough. It was a brand-new development; the first single furrow reversible plough that went through 180 degrees in a vertical plain rather than tipping over through ninety degrees.[71]

In 1958 the College purchased its first potato and sugar beet harvester, a Whitsed harvester.[27] A purchase, as opposed to a loan, was a rare event.[58] By the late 1950s sixty-two percent of sugar beet was harvested by machine nationally. Interestingly, in October 1952, the National Potato Harvesting Demonstrations organized by the Royal Agricultural Society of England (RASE) were held in association with the College.[72] A competition to find the best potato harvesters had first been announced by the RASE in 1949. In 1952 twenty-three entrants, from places as far away as Denmark and France,[58] took part over a three-week period, using land in the Fens and Shropshire.[73] A first prize was awarded to the Globe Harvester Company of London, but no machine was thought sufficiently good to win a medal.[58]

Many of the people who worked on the Farm were real characters and remembered with great fondness by the students and staff. Charlie Spencer at one time worked with the sheep and 'what he lacked in this field [scientific knowledge] he made up by knowing every countryman's trick in the book, and when it came to lambing a difficult ewe there were few that were his equal'.[58] On one occasion the sheep were being fed with steamed potatoes. In 1952 the Farm, seeking to find an outlet for small potatoes after riddling, had purchased a cheap steam traction engine for sixty pounds to steam the potatoes in bulk. However, initially the sheep would not eat the steamed potatoes: *'I came back about the middle of the day and the troughs were clear. I said "What have you done Charlie. Have you taken the potatoes out?" He said "No, they've eaten them. [...] I put some salt on them!".'* [58]

The College Farm of the 1950s has one last claim to fame. In the late 1950s it was used by the BBC for recordings of farm sounds for 'The Archers' radio series. Thus recording crews came on several occasions to capture the noises of animals and machines, discovering very quickly that neither animal nor agricultural worker would perform on cue. The sheep would not bleat and the tractor did not sound rough enough for Walter Gabriel's machine. Perhaps, the greatest resources of patience were needed for the session with the pigs, or rather the pigman:

> *[The pigman] was a very nice chap except that he used to get excited about things all the time and whatever he was doing he used to sort of swear and curse everything that didn't go right. So they came on to the loading ramp to load pigs, the pigs got out [...] and he kept saying 'Come on you buggers. Come on you buggers'. Then they'd switch off and say 'No, no, no, no. This is being recorded for the Archers. Would you please be quiet' [...] and he never did it. In the end they had to tell him 'Well, we'll go and record somewhere else'. They took him round the corner with a tape recorder which wasn't going [...], whilst somebody else loaded the pigs into the lorry. It was the only way after about an hour and a half that they actually recorded the pig noises.* [71]

It was fitting that as Principal Price's days at the College grew to a close, his achievement was recognized. In January 1960 he was awarded a CBE.[29] Thus with honours and praise, and no doubt much regret, Price retired as Principal in August 1962. His last years were spent in Leamington Spa and it was here that he died in January 1982.[74] Perhaps, of all the achievements which occurred during his time at the College, he would have liked to have been remembered for the sheer joyousness of those years. Luscombe portrayed them as 'halcyon days' and there seems little doubt that the golden memories always outshone any difficulties.[49] Many of the students of the 1950s have maintained close ties and today still testify to 'a tremendous feeling of mutual respect and mutual friendship which [has] persisted over the years'.[75]

Emphasis on the friendliness and fun of the period is not to deny that there was always a degree of 'accepted regimentation' and that there were 'rules of attendance, dress, eating etc.'.[76] Anthony Warner remembered the College still being 'like an extension to school' in the late 1950s. Students called the lecturers 'Sir' and asked their permission before they dared to do anything. Standards were there to be maintained. Don Findlay, arriving as a student in 1956, remembered: *'All students wore a collar and tie with jackets for lectures, until the summer months when the word went out we could dispense with jackets. The standard college dress was brown brogue shoes, cavalry twill trousers, sports jacket, check shirt, cravat or tie.'* Ian Gedye, a contemporary of Findlay, found that the Warden viewed his growing of a beard as *'appalling behaviour, [...] worse than breaking the hostel rules'*. Regulations had to be obeyed. However, there was not 'unnecessary

11.16 Lecturer Peter Missen and students, probably early 1950s. Private Collection.

discipline'. Although gates were no longer literally locked,[11] there was a deadline to return to College, namely 10.30 p.m.[76] Students had to be in their rooms by 11.00 [69] and the Warden, Dr Norman Gill, Head of Biology, would appear 'magically' at any hour of night to confront the errant.[76] Other regulations banned students from playing table tennis, billiards, musical instruments and radios 'during instructional hours'. Games were not permitted on College grounds before 1.00 p.m. on Sundays.[69] *'Doc'* was 'a firm advocate of maintaining standards without being bombastic, [and possessed] the gift to maintain discipline without being pedantic and the talent to instil in students the need for formality but at the same time accepting the inevitability of change'. He had a good sense of humour and was a great supporter of sport at the College, notably cricket.[77] Gill's counterpart was Dr Janet MacLagan, Assistant Lecturer in Agricultural Botany: *'He was the bachelor who looked after the men and she was the spinster who looked after the women's hostel down at Ancellor.'* [75] Her 'unassuming but always competent approach coupled with a deep knowledge of her subject' gave her the ability to win the students over 'from an attitude of indifference to one of genuine enthusiasm'.[77]

Student high spirits abounded, but pranks were always kept within bounds.[44] Serious misdemeanours could range from breaking windows to throwing water at staff.[3] A favourite recreation was water fights: *'If you want a full scale battle pick some staunch allies, some waste paper canisters and some bicycle pumps and get up to the top of Boughey Hall from whence you have a grand view of your prospective enemies trotting [...] unsuspectingly between the two hostels.'* [17] Warner can remember Old Boughey Hostel being awash with water to about the depth of one foot at a time when there was a water shortage. The ultimate punishment for anybody 'considered too big for his boots',[11] or for someone who simply had a birthday, was the trough. Those who failed to provide entertainment of a sufficiently high calibre at Freshers Night were front line candidates for a watery fate:

> *We [the Freshers] all had to sing and dance and generally make a complete fool of ourselves, purely to entertain the second years, and after each act, the thumbs were then turned up or down, like the old Roman circus for the poor Christians. Those who did not entertain to the satisfaction of the enraged crowd were then sentenced to 'the trough'! The general rule of the jury, was that they chant first slowly and quite low, 'trough, trough, trough!!!' getting louder and louder until the Student President acting as Master of Ceremonies thought that the masses wished to see the poor first-year student enter the muddiest horse trough that could be imagined.*[78]

This large trough, regularly used by the animals, was situated close to the gardens, now the site of the staff & visitors' car park and had been used for its entertainment value since the early years of the College.[79] By the late 1950s the ducking had transferred to a 'very stagnant, revolting looking pond' in the Principal's garden[11] *(see photo page 146).* Brian Lloyd-James, student 1957-59, had to contend with a paratyphoid infection as a result of a ducking. In 1971 the Student Union banned ducking in a new atmosphere centred on belief in 'free speech and free thought'. The Cat-A-Mountain of 1971 declared: *'Tradition has many benefits, but the Troffe is the epitome of tradition at its worse.'* Today, the pond is preserved as a wildlife habitat.

As ever, sport was very important and flourished at the College. The participation level was excellent and events were well supported.[67] In the season 1959-60 it was estimated that often more than one hundred students took part in sport on one day.[29] This figure represented virtually half of the College. Sport was given a tremendous boost when a new sports field was opened ready for the session 1947-48. It was situated across the Caynton Road, on the site of the old laying trials.[80] Prior to the move sport had been played in front of the College or around the fields.[35] By 1957 the new site was proving uneven and adjacent fields to the south, altogether ten acres, were ploughed up, reseeded and henceforth used.[25]

Rugby remained the leading sport at the College and the late 1950s saw the College fielding three teams and in the season 1958-59, for the first time, winning the Shropshire Cup outright.[27] Cricket continued to be played in front of the College, although the pitch was not good.[19] In autumn 1953 two hard tennis courts, situated beside the car park opposite the main gates,[35] were added to the four grass courts available to students.[19] Staff already had use of a hard court at the NIPH.[70]

With the growing emphasis on engineering, ploughing matches became very much a feature of College life. The first ever contest, that between the College and the Royal Agricultural College, took place in November 1952 and the College won.[72] For many years

11.17 Sports Day 1957. Back row from left: David le Marquand second, Russ Dorrell third, Ian Gedye fourth, Jim Duffy fifth, Rodney Jones sixth, Bob Horton seventh. Front row from left: Mrs Price first, John Wood SU President second, Bob Wake sixth, Principal Price eighth: Wellington Journal & Shrewsbury News.

in the 1950s there was an annual triangular competition between the College, the Royal Agricultural College and Seale Hayne. Companies such as Ferguson would send tractors on loan for the event, held in rota at the three colleges.[70] In 1954 the Colleges competed for the newly donated Massey-Harris-Ferguson Cup.[43]

June 1948 saw the first sports day to be held on the new site across the Caynton Road.[15] The College Sports Day continued to be as popular as ever and all associated with the College in any way were invited, with tea on the cricket pitch and ice cream for all the children.[58] Almost half of Edgmond attended and it was very much part of the village social scene.[81] Parents of students would also be invited and given the opportunity to view the College. It was a day to remember, 'more like Ascot', with 'floral dresses and big hats' very much in evidence.[58] On one occasion in the early 1950s none other than Mrs Dugdale, wife of the Minister of Agriculture, came to present the prizes.[3]

The 1950s was a time when the social life of the College blossomed. At various times there were thriving groups for everything from Scottish and English country dancing to skiffle, from field sports to athletics and from chess to debates and discussions.[6] The Field Sports Club was a new venture, started in 1950. An early request from the Club to hold firearms was met with a suggestion from the Governors that archery would be more acceptable.[17]

11.18 Skiffle group, c. 1957. Front from left: Nigel Parker, John Usher, Charlie Parker, Bob Hill, Roy Beaven, Mike Czarnota-Bojarski: Private Collection.

Members followed the Shropshire Beagles, attended the Albrighton Point to Point, fished at Caynton Mill and enjoyed rabbiting with the Club's one ferret.[17] This ferret may well have been the one which made a successful appearance in a revue sketch with Peter Missen, Harold Temperton and Peter Luscombe. It was brought out from under the latter's jumper, where it had been hidden from all including the participants, and held against Missen's ear to the audience's considerable delight.[54] The Music Club was still meeting on Sunday evenings in the library and venturing,[25] with a degree of boldness, into 'light music' with 'The King and I' and Eartha Kitt.[27]

Dances, great and small, proved continually popular. For the Saturday hops, girls were brought from Radbrook College, Shrewsbury, and from Nelson Hall, a teacher training college near Eccleshall.[19] The regular dance on the first Saturday of Term could be expensive. Irene Doody remembered admission cost sixpence. There was a live band, but no bar, and the dance would stop at eleven. At the major dances the students would often decorate the College. They could be ingenious, one year making a pool from a stack sheet and filling it with fish.[58] Robin Harris, appointed Farm Recorder in 1959, remembered the End of Session Ball: *'[It] was a very big occasion. The Lord Lieutenant and other dignitaries of the County came and it was an event everyone attended, whether you wanted to or not.'* Plays continued to be produced at the College and these farces were sell outs, drawing an audience from College and Newport alike *(see photo page 147)*. Don Findlay, minus his trousers, was well remembered as a star of some calibre.[50] The year 1959 saw the first production to be entirely in the hands of students. The team of Gill and MacLagan graciously stood down after nearly twenty years.[29]

In August 1962 Reginald Kenney became Principal of Harper Adams. The essence of the Price years did not disappear overnight. However, from the very first months it was clear that Kenney was a man of very different character to Price and a man determined to stamp his ideas on the College. Despite the bounty of the inheritance passed down by his predecessor in the form of an impressive building programme, signed, sealed and shortly to be delivered, Kenney was to encounter many problems. It was an era of rapid and significant changes, which all had to be accommodated. To these inevitable difficulties must be added the human element. Both staff and students felt little affinity with either the Principal or his ideas and discontent, conceived of misunderstanding, was born.

Rag Stunts of the 1950s

The students of Harper Adams could never be accused of a lack of ingenuity when it came to the annual Rag. After all, as the Cat-A-Mountain of 1961 declared, the Rag 'encouraged initiative and enterprise'. The Price Era was a time when many considered the rag stunts were at their best. In May 1955, with the General Election about to take place, the students created the English Nationalist Party, distributing posters around the local towns, sending out speakers and organizing rallies. The fictitious candidate was Tommy Bates alias Keith Stevens.[a] It was not until the day before the Election that the trick was exposed. On Election Day the College made national news in the Daily Sketch with the headline 'The Party that never was (before) peps up the fight'.[b]

11.19 Electioneering for Tommy Bates, Rag 1955: Dan Arden.

In 1959 Student Brian Lloyd-James remembered:

One ingenious event was the diversion of traffic leaving Newport along the Edgmond Road, through the college farm and via the back lane to the A41. This was achieved with a group of students completely kitted out as Shropshire County Council workmen, with pickaxes, shovels and road signs effectively blocking the A41 and demanding money. It worked very well until a lorry loaded with timber got stuck on the sharp bends outside Newport. All signs, and evidence of the diversion were gone in a few seconds.

The stunt that will forever live in the memory is the kidnapping of Ferdinand the Bull in October 1959. Maurice Lancaster, Lecturer at the time, set the scene: '*In those days bulls to be registered had to look masculine. [...] It was a Friesian bull which had rather a feminine*

11.20 The kidnappers of Ferdinand the Bull, October 1959. From left: Chris Milne-Day, Marcus Barker, Nick Hoskins, unknown, Iwan Roberts, John McBride, Ferdinand, John Holm, Tony Warner, Mike Blakemore, Jeff Chalk. Private Collection.

appearance and the authorities wouldn't register it and it made headlines in the press.'

Anthony Warner, one of the first-year poultry students involved, recalled the events: *'A couple of students went down a couple of days before hand and spied out the land for us. Then it was my job to borrow Father's Landrover. [...] So, I was the driver. We went down a couple of days later. I think it was a Saturday night we went down. [Ferdinand was being kept by the Mirror in an animal sanctuary at Henley on Thames ᶜ] [..] Suddenly, the bull was there, went in [the trailer] quite happily and off we went.'* Initially, the bull was hidden on a local farm in Edgmond, but as students got to hear the news, the location became too hot and a move was made to Buildwas.

Warner continued: *'On the Sunday, gradually word got round the College and everything was a buzz. We couldn't wait for the Monday morning with all the newspapers to come out. I think it was on the television news the Sunday night. Monday Morning we wanted the daily papers. The Daily Mirror, who had sort of rescued the bull from its fate, they followed it and so did all the other newspapers. They were poking fun at the Daily Mirror. They all thought it a great joke. By then, Monday morning, the Police were involved.'* Dr Gill, receiving word that the College was implicated, got the culprits to confess: *'The Police were very good about it. They thought it was a huge joke, but they had to go through the motions. [...] The College, secretly, were highly chuffed with it all.'* The students had demanded a ransom of *'something like £100'*, which was not paid, but as news of Harper Adams' involvement spread many ex-students contacted the College to offer congratulations and to make a contribution to the Rag fund. The main carnival float that November was not surprisingly on the theme of the nation's most famous kidnap victim.

Chapter Twelve —

All Change: the DES and the HND 1962-77

The 1960s was a time of immense change for Harper Adams. There had not been a period when so much happened so quickly. The College not only grew but expanded beyond its original site. The number of students increased by a third, from 232 in 1962 to 310 in 1976. The Department of Education and Science (DES) took over responsibility for the funding of the College from the Ministry of Agriculture. Furthermore, the College was being drawn into a more rigidly structured national picture for agricultural education. This was the time of important education reports, such as that of Robbins and Pilkington. A new interest in teaching methods was emerging. More significantly, the NDA was replaced by the HND with its sandwich period and internally set and marked examinations. Adding to this drive for the new was the recently appointed Principal keenly enthusiastic about the challenge which lay ahead.

Reginald Kenney arrived at Harper Adams with an impressive background. He was a graduate of Leeds University and possessed the NDD from the West of Scotland Agricultural College.[1] His special interest was in dairy cows. He had begun work as a lecturer at the Staffordshire Farm Institute before being appointed Assistant County Agricultural Education Officer for Bedfordshire County Council and later becoming involved in work for the War Ag. In 1942 he was appointed Lecturer in Farm Management and Animal Husbandry at the University of Reading. Six years later he was made Principal of the newly created Dorset Farm Institute, undertaking sterling work building up that establishment.[2] His industrious and conscientious approach to his work was never denied, but it soon became apparent that the new Principal was a very different man to his predecessor: *'Whereas Price was outgoing, social, amiable. Kenney was rather diffident, shy, rather withdrawn.'* [3] His personality was at odds with his role. Students considered him at best 'a remote figure',[4] who they would pass on the rare occasion and who would not speak.[5] He rarely attended student events

12.1 Principal Kenney, students and cattle.

and gave a poor impression when lecturing.[6] Perhaps, only too aware of his unease with people, he could be clumsy in his relationship with students.[3] At times he was seen to be unnecessarily crossing the students when he should have 'gone with the flow'.[7] At other times, when a firm hand was called for, he could appear indecisive [6] to the point of avoiding confrontation.[8] The 1960s was not the time to falter in the face of the growing rebelliousness of the young.

Kenney soon encountered difficulties managing a College growing rapidly in size and in the expectations made of it. His experience lay in dealing with 'a very small county college with a very small number of students'.[9] Arriving in 1962, he was ambitious for the College and sought to make improvements. Indeed, a spring clean was in order.[10] He clearly had the best interests of the College at heart. However, he had little understanding of its people and his new broom approach sought too much too quickly. In his position in Dorset, as head of a new institution, he had not encountered the same need for diplomacy as that required with staff established in their posts. His first address to the staff at Harper Adams was a long list of one new idea after another, giving the impression that all that had gone before had been wrong.[11] His secretary, Irene Doody got a very strong message: *'The first thing he did when he came, everything Mr Price had done had got to be altered. We'll do it differently, the opposite, no matter what.'* As Principal it was essential for him to take his staff on board, but sadly, the strength of his enthusiasm just gave the impression that he wanted his own way regardless of all else. Staff feathers were increasingly ruffled and resentment built up. In defence he stood his ground and became difficult and dogmatic.[3] It was a breakdown in communications clearly springing from gross misunderstanding on both sides.

By the 1970s the situation had deteriorated still further. Staff felt that they were not consulted, that their views and experience were not respected and even that they were seen as 'stirrers, stirring up difficulties'.[8] More importantly, Kenney was not a risk taker [12] and consequently, new ideas were not supported.[8] When as Principal he should have been encouraging and pushing for innovation, he gave the impression of being wary of progress.[7] There was little impetus for change: *'It [The College] was just going along as before and [we were] reacting to changes rather than developing them ourselves.'* [7] Kenney found himself increasingly isolated and without support and his earlier enthusiasm and drive were clearly drained. The difficult situation only served to strengthen the accusations that he was out of his depth.

With hindsight, the opinion has been expressed that, in a different age with different expectations, Kenney had been appointed primarily for his skills in dairying. Few, knowing the man, would deny that it was here that his true enthusiasm lay. However, a principal of the 1960s had to be a multi-faceted leader, for as Bob Bradley so succinctly said: *'The days of being timeless were disappearing rapidly'*. Kenney, unenviably, *'was in the heat of the battle in the kitchen'*.[13] Everything from the very appearance and size of the College to the actual teaching and

12.2 Dining room, Queen Mother Hall: Shropshire Star & Journal.

courses was to undergo dramatic change under the auspices of the DES. As for the problem of the Swinging Sixties, perhaps even the charisma of a Price would have been insufficient. Principal Kenney was a man of talent and good intentions, but by the time of his retirement in 1977, few felt that he was the man for the job. Yet, for all the difficulties during the Kenney Era, the College 'seemed to run despite him'.[6] The College was growing in size, with a maturing structure increasingly tied to national policy. Thus, against this troubled background a story of progress unfolds and the first very visual sign of this was the physical expansion of the College.

The new buildings, first proposed by Price, meant that the College was now expanding across the Caynton Road *(see photo page 148)*. Until the building work of the 1960s there had been no major structure on the eastern side of the road.[9] After years when little significant building activity had taken place, the day in May 1961 when the diggers began work was indeed an exciting one. Just this small move brought about significant changes in the psychology of the College. Inevitably, there was a new atmosphere and some loss of intimacy. The building programme was impressive. The Queen Mother Hall included on the first floor an assembly room for 400, with fully equipped stage, and on the ground floor there was a kitchen and a dining room which could seat 350.[14] Here, from early 1963, meals were served by waitresses, with a menu ranging from fried breakfast and porridge to salad, braised steak and pineapple trifle.[15] Two new hostels were also built: Bradford Hall for fifty-eight men, in use from late September 1962, and Ward Hall for thirty-five women.[14] The firm of architects was Holland W. Hobbiss and Partners of Birmingham.[16]

In parallel with the erection of these new buildings went an extensive reorganization of the Main Building, now free of the need to provide catering and dining facilities. The initial plans were for the building to become the administrative centre of the College, but the continuing need for the building to house science and student recreational facilities restricted these plans.[17] At one point, amid the passion to modernize, the idea was raised that the Main Building itself should be demolished along with various outbuildings to the rear.[18] Hopes were expressed that the facade at least should be spared.[19] In the event the only building to be demolished was Boughey Hall in August 1968. The Hall had long been recognized as inadequate for students and with the completion of New Boughey Hall that year, the opportunity to remove it was taken.

With the modernization, opportunity was taken to improve the heating arrangements, which were by now antiquated and extremely costly to run. In the harsh winter of 1963 three tons of coal a day were used to heat the old buildings.[20] By 1967 coal had been replaced by oil.[21] Restructuring of the Main Building also meant that the Principal was no longer a resident. As long ago as 1944 Crowther had emphasized the unsuitability of the Principal's accommodation. After the transfer of the women students to Ward Hall, Ancellor House was used until 1964 as a hostel for men. By 1966 hopes of funding a new detached house for the Principal were dashed and the Principal and his family took up residence in Ancellor House.[22] It remains the home of the Principal to this day.

The very respectable sum of approximately £200,000 had been awarded to the College by the Ministry of Agriculture to cover the majority of the building programme.[23] However, the Ministry was not renowned for speedy payment. All of the bills initially had to be met by the College and, inevitably, this would result in a large overdraft.[24] Furthermore, the College had to contribute towards the building costs and the Governors had also taken upon themselves to use profits from the Farm and piggery to finance projects outside Ministry approval.[25]

No allowance was made by the Ministry to cover the rising running costs of the new buildings or the wages for the increased numbers of domestic and estate staff. For the session 1962-63 it was calculated that revenue expenditure had increased by nearly £25,000 on the previous year. That same session coincided with an unexplained drop in grant from the Ministry of Agriculture of over £5,000, making the grant £17,999.[25] When it is noted that Kenney inherited a sizeable overdraft from his predecessor, the College was clearly on a rocky footing financially.

By the autumn of 1964 the bank overdraft had risen dramatically from £50,000 in 1962 [22] to £101,000.[25] Crisis loomed, with talk of the financial plight of the College bringing about its closure. However, in April 1964 responsibility for Harper Adams along with its fellow national agricultural colleges passed from the Ministry of Agriculture to the DES.[26] The creation of Harper Adams as a direct grant college proved its salvation. Kenney declared *'a tremendous burden and a constant worry'* had been removed.[27] By 1969 he could write: *'The overall financial position today is very satisfactory and certainly a vast improvement on the period 1950-64.'* [28] In September 1964, as a result of a crisis meeting between the College, the DES and the Ministry of Agriculture, the DES agreed to liquidate the overdraft,[25] an action which Kenney described as generous.[29] The College was to contribute £11,000 and to undertake no capital project, whether it was funded by the DES or not, without DES clearance.[30] Capital projects approved by the DES would be fully grant aided. There would also be a deficit grant paid on approved annual revenue estimates.[29] The annual maintenance grant was increased throughout this time.[27] For the session 1965-66 the grant stood at £68,894 and by 1974-75 it had risen to £292,000.[30]

The DES promoted a positive approach to the funding of capital projects, as part of its overall programme to modernize and expand agricultural education. The occasional tightening of purse strings in the 1960s led to 'standstill orders' rather than cancellations of plans.[31] This was just as well for the College, for, no sooner had the new buildings opened, than it became apparent that rising student numbers called for a second phase of development. In 1964 the need was declared for three further halls of residence, although funding for all three was never thought an immediate prospect.[32] There was also a primary need for a new teaching block.[33] The old lecture room in the Main Building was no longer adequate for 105 first-year agricultural students. By 1963 the new assembly hall at the NIPH was already in use as a lecture hall for agricultural students and there were plans to use the Queen Mother Hall.

After an inspection of Harper Adams by the DES in 1964, a long-term plan was drawn up.[28] In 1965 it was announced that the College would receive £180,000 to cover the next stage of development. By 1968, as a result of this allocation, New Boughey Hall had been built, providing accommodation for sixty.[34] Government cutbacks in the 1970s meant that additional hostels were never forthcoming during the Kenney years, although the need was there. In the session 1969-70 the student population turned the three hundred mark for the first time in the College's history. In the session 1976-77 the severe shortage of accommodation led to the suggestion that bunk beds and caravans might have to be utilized.[35] The remaining grant went towards a new teaching block, containing two lecture rooms with tiered seating, capable of holding 120 each,[36] tutorial rooms, a library and staff offices.[37] The state-of-the-art building was in use for the session 1968-69,[36] allowing students wryly to comment: *'The attempts of some [lecturers] to operate the multitude of electrical gadgets at their command have been jubilantly observed by a critical audience on many occasions.'* [38] The creation of a larger modern library was a

12.3 A student study bedroom, 1969: Crown copyright.

12.4 The new lecture theatre, 1969: Crown copyright.

significant step forward for the College. For the year 1961-62 book loans had increased by sixty percent over those for 1959-60. However, the old library had remained very limiting and was described by one member of staff as 'next to useless'. On one occasion he had discovered old journals, removed from the library at some stage, stuffed in a chimney. The new library became fully functional in the spring of 1969, after the appointment in 1968 of the College's first full-time Librarian, Olga Beaumont. She 'strolled about in brogues and tweeds'.[6] The old library became a committee room.

There was also a requirement for a new science block,[33] but this was not realized. However, in the mid-1960s plans went ahead to reorganize and extend laboratory space in the Main Building.[14] Although the chemistry laboratory had been doubled in size in the early 1960s, space was still very much at a premium and the situation bordered on 'absolute mayhem'.[6] By the autumn of 1971 final improvements had been made to the science facilities in the Main Building, so that the whole eastern wing could in effect be considered a science block.[39] However, conditions were still far from perfect: *'The fume cupboards were the extraordinary thing. They were dark brown wood which didn't really extract any fumes at all. There was a bit of a hole near the ceiling and you'd put something in with boiling sulphuric acid and half the fumes came back in the laboratory and not many went outside. [...] Students coughing and spluttering as you did that sort of thing.'* [40] It has been said that it was the continued use of the Main Building for science, as a result of the delay in the erection of a purpose-built science block, which saved the Main Building from demolition.[6] Those at Harper Adams today have much to be grateful for.

The College's last remaining requirements were a machinery teaching block to replace the old Boutflour building at Ancellor and a student union building.[41] The machinery

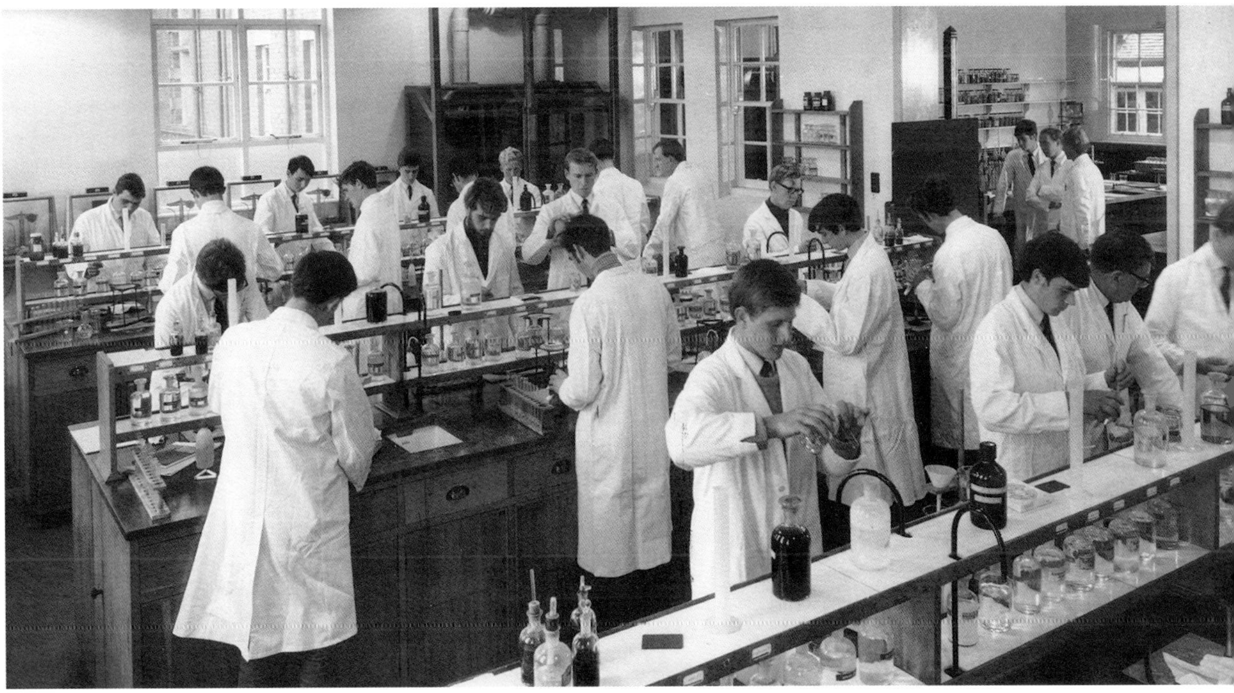

12.5 Improved chemistry laboratory, Main Building, early 1970s: West Midland Photo Services Ltd.

teaching block, situated on the eastern side of the Caynton Road, opposite the farm buildings, was first in use in the autumn of 1969. The next project to be completed was the livestock demonstration hall. It was built on the western side of the Dutch barns and completed in the autumn of 1970.[42] Here live animals and carcasses were displayed and a small slaughter-house was included on site.[22] The need for a building providing student facilities had been recognized for some time, but once agreed the wait was not at an end. The student union building suffered delays and mounting costs due to the bankruptcy of the contractors.[43] Students had been asked what they wanted in the building and on completion in late 1971 it soon became a focal point: *'Most nights of the week the majority of students would end up in the bar at some point in time'*.[44] At last the students could say: *'Today we can claim that our overall facilities are at least equal to those of any Agricultural College in Britain.'* The sports hall could be adapted to accommodate 'virtually any kind of ball game including cricket and golf'.[45] The ever popular game of billiards was naturally not forgotten and a room was dedicated to the pastime. Those students less sports minded could enjoy the facilities of the bar, canteen and common room.[46]

By the 1970s it was clear that Government cutbacks in the face of sharp inflation would limit future developments at the College.[28] Even minor projects had to be discontinued. It was a period of severe financial stricture for the College. Fees were increased quite steeply during this period, but this act could not compensate for the 'tremendous increases in revenue costs'.[22] In 1964 fees had been raised from £315 to £350 a year for a residential student. By the autumn of 1973 the same fees were £528 [47] and three years later £696.[48] In the early 1970s, the financial situation was aggravated further by the decline in student numbers, resulting from the uncertain situation in farming and poor employment prospects. In the session 1973-74 numbers reached the lowest point of 244.

At the beginning of the Kenney Era there was just one agricultural course. It ran for two years and led to the College Diploma and the NDA. The option of studying for an extra year to

12.6 The Student Union Building close to completion, c. 1971: Shropshire Star & Journal.

obtain an honours qualification remained, but it was an option taken by very few.[49] By the mid-1960s the post-diploma subjects offered were pig husbandry, crop technology and marketing.[50] The situation was very different by the end of Kenney's principalship. The College was clearly moving into the realms of Higher Education. Gone were both the College Diploma and the NDA and in their place was the Higher National Diploma (HND), a three-year course with sandwich training. Accompanying the HND were three specialist one-year post-diploma courses. With these developments had come a very new approach to learning. As Kenney himself wrote on his retirement: *'Continuous assessment, involving major and minor projects and practical tests as well as written examinations coupled with a period of industrial training provide in my opinion, a more accurate picture of a student's real ability and a better all-round training for them.'* [37]

The HND was first offered at Harper Adams in 1969, [22] but the idea of raising the standard of the NDA had been discussed for over a decade. The De la Warr Committee of 1958 concerned itself with farm institutes.[51] However, the Committee felt that in the face of its recommendations for improved standards at farm institutes, standards must also be raised at college level. The suggestion was made that college courses should be lengthened to three years with the possible adoption of a sandwich period for practical experience.[52] The Robbins Report appeared in 1963.[53] Described as 'the most comprehensive account of higher education in this country ever attempted', it called for a great expansion in Higher Education. By 1973-74 it envisaged 390,000 students as opposed to the 216,000 at the time of publication.[52] Not only were new universities to be built, but existing regional centres were to be encouraged to offer advanced courses. Special reference was made to the five independent agricultural colleges, namely Harper Adams, Seale Hayne, the Royal, Shuttleworth and Studley. Here the suggestion was made that entry standards should rise to include A-Levels and that courses should be lengthened to three years with a view to offering degrees at a later date.

In 1966 the Pilkington Report followed.[54] The Pilkington Report sought to bring agricultural education into line with thinking on education in other industries. Qualifications were to have national currency and courses were to be streamlined. The Report recommended the introduction of the Higher National Diploma and the Ordinary National Diploma (OND). Both were to be three-year sandwich courses, tied to practical experience.[55] The HND was a qualification of higher standing than the NDA. It was to be a science-based course, but with science integrated with husbandry throughout. The level of the OND was to be higher than the National Certificate taught at farm institutes but lower than the old NDA. Plans for teaching the OND in agriculture at Harper Adams never materialized in the face of competition from the farm institutes.[56] Many welcomed the impending change. Peter Watson Jones, a local farmer and a young Governor at Harper Adams at this time, was one: *'I didn't think the NDA was quite honestly the right sort of course for the future top farmers and top farming. [...] The NDA course was coming to the end of its time.'* Quite simply the NDA was in many people's minds not in touch with farming. The HND in contrast was a balance of science and practice, with the notable addition of the sandwich period.

Entry requirements for the HND were higher than those needed to study the NDA. College prospectuses show that entry requirements had been slowly rising at Harper Adams in line with Government thinking on improved standards. Entry for the HND required at least one pass in a science subject at A-Level and evidence that another subject had been studied to that standard.[57] From September 1972 Harper Adams offered the Pre-Entry course, for those with an insufficient science background who wished to study for the HND.[58] It was a one-year course, aimed primarily at those with good O-Levels but without an A-Level. For its first year ten students were enrolled.[59] The course was the idea of Dr Boyns, Head of Science, who realized that higher scientific requirements were losing many students for the HND course.[6] The Pre-Entry course was the only one of its kind and a pass allowed entry to any HND course in agriculture in the country.

The first thirty-four HND students enrolled at Harper Adams in 1969.[46] In 1972 Harper Adams had the highest number of HND entrants, over seventy, for any centre in Britain.[59] By 1975 the annual intake was between seventy-five and eighty students.[22] Standards were good. From 1969 the higher standard of student had been evident to many staff: *'The students had had more experience of education by going on to sixth form and were more in touch with learning.'* [40] In 1973 the Educational Assessor declared of the College's HND students: *'The top end of the group compared very favourably with the standard at university.'* [60] In 1976 three Harper Adams students reached distinction standard, the highest number for any agricultural college in England and Wales. This was the first time any agricultural college had achieved three distinctions in one year.[61] Significantly, women were included among enrolments for the HND.[6] The Pilkington Report had specified that opportunities in agricultural education should be open to women and the Governors had consequently passed a special resolution allowing the admission of women to the HND course. Thus, in 1970 after a gap of twenty-five years, the first women, two in number, enrolled to study agriculture at Harper Adams.[62] In 1974 the first woman to obtain a distinction in the HND in agriculture, Sheila Driver, was a Harper Adams student.[63] Initially the NDA ran alongside the HND, but the last entry for the NDA occurred in 1971.[64] Many regretted the passing of a qualification which had earned high respect. Indeed, between 1965 and 1973 Harper Adams was well above the national average as regards results in overall and subject passes. The Queen's Prize, awarded to the top student in the NDA examinations, was won three times by students from Harper Adams during this period.[22] Students needed to cover the syllabus, learning as many facts as possible to pass the examination. The NDA examinations were real trials of strength and stamina. However, thinking on educational methods was changing.

The Pilkington Report advised that, in future, examinations should suit the courses rather than vice versa.[65] The Report recommended that responsibility for syllabus and teaching methods should belong with the College. Examinations for the NDA had been set and marked externally. In contrast, the HND was to be internally set and marked, but with external assessment and validation by the new Joint Committee for National Awards in Agricultural Subjects. It was a system which encouraged enthusiasm rather than tied students and staff to a list of topics which had been decided externally.[40] All this had to be taken on board and involved a great deal of extra work on the part of the staff. Furthermore, staff had to keep abreast of new developments in teaching methods, with new concepts such as continuous assessment and sandwich placements.[7] Interestingly, in 1974 Harper Adams organized jointly with Wolverhampton Technical Teachers College a teacher training course for staff, the first of its kind to be offered by an agricultural college. Fourteen members of staff joined the course, along with staff from three county colleges.[63]

Changes in teaching methods had been encouraged by Kenney since the early 1960s. He had not been happy with a traditional

12.7 1971-72 staff and students.

system which was based on notes dictated at top speed and remembered parrot fashion. The Cat-A-Mountain of 1963 declared that students were no longer 'note-quoting automatons'. Earlier students had often accepted the situation knowing little different: *'Nobody objected to that at all. We just sat there and took notes for fifty minutes. That's the way it was. That's the way you'd been taught at school.'* [9] Adopting contemporary thinking on education, by the mid-1960s Kenney was urging the need to stimulate students, encouraging them to broaden their knowledge through their own initiative and efforts under the guiding hand of the lecturer. New emphasis was to be placed on discussion and reading with tutorials and private study periods.[66] Formal lecturing and note taking was no longer to dominate. The first tentative stages of a tutorial system appeared at Harper Adams late in 1964, with the arrival of personal tutors. Every member of staff was allocated a group of students and a student could turn to his tutor for help and advice concerning his work or personal life.[67] This was especially important as a degree of impersonality [6] had arrived with student numbers attending lectures rocketing to around 120.[12] In the session 1966-67 the system was extended to academic life with the arrival of tutorials and more project work for the final-year students, including a thesis which was to count towards the College Diploma.[66] Prior to this time there had been some instruction in small groups of around six people, often practical sessions involving wider discussion, but these had never been known as tutorials.[68] As Kenney declared, the new system was *'a far better test of "real" ability'*.[66]

These developments were accelerated by the arrival of the HND course. With the arrival

of the HND in 1969, fifty percent of marks rested on continuous assessment, with modules undertaken throughout the three years.[64] Continuous assessment did change student life in a significant way, as Ros Barsley, nee Forrester, student 1970-73, noted: *'You have to work through the three years. You can't just do nothing and then swot furiously for the final exam.'* However, it would be wrong to presume that change was quick and dramatic.
The student timetable was still very traditional, filled with lectures every morning from nine until one and four afternoons a week from two until five, when there would be tutorials and laboratory work. There was no time set aside for private study for minor projects (assignments) beyond evenings and weekends.

A very important part of the HND was the emphasis on the practical with sandwich training, described by Alan Costley, Head of Education Development and Quality Assurance at the College, 'as the single most important development in vocational higher education since the 1960s'. The sandwich period assisted the employability of students, cemented links with the farming world and put flesh on the bones of academic learning, being of

12.8 Cartoon of staff from Cat-A-Mountain 1973. Back row from left: unknown, Ray Smith, Henry Hollinshead, Ian Gedye. Front row from left: Peter Dickie, Derek Sadler, Colonel Nixon and Principal Kenney, Maurice Eddowes, Josie Hutchinson.

considerable importance placed as it was in the middle of the course.[7] The sandwich period accounted for fifteen percent of the overall assessment in 1971 and, if the farmer's report was completely adverse, the student could fail the course. Students were paid for their work and in the early years the farmer received a grant after the training period had been completed. With the introduction of sandwich training the staff had the novel experience of travelling to meet students at their work placements and to discuss their progress with their employers.[69] Such visits were invariably worthwhile for all parties, but could result in some memorable experiences for the visiting member of staff: *'I went across and there he [the student] was carrying two dead little pigs, filthy hands. [...] They were feeding the pigs on manufacturing by-products including waste sweets from the big sweet factory. He took me into a big room and there were piles of Wrigley's chewing gum, all in their paper as well which they were feeding to the pigs. He said, "These are quite good, have one", and he opened one, his hands were black.'* Coffee was subsequently 'very politely refused'. Be it known that the student has since declared he had every intention of washing his hands. Not all students gained 'hands on' experience, whether of this colourful kind or not. Women students, especially, could find themselves working as nursemaids and domestics.[6] Prejudice against women being fully involved in agriculture still persisted. Ros Forrester, joining the College in 1970, gained her practical experience before enrolment on a farm in Staffordshire: *'I did mainly poultry. Just occasionally I was allowed to touch the calves, but [I had to] keep well away from cows or machinery.'*

Increasingly, it was recognized that farming needed specialized expertise and the Pilkington Report highlighted the need for supplementary post-diploma courses 'for specialist aspects of production which involved new and advanced techniques'. Harper Adams did not neglect post-diploma qualifications. In 1966 the first students enrolled on a one-year post-diploma course in pig technology,[36] the idea of Peter Luscombe.[70] Its ancestry lay in the one-year honours course in pig husbandry.[67] The course was aimed at those seeking to become managers of large-scale herds or wishing to take up posts in advisory work or with the Agricultural Development and Advisory Service (ADAS) and the Meat and Livestock Commission.[71] The course had a chequered history and lost popularity as the HND took hold. Few students were happy to study for an extra year after taking three years for the HND and the course ended with the session 1974-75.

A post-diploma course in Agricultural Marketing and Business Administration (AMBA) was started in 1969.[58] The Pilkington Report had highlighted the need to study the business aspects of farming. Business management was very much the in-word. The new course aimed to produce people knowledgeable in marketing and business techniques for posts at managerial level in commercial organisations, marketing boards and large-scale farming organisations.[72] It was a one-year course and was organized in association with North Staffordshire Polytechnic.[57] For the session 1972-73 there were twelve to fourteen students.[73] By 1974, the 'tremendous demand' for AMBA students led the College to discuss the possibility of setting up a HND in business studies, a course which in its strong agricultural bias would be unique to Harper Adams.[74] The HND AMBA started in September 1976 with fifteen students.[37] However, in 1978 the Post-Diploma AMBA was suspended due to insufficient numbers.[75]

Ray Mortimer, by the 1970s designated Head of Agriculture and Director of Studies,[71] was the man behind the AMBA courses. Indeed, he made a great contribution to course development as a whole and was highly respected for his involvement in all aspects of life at College. On the retirement of Dr Temperton he acted as Deputy Principal and in 1977 his role was recognized by his appointment as Vice Principal.[35] His 'ability to fit in' [8] ensured that he worked well with the Principal. Kenney's successor as Principal, Tony Harris, declared: *'He carried responsibility with ease and distinction.'* [76] Mortimer was especially good at advancing the cause of the College nationally: *'[He was] the upfront man and did it very well. He was very good saying all the right things at the right time and being knowledgeable in a wide range of areas.'* [8] In this capacity he wrote books and articles and was a regular contributor to the Shropshire newspapers. He also attended numerous conferences and took part in radio and television broadcasts. Tragically, in 1982 he died aged fifty-eight.[76]

It was the age when commerce was making increased use of the consultant, whose specialist knowledge of the sophisticated services and products on offer was in demand from farmers with neither the time nor knowledge to keep abreast of developments. Against this background and with a move away from broad definitions of animal or crop husbandry, crop production was gradually being built into the framework of the HND as a subject in its own right drawing on a wide range of expertise.[3] Thus, in 1972 [62] the number of post-diploma courses was enlarged to three with the inclusion of crop protection under the supervision of Maurice Eddowes.[77]
The course was a year in length and sought to produce specialists for ADAS, education and research, the agrochemical industry and commerce.[71] Success was by no means certain, as Eddowes related: *'We started our crop protection course more in hope than expectation in 1972. Our initial intake was only three students. [...] We progressed with it and gradually built this up.'* Other courses offered by fellow institutions did not survive.[3] By the session 1975-76 fourteen students were studying on the course.[78] Part of the success was due to the course's ability to attract considerable interest from overseas students. Of the fourteen students in 1975-76 six came from abroad, namely the Sudan, Sri Lanka and Guyana.[3]

There had been a small number of overseas students at Harper Adams from the very early days. Very often they were on the poultry courses. Of the average number of poultry students in the early 1960s, namely sixty-five to seventy, some twelve to fifteen were usually from overseas. The Diploma in Poultry Technology was especially popular with overseas students.[49] In the mid-1970s specialist twelve-week courses for trainee poultry farm managers from developing countries were offered, financed by the Ministry for Overseas Development.[79] An overseas club was established in 1963 to bring foreign students together and to stimulate interest in foreign lands.[80] Often the ways of the students were new to the College, but the staff did their utmost to accommodate them.
Bob Bradley remembered one particular trip

12.9 A farm management exercise, with Bruce Cox, 1969: Crown copyright.

to Silsoe in the 1970s and the problem of finding Mecca for the accompanying Muslim students: *'As we progressed on, the fog came down and when it was time for them [to pray] nobody knew where they were supposed to put their prayer mats. So, by now we'd got into Silsoe village and Guy Whitley, who was a Red Arrows pilot, said "Don't worry. No problem". [...] We rushed into the church, right up the bell tower and stood on the top of the parapet of the church in Silsoe yelling down, people watching around. We were saying to these guys "Left a bit. Left a bit. That's it. Now you're right" and he was giving the directions off the weather vane.'*

The great changes in agricultural education in the 1960s did not ignore poultry husbandry. As a result of the Pilkington Report the NIPH sought to introduce both the HND and the OND.[81] The DPT was to continue to be offered as it fulfilled the Report's requirement for technological degree type courses and there were hopes that it would be transformed into a degree in its own right. However, enrolments for the DPT ceased in 1968 due to a lack of applicants.[82] This disappointment was, however, balanced by the fact that in July 1967 the DES had given approval for a degree in poultry science and technology to be jointly organized by the NIPH and the University of Nottingham.[83] In 1974 this course also failed, with the University of Nottingham ending the arrangements due to poor recruitment. The prospect of coming 'to the outback' for the final year had not appealed.[68] Ironically, the first degree students, two in number,[82] had only just arrived at the NIPH in the previous session 1972-73, after two years studying general agriculture at Nottingham.[59]

Plans for the HND and OND were in place, but the plans did not materialize as envisaged. Despite earnest efforts on the part of the NIPH, numbers of applicants for the HND in poultry husbandry were never sufficient for the course to run.[84] Plans to remodel the HND as an HND in agriculture with a poultry option were approved by the DES in 1975, but again met with little support.[85] The option was dropped for the 1978-79 session.[86] The recruitment problem had been growing since the mid-1960s. For the session 1964-65 there had been a record number of poultry students, with sixty-seven students studying for the College Diploma and the NDP and a further twenty-eight on the DPT course.[54] The total number of poultry students represented a third of the College. However, numbers declined thereafter. For the session 1971-72, the last before major course changes, NDP students numbered just thirty-three. The reasons for poor recruitment were many and various, but in essence stemmed from the fact that the Poultry Industry was becoming a factory process with larger and larger units and consequently with fewer requirements for employment.[59] Training was often undertaken internally and there was less need for formal poultry education at specialized centres. In many circles it was perceived that academic institutions were still embedded in a traditional view of poultry husbandry linked with general farming and did not take sufficient account of the growing movement towards specialization, the factory process and the business ethic. In the opinion of some, research and academic orientation were now labels which no longer worked to the advantage of the NIPH.[68] The NIPH was becoming 'isolated' from Industry. As Rowley remarked: *'They didn't understand what each other was doing. [..] They weren't talking the same language.'* The poor image of an intensive industry and the economic recession, with the accompanying uncertainty as to employment, also contributed to make a bad recruitment situation considerably worse.[87]

There was also much mistrust of the new qualifications. The NDP, as with the NDA, had been tied to external examinations, 'each question paper was set by an industrial expert', but the new qualifications relied upon internally set and marked examinations.[88] This resulted in some loss of industrial

12.10 1963 NIPH staff and students: Private Collection.

credibility.[68] Furthermore, courses covering four years, if the required one year's practical before enrolment was taken into account, had little appeal. In the case of the HND the high entry requirements severely restricted those eligible to enrol.[81] The OND similarly ran into difficulties because of competition from similar but shorter courses at county college level. There was also the feeling in some quarters that the very title with the inclusion of the word 'ordinary' was not conducive to recruitment.[68]

The first OND course, unlike that for the HND, did get underway in 1972.[89] However, enrolment was never high and when NDP students ceased at the end of the session 1972-73, a crisis in poultry student numbers was reached. For the session 1973-74 poultry students, all studying for the OND, were just eight in number. Everything was done to encourage recruitment. In July 1972 the NIPH organized taster courses for school-leavers, consisting of poultry projects, a barbecue and games. Twenty-three children attended from Shropshire and neighbouring counties. In 1974 the Governors established the Poultry Education Consultative Committee to investigate the training requirements of Industry and ways of attracting students.[74] It was composed of representatives of Harper Adams and the Poultry Industry and one of its first objectives was to encourage commercial firms to sponsor students. Students were to spend part of their time at the NIPH and part with a commercial firm.[90] With the support of the Poultry Industry, plans materialized to introduce a shortened two-year OND [91] which was to include a sandwich period of just six months.[85] The revamped course began in September 1976 with nineteen students.[48] The crisis had been averted for the time being at least.

With the increase in course diversification in the Kenney Era, improved management was essential. Thus, in November 1970, on the recommendation of the DES, Harper Adams established the Academic Board.[92] Initially, the Committee met twice a year and included representation across the College, including students. Its task was to overview the structure, content and implementation of all courses, methods of assessment and examination and teaching facilities.[93] However, it was to have little real role in this area until fully utilized by Kenney's successor.

It was also the DES which instigated changes in the Foundation Scheme at this time.
In 1967 there was a reduction in the number of Governors from thirty-one to twenty-eight. Representation from the County Councils was reduced to take account of the fact that financial contribution from that quarter was at an end. In future there would be just one representative from Shropshire, Staffordshire, Warwickshire, Worcestershire, Herefordshire and Cheshire. Governors now came from a wider spectrum, with representation invited from the West Midlands Advisory Council for Further Education, the National Farmers Union, the DES, the Ministry of Agriculture, the British Egg Marketing Board, the University of Nottingham, the RASE, the National Union of Agricultural Workers and the Country Landowners' Association. Most significantly, for the first time, two representatives were invited from and chosen by the academic staff. The Principal, as before, attended as an ex-officio member. In guide-lines issued by Local Education Authorities and the National Union of Students concerning relations between staff and colleges the matter of student representation on the Governing Body was left to the individual institutions.[42] In the case of Harper Adams the Governors refused student representation,[94] despite student requests for the right to attend.[95] It was felt that existing committees, such as the Joint Staff Student Committee and the Academic Board, allowed the student voice to be heard and that in special circumstances a request for student representation at a meeting of the Governors would suffice.

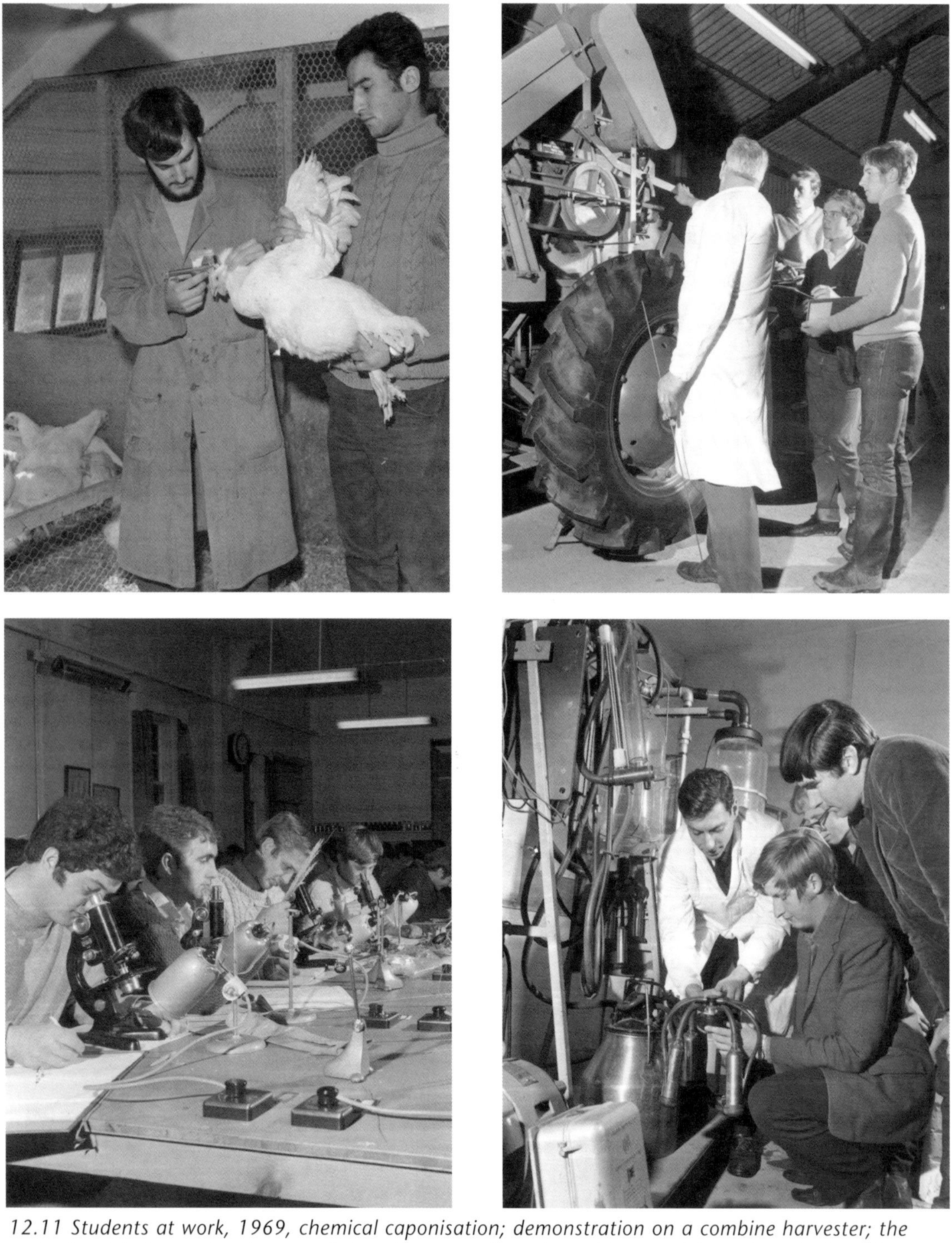

12.11 Students at work, 1969, chemical caponisation; demonstration on a combine harvester; the agricultural botany laboratory; workshop demonstration of a milking machine unit and plant: Crown copyright.

Royal Visits

12.12 Above and overleaf - the visit of H.R.H. The Princess Elizabeth on the occasion of the College's Jubilee and the opening of the Jubilee Hostel, 8 June 1951: Dan Arden, Wolverhampton Express & Star.

12.13 Below - the visit of H.M. Queen Elizabeth the Queen Mother on the occasion of the opening of Queen Mother Hall, 1 May 1963: Wellington Journal & Shrewsbury News.

Chapter Thirteen —

People, Pigs and Poultry 1962-77

The students of the Kenney Era relished their time at Harper Adams every bit as much as their predecessors. Christine Justice, student 1964-67, 'enjoyed every minute of it' and her future husband, Trevor Parnham, student 1965-67, 'had a great time', relishing the 'wonderful shock to the system' with 'no responsibility whatsoever'. They remember their Harper years as the happiest time of their lives. Desmond Ennis, on arriving as a student in 1966, felt he was checking into 'a hotel for two years' with 'magnificent food, swimming pool, in-house entertainment and organised social events'.[1] Student tales of lectures, dances, sport and the pleasures of youth are interwoven to form indissoluble memories. The College was rapidly growing in size, but everyone still knew everyone and the majority of students lived on campus. When Ros Forrester arrived in 1970, she remembered that it was like acquiring three hundred brothers overnight. The 'terrific community spirit' still lives on in the frequent reunions that take place.[2]

Sport remained the bed-rock of College life beyond the lecture room and Harper Adams had a considerable reputation. For three consecutive years from 1967 the College won the sporting event organized by the United Kingdom and Ireland Agricultural Students Association (UKIASA), which had been set up in 1963 to promote liaison between students.[3] The competition was stiff and in 1969 Harper Adams won on home ground against eleven universities and colleges.[4] Rugby was the dominant sport, blessed with the enthusiasm of a principal with a passionate interest and considerable skill in the game.[5] At the end of the 1971 Season the College First XV had beaten every other college team both at home and away, the Principal adding with pride 'including the Royal'. Long-standing rivalries were as strong as ever.

During the 1960s, running captured the imagination of the College with some outstanding students. In 1962 Seale Hayne organized the first run between the two colleges *(see photo page 149)*. Its team took thirty-nine hours and thirty-four minutes to cover the 332 miles. The route took them

13.1 1966 First XI cricket team. Back row from left: D.M. Wodfold, J.P. Franklin, F. Maxwell, J. Anderson, A. Waterworth, C. Jones. Front row from left: J. Timms, R. Blades, M.C. A'Bear, J. Old (Capt.), H. Morgan, D. Muscamp: Dan Arden.

via Shuttleworth and the Royal.[6] Harper Adams took up the challenge in 1965, seizing the record with a time of thirty-six hours and forty-five minutes.[7] In the spring of 1966 the College lost its title to Shuttleworth,[8] but in October the College team fought its way back with a vengeance [9] and took just thirty-three hours and forty minutes to reach Harper Adams.[10] Trevor Parnham recalled that they received a hero's welcome: *'Word got passed up the road that we were actually well ahead of the record time and from the A5 back to Harper, the road was lined all the way with students and staff cheering us on.'* The celebratory meal cooked for the runners and their supporting team by the staff must have tasted exceptionally good that day. Despite the continuing popularity of sport at the College, the idea of a formal sports day declined in popularity and by the early 1970s there was a decided lack of entries for the event.[11] Sports day died out in the 1970s. The fact that there is no record of the exact date is evidence of loss of interest, despite its former glory.

As in earlier years there were clubs to join. The drama group was still popular. Revues

13.2 Student play 'The French Mistress', January 1976. From left: Dave Blair, Colin Maddison, Tim Nelson, Colin Rayner, Geoff Gorton, Janet de Souza, Nick Burton, Janet Royce, Chris Neave, Charlie Shepard.

were still performed and enjoyed. The revue of 1968 'came to a hilarious conclusion with Graham Solari doing an impersonation of "Doc Boyns". The audience was lectured on the distribution, habits and husbandry of Bactrian and Dromedian camels from S.E. Wales and the warmer regions of the Arctic.' [12]

The 'victim', renowned for his tales of days in the Sudan, was the first to applaud the talent displayed. The key element of social life remained the dance, whether formal or informal. By the mid-1960s there was, however, a clash of musical tastes. The Cat-A-Mountain of 1965 published the report of the Music Club: *'As has always been the policy of the Club, the music chosen was of a serious nature for the benefit of those not addicted to the meaningless gibberings of otherwise unemployable, drooling Liverpudlians.'* It is worth noting that the Fab Four clearly took heed of this criticism and have since taken up gainful employment as professional millionaires. The 'difficult problem of successfully catering for both staff and students in a day of electric guitars playing havoc with the traditional dance steps and polite conversation' had to be confronted.[6] Thus, there was a diplomatic division of the Queen Mother Hall for the End of Session Ball in the mid-1960s: *'Two-level dancing was available, upstairs for the sedate or when a rest was needed, downstairs for rock and rollers of all ages.'* [1] For a time from 1965 the College even had its own rock group, the Moogees, originally starting life under the less fashionable but equally memorable name, the Sugarbeats.[8]

With a growing number of students the choice of clubs also increased, although the life-span of a club could be short.[13] Very often a club would start, encouraged by the interest of a particular body of students, only to fade as those students left and then, later revive as fresh interest awoke.[14] In the early 1970s the Riding Club appeared. The Judo Club started in 1964, only to fold as karate gained in popularity and a club for that was set up in 1973. There was the Mountaineering Club, which started in the session 1963-64, where less adventurous members could choose walking rather than climbing on a weekend away.[15] The Canoe Club opted for their excitement in the swimming pool. The Club opened in 1972, the pool in 1963. The pool became especially popular as a ducking arena after dances.[15] A pool had been a long-held hope since Price's years. From pre-war days students had frequented Wright's Picture House in Newport. Maurice Lancaster as a student in the early 1950s remembered the queues way down Stafford Street and the fact that students were banned from attending on Thursdays. However, the cinema closed in the early 1960s and the students duly started the Film Society in the session 1962-63. This proved very popular with average attendances of 150 in the late 1960s. The intrepid organizers had to deal with everything from projector bulbs exploding in mid-performance to 'Tom Jones' getting lost on British Rail.

In essence student life was no different from that of earlier years, but there is no doubt that currents of change were flowing through the nation and Harper Adams was not isolated. The issue of student cars illustrates the point. By the last years of Principal Price it was noticeable that the number of student cars was growing. This led to problems of parking and also a genuine fear that greater mobility would undermine the corporate spirit of the College.[16] Complaints of poor participation in College activities were common and blamed on the fact that the cost of a car was within the grasp of most students. It was quite common for a group of friends to club together and buy a share in an old banger.[17] In 1961 the College banned first-year students from keeping cars.[18] However, those students affected by the ban simply used the cars of their colleagues, or with due initiative parked their cars in the village.[2] One local farmer made a respectable sideline offering parking facilities.[19] The College could not hold back the tide of change and in 1966, despite misgivings, the ban was ended.[20] By the early 1970s the majority of students had cars.[21]

The arrival of cars has been regarded ever since as the main cause of a decline in College life. Even with the availability of cars, students were not unduly tempted away from the College during the weekdays. In 1966 a permanent bar run by and for students was set up in the old laboratory on the first floor of the Main Building.[8] This and the opening of a purpose-built bar in the new student union building in 1971 encouraged many students to stay on site. However, as many students could still be regarded as being local, a car did present the opportunity to go home at weekends. In 1970 the end to Saturday morning lectures as a result of student pressure eradicated the last hurdle to the 'five-day week'.[22] However, it should be stressed that student apathy towards College events and accusations of the College being dead at weekends had been common before allowing all students the freedom to have cars.[23] Furthermore, the Cat-A-Mountains of the 1970s confirm that there was a broad spectrum of entertainment on offer throughout the week and weekends. In all probability cars were not the cause of a change in the social life of the College, but rather an indication that change was taking place.

This was the time of greater freedom for the young and not even the College with the conservatism of its students could avoid the

13.3 New student bar, Student Union Building, 1973.

changing climate. Opportunities for education beyond school were available to ever larger numbers and, as a result, the younger generation was gaining in confidence and their expectations of life were the greater.[21] In 1969 the Representation of the People Act had lowered the age of majority to eighteen. Ros Forrester, considered 'the speed of change from the 50s through to the 70s was absolutely horrendous'. Meanwhile, Harper Adams in the early 1960s had remained a traditional environment. The College was still quite formal in character. The public school atmosphere lingered, although few considered the rules stifling in an age where rules were expected by many straight from school and home.[2] In 1962 there was still one formal meal a day with grace, at which all male students were required to wear jackets and ties and the ladies skirts or dresses.[24] Nine years later, Sunday dinner was the one remaining meal where grace was said and smart dress applied.[25] Throughout the country change was afoot. Focussing on the College, with the expansion of student numbers had come an expansion in staff and the new lecturers were young and fresh out of university, with a more modern attitude to life. Rapport with the students thus came more easily and barriers began to fall.[14]

In the 1960s the Governors, Principal and traditional elements amongst the staff found themselves with little experience of how to deal with the growing student demands. Thus, on one occasion Kenney, sounding more like a beleaguered parent than Principal, declared he was surprised at the resentment of the students to anything like discussion, 'sometimes at the slightest hint of offering help'.[26] There were complaints of 'tremendous noise' from the visiting bands.[27] The mid-1960s was the time when the first big bands came to Harper Adams, bands such as Unit Four Plus Two with their hit 'Concrete and Clay'.[2] There were complaints concerning 'the display of bad taste and discourtesy' by some students [28] and of a decline in respect by students for staff.[29] Student dress was another cause for concern. As Dr Boyns succinctly put it: *'We had a period when some of them were dressing in king's robes, Napoleonic dress and things like that.'* On occasion a clash of ideas did have its lighter moments. Bob Bradley remembered one particular trip to Crudgington Creamery shortly after the students had been reprimanded for being shabbily dressed:

> *The students that day had been to Shrewsbury, to Moss Bros. and hired sort of evening dress and morning suits and bowler hats and umbrellas and they all sat on the coach, sort of bolt upright, as if they were going to a funeral. Ray [Mortimer] and I, in our sports jackets and trousers, were totally speechless. [...] Nobody said a word. It was just one of those lovely things that you just carry straight the way through. [...] If anybody was scruffy that afternoon, it was Ray and myself.*

Of course, the issues were more often than not quite minor in nature, but there is no doubt that there was a gulf in understanding opening up. Perhaps, nowhere was the gulf between the old and the new better illustrated than in one talk given to the students late in 1972 as part of the general studies requirement for the HND.[30] This very sensibly was a talk on drug addiction. The choice of speaker was a little more in question. Sister Patricia of St Mary's Convent, Wantage, however, was described as possessing a wonderful personality 'in spite of being 80 years' young'.

There were attempts to meet in the middle ground. In 1966 Kenney instigated a series of meetings to promote an interchange of ideas between the SU, the Governors and the staff.[31] By 1970 this had led to the creation of a Joint Staff Student Committee, which was to meet regularly throughout the year.[32] Some concessions in rules resulted. In 1971, for example, it was agreed that students were no longer required to sign a leave book when absent from the College for any length of time.[33] Signing out if returning late at night was also made optional.[34] The Cat-A-Mountain of 1968 commented upon the fair and easy exchange of views with Kenney. Similarly, student good behaviour and 'mutual understanding and respect' were underlined by Kenney in the Principal's Reports of the same period. The Academic Board with student representation was also used very much as a mouthpiece for student issues during its early years in the 1970s.[14] Yet despite some progress, little time would go by without some issue or another giving cause for concern. From the student point of view, Trevor Parnham remembered 'there was always a continual battle with the Principal to get change'. Bill Wilder, SU President 1967, declared:

> *[These were the days of] Student Unrest, when we Harper Agrics were dragged screaming to the world of student politics. [...] There was just enough activists amongst us to stir the pot and force us to confront the authorities. Reg Kenney, the Principal of the day, was not the sort of man to relax his guard to such a threat. There was endless issues over which we scrapped; student grants, student representation on the Governing Body and at disciplinary hearings, staff/student committees, Union funds and colour television, and even the laundries.*[9]

Rag stunts throughout this period got progressively more out of hand and were frequently a bone of contention. The habit of kidnapping shop assistants and auctioning them at a slave market on the balcony of the Vine Vaults in Newport Square usually caused much merriment.[2] However, on one occasion both assistants in Mr Cawthorne's shop were

13.4 Piano smashing in the Square, Newport, Rag 1965: Private Collection.

13.5 Rag procession, 1960s: Wellington Journal & Shrewsbury News.

seized and the shop was left unattended. The fact that it was a jeweller's shop did not help the cause of the students.[35] In 1965 the cannon outside the Headquarters of the Territorial Army at Trench was borrowed by the students and police prosecution was threatened.[36] In 1967 a realistic street 'shooting' and kidnapping in Stoke-on-Trent on a Saturday afternoon led to police road blocks. The bursting of a bag of tomato sauce resulted in a hospital alert.[37] The College's decision to end all further rag events that year was not a surprising outcome.[9] In 1972 students in the Rag procession threw water bombs filled with slurry about Newport High Street and a bucket of slurry was tipped over a private car.[21] All future processions were banned. It should be stressed that the Rag events were overall well organized and primarily good fun, with great benefits going to charities. The 1973 Rag, for example, saw the sum of £4,720 being raised. This represented £23.50 per student, a figure which was 'better than [that for] any other group of students in the country'. Instead of the procession there was a Rag Fayre and a re-enactment of Dick Turpin's famous ride.[38] The students could not, however, resist a little mischief making and carried out the kidnapping of Miss Great Britain on her visit to Newport as part of a promotional tour for Tesco.

One of the most contentious issues between the students and the College was that of the integration of the sexes and whereas many issues had been resolved by the 1970s, this showed no signs of abating. With the opening of Ward Hall the women students came in closer contact with the main site. However, their numbers were never large.

13.6 Womens study bedroom: Farmer & Stockbreeder.

In the session 1964-65 there were twenty-one women at Harper Adams, all poultry students. A little over ten years later, in the session 1976-77, the number was thirty-eight due primarily to the addition of women studying agriculture on the HND course. Many found it a charmed life with so many men. However, there were hurdles to overcome. The very first obstacle to confront any female student 'was tea that first day with over two hundred pairs of eyes wanting to see what the new girls would be like'. At times the allocation of feminine duties, with so few women to take them up, was little short of a chore: *'[There are] dances when all we see are buttered rolls, dirty coffee cups and tables to be laid for breakfast at 1.00 in the morning'.*[39] By the 1970s women were assuming responsibility for far more than catering. In 1972-73 women were given charge of that all-important male commodity beer, when they were appointed managers of the student bar in the new union building.[21] Within months, the Cat-A-Mountain of 1974 was celebrating for the first time the appointment of women to the key positions of SU Secretary and Treasurer.[38]

The unrest of the mid-1960s saw a call for a relaxation of the segregation rules. In 1966, for the first time, the sexes were allowed in each other's rooms, although only for a four-hour period on Saturday and Sunday

13.7 Student Union Committee, 1973-74. Back row from left: Miss M. Spratt (Ladies Rep.), B. Montocchio (NIPH), R. Brown (First Year Rep.), D. Walker (Third Year Rep.), N. Sprackling (Sports Secretary), A. Buchanan (Bar Manager), N. Simkin (Social Secretary). Front row from left: T. Benson (NUS Rep.), Miss S. Rooney (Secretary), P. Heydon (President), Miss S. Robinson (Treasurer), P. Shaw (Senior Treasurer), P. Lacey (Canteen Manager).

afternoons.[8] Christine Justice described the concession as 'a little tiny step', but nonetheless indicative that 'a change was coming'. That change was not quick to come. As late as 1970 the female students were so heavily guarded by locks and wardens that, as Ros Forrester reported: *'We could neither escape nor let anyone in.'* Things got progressively better as women grew in numbers and ground-floor rooms with their obliging windows meant that locked doors had little meaning.[21] By this time male visitors were meant to leave the women's hostel by 10.30 p.m.[40] However, in 1975 the decision to suspend temporarily two students for spending the night together resulted in an orderly overnight sit-in in Ward Hall. Two years later, in the last year of Kenney's principalship, the issue of unrestricted visiting hours had still not been resolved. The students considered existing rules 'imposed an unwarranted restriction on the freedom of individuals, implied a form of moral censorship and could be regarded as treating students as minors'.[41] Perhaps, Peter Luscombe should have the last word on the matter. At one meeting of the Governors, when the issue of unrestricted visiting hours was discussed long and hard, he quipped: *'It seems to me all the business about a rather futile thing comes down to one point, that, if they [the visiting male students] can't get done what they want by 11 o'clock, then they should not have been there in the first place.'*

If Kenney encountered endless difficulties in his dealings with students, then it was as Farm Director that he came into his element. It has been said that his ambition lay more in farming rather than in education.[42] No one can deny he was at his happiest issuing daily orders concerning the Farm [24] and walking the fields every day. As Principal of the Dorset Farm Institute, he had been closely involved with farming policy there and he saw no reason why his role should change at Harper Adams. Indeed, the Principal's Reports often contain more on farming matters than issues educational. Parties of farmers, dignitaries and members of the Women's Institute visiting the College became very much a feature of the Kenney years as he showed off the Farm as a fine example of modern enterprise.[43] He was very keen that the Farm should do well and represent progressive ideas. However, his plans to modernize often rode rough-shod over the feelings of others with longer experience of the College. At the very heart of the problem was the question of who had responsibility for the day to day running of the Farm. Increasingly the Principal asserted his authority [44] and there was conflict.[43] Doubts were expressed as to the wisdom of such a policy, but it was not until Kenney's retirement that the Farm Manager was once again given full authority.[45]

Nationally, throughout the 1960s, farming was moving away from subsidies. Criticism of the Government's mounting subsidy bill had grown in the 1950s in the face of world-wide glut. The Agriculture Act of 1957 had in effect put the brakes on subsidization without abandoning the 1947 Act. The 1957 Act introduced the principle of standard quantities. The Government would pay subsidies in full up to a certain tonnage or volume, but anything over that amount would receive a much-reduced payment. As a concession to farmers the Government undertook a three-year basis for fixing prices and agreed that the guaranteed price of a commodity would not be reduced by more than four percent in any one year. Allied to this move away from straightforward price guarantees was the increased availability of improvement grants. Thus a favourable climate for farming was guaranteed for the foreseeable future and this policy continued, with minor modifications, throughout the 1960s.

Farming in the early 1970s was not in such optimistic vein. On 1 January 1973 Britain entered the European Economic Community

(EEC). There was a good deal of uncertainty as to how entry would affect British farmers. Control of prices and policy passed to Brussels and a completely new system of support was introduced. In the past, food costs had been kept down by paying low prices to farmers, who in return were assisted by subsidies. In contrast, within the EEC, support for farmers came from the consumers by means of high market prices. Intervention Boards took the surplus off the market when prices dropped below a certain level. The EEC's target prices were much higher than the minimum guaranteed prices in Britain. During the five-year transitionary period British food prices gradually had to be brought into line with those prevailing in the existing member countries. Thus, in the event, entry to the EEC worked initially to the advantage of British farmers with higher prices for cereals and small increases in other commodities. However, not all was smooth sailing. The early 1970s was also the period of the Oil Crisis, when Arab oil producing countries forced dramatic rises in the price of oil, which consequently had a knock-on effect on a wide range of items used by farmers from fertilisers to machinery. Amid high rising inflation the costs of production escalated.

Kenney was very much caught up in the optimistic mood current in farming in the 1960s. Expansion and progress were to the fore. Within months of his arrival he had drawn up an all-embracing plan for the Farm, one which would change the appearance of the land considerably. He viewed the Farm as being in two parts. South of the Shrewsbury Road he considered most suitable for arable production.[46] The fields here were to be rationalized to increase the use of mechanization. By means of hedge clearance, the relocation of field boundaries, drainage and the installation of access roads the land would be turned into a workable block of around 120 acres, subdivided into fields of around twenty acres.[47] The skills and experience Kenney had gained working for the War Ag, when he had responsibility for draining and farming 3,000 acres in Bedfordshire taken over for reclamation and improvement, were put to good use.[48] As Fordham said: *'Opening up the farm was a major achievement. He did put roads all round the farm so that you could drive all round the farm on firm roads.'* At the same time the Farm received a logical numbering system for its fields.[49]

To the north of the Shrewsbury Road the land was given over to grazing. Here the fields were considered of a workable size, but there was a need for drainage [46] and some realignment of boundaries.[50] The major programme got underway in 1963 with a grant from the DES and was still in full progress two years later.[51] Proudly, Kenney stated that, despite much land being taken up for new buildings and roadways, altogether some eight acres of land had come into cultivation as a result of the programme. By 1966 Kenney was declaring 'the new road system and the replanning of many fields including the elimination of hedges and ditches on the main arable block has really paid dividends this season'. In the drive towards modern efficiency, many looked back with regret, the new-look fields receiving the nickname 'Kenney's prairies'.

Kenney did not falter in the march on improvement. He also began an extensive programme of irrigation. By 1965 an irrigation system provided water to every field from a surface reservoir holding over two and a half million gallons.[51] Kenney even proposed that the reservoir should be stocked with fish and a staff fishing club formed.[27] A centralized storage and mixing and milling unit was in operation by the autumn of 1966.[52] In 1976 further improvements to the grain storage took place to the sum of £10,000. With the improvements in field layout came a resulting increase in mechanization, often by means of contract. There was, however, the need for a better machinery store and an umbrella-type building was proposed in 1969 as an extension to the farm machinery workshop.[53] This drive to efficiency was not without its casualties. Employment on the Farm was drastically reduced during the Kenney Era. In 1963 one manager, three stockmen, three tractor drivers, one general worker and one mechanic were employed. In 1971 three full-time workers including the Farm Manager and one part-timer were employed and supplemented with casual workers and students during the vacation.

In the interests of 'modern farming techniques and teaching requirements' Kenney urged the Governors to acquire more land.[50] Unfortunately, limitations on finances meant that not all bids were successful, but slowly the acreage grew as land adjacent to the College fields became available and was bought. There was the purchase of three and a half acres of Flatt Pit Farm in 1969 [54] and of twenty-nine acres to the north of Black Brook in 1970. In 1971 the Farm amounted to 336 acres.[55] The purchase of the land beyond Black Brook led to an extensive drainage programme of 150 acres of land, accompanied by an improvement scheme widening and straightening the brook, financed by a grant from the DES.

In 1967-68 the Farm was confronted by the serious threat of foot and mouth disease. These were bleak days when the local countryside was lit up with bonfires of carcases. Shropshire was one of the worst affected counties with the value of animals lost to disease or slaughter in the county amounting to around ten million pounds.[56] Fordham remembered: *'It was terrible. I remember coming back from Shrewsbury and seeing the fires on the farms round about where they had been burning cattle. You could see them blazing in the night. [...] It was heartbreaking.'* The College escaped the outbreak, but all around neighbouring farms were not so lucky. The first outbreak in Shropshire was registered on 25 October 1967.[57] College life was disrupted with the Autumn Term of 1967 ending sixteen days earlier than planned to ensure that students could spend the holidays at home.[58] Farm visits, livestock demonstration classes, the Easter vacation trip and the Harper Adams Club weekend were all cancelled. Sports and social events were seriously curtailed [58] and students living off campus were accommodated on site.[43] The disruption lasted throughout the first six months of 1968.[12]

If the Principal's main interest lay in the College Farm then his passion was for the dairy herd. Although Price had devoted attention to the herd, there was room for considerable improvement. This was the age of the large herd and parlours and associated automation. Kenney, with his background in the West Country, wanted to make the College herd the profit-making core of the Farm and indeed, he did build up 'a formidable herd'.[42] In Kenney's first year at the College the dairy cows numbered fifty-three. In the year of his retirement, 1977, the number had risen to ninety-two. By 1977 capacity had been reached in the cubicle housing and milking parlour and there were severe problems with slurry storage despite the introduction of a new slurry system in the early 1970s.[59] In early 1977 the decision was made to undertake a cull of the cattle, which with a total dairy herd of 159 and twenty beef steers stood at 179. In 1967 a new calf rearing, young stock and beef unit had been completed as a result of conversion work. It had the capacity to handle 140 head of cattle from calf to yearling stage. As a consequence beef numbers had risen from twenty-five in 1963 to a peak of seventy-six at the close of December 1974.

On arrival Kenney saw that the dairy herd was desperately in need of a new yard and parlour system. The old cowshed he condemned as out-of-date and 'grossly wasteful of labour'. As a consequence the dairy enterprise was

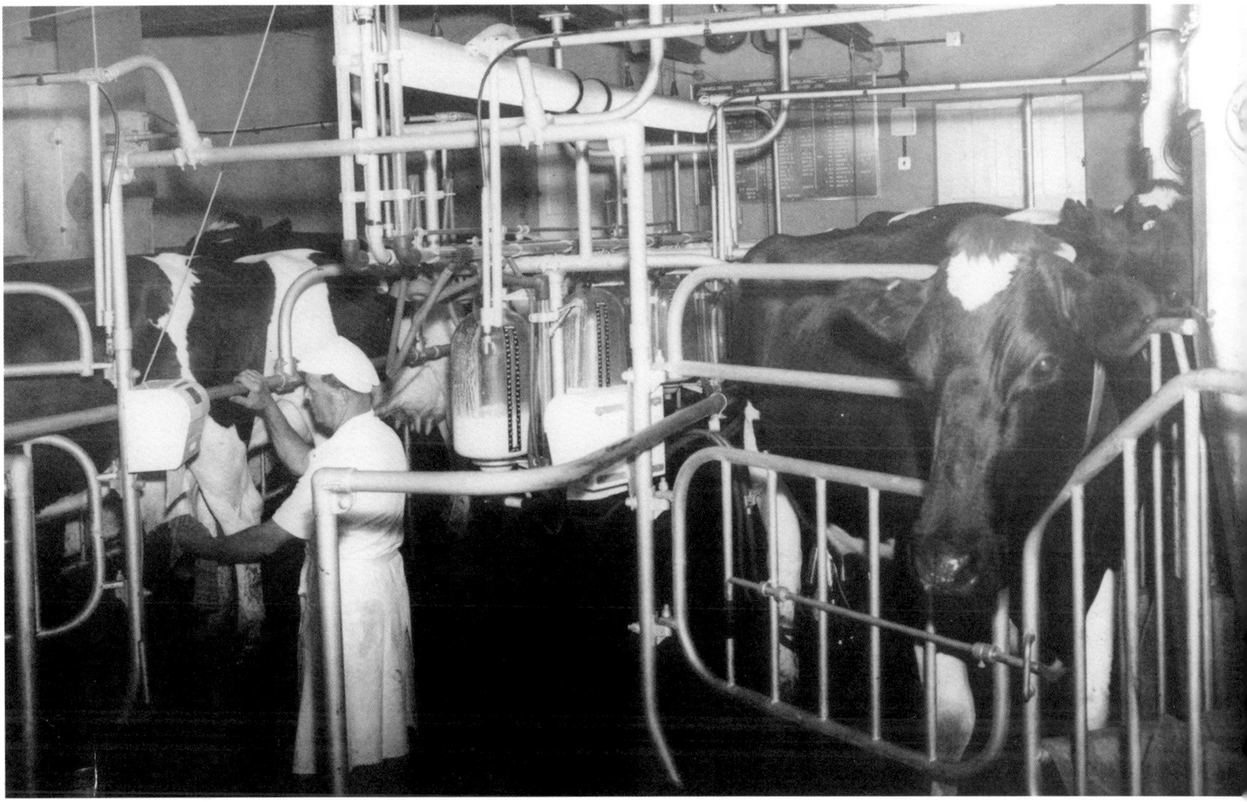

13.8 Milking in the new parlour: Shropshire Star & Journal.

barely paying its way. The Governors demurred at the costs involved, but Kenney's plans for a single breed herd were accepted.[50] By 1964 most of the Ayrshires and cull Friesians had been sold and a respectable Friesian herd remained. That year Kenney was granted his wish for a new parlour. Milking first took place in the new herringbone parlour on 5 April 1966. With the new parlour came greater efficiency in labour, as milking could be carried out by one man. Unfortunately, progress was jeopardized by the health of the herd. Brucellosis was diagnosed in the herd over the summer of 1965. The problem of loss of milk sales was made worse by the fact that the College had now to purchase milk for domestic consumption. It was to be several years before brucellosis was totally eradicated, but in 1969 the College was at last registered with the Accredited (Brucellosis) Herds Scheme.[60] Milk production was to increase impressively. In 1969-70 the figure was 60,170 gallons and in 1974-75 it rose to 92,084 gallons. In 1972, for the first time, annual sales per cow and heifer had reached over 1,000 gallons. This had been achieved with no additional acreage devoted to the dairy enterprise.

By the early 1970s Kenney was seeking to expand the dairy herd and consequently needed improvements to both milking and accommodation arrangements. In 1973 the parlour was altered to a 5/10 system [61] and more importantly, the old straw-yard system operating in three separate covered yards was to be abandoned in favour of a new cubicle house for ninety cows, completed late in 1974.[62] The additional animals put pressure on the slurry system and subsequent improvements took place in the form of a Howard slurry store unit and a new lagoon.[61]

Field experiments continued at Harper Adams throughout the Kenney years and grew in number and range. Most were undertaken by NIAB and in 1972 the Governors agreed that the acreage under the control of NIAB could be increased from twenty to twenty-three acres with the establishment of the College as one of five regional centres in England and Wales for extensive grassland and clover trials.[38] Other pioneering work was undertaken by Maurice Eddowes in the area of herbicide use. Originally the work had concentrated on maize which required special considerations, but in 1962 Eddowes began trials on the potato.[42] Over a period of several years, comparisons were made between the use of contact and residual herbicides as compared with traditional cultivation methods. These were exciting but taxing days, as Eddowes recalled:

> *Our new method was to put the potatoes in ridges, wait until they were just about to emerge and then we put on these pre-emergence sprays. [...] One operation spray was all you need do. But, of course, there were snags, because at that time the chemicals were relatively expensive, whereas farmers were arguing that cultivation was relatively cheap. [...] We pioneered the concept that it wasn't necessarily the physical effort. It was the cost and convenience and it was very difficult to quantify in financial terms convenience, but everybody knows what convenience is. [...] Again, we were treated at first with suspicion. Some people thought that the chemicals would be toxic and harm the potatoes. Others simply said it was unnatural.*

Chemical companies, such as Fisons Pest Control, would supply free chemicals and basic knowledge of the product and the College would provide the test patch and the labour.[42] The work was carried out in association with ADAS and the Potato Marketing Board and it did much to promote the College nationally. Television appearances, articles and conference talks all got the College's name before the public. By 1970, as a result of this pioneering work at Harper Adams and other centres around the country, the use of herbicides on potatoes was universally accepted 'on the grounds of costs, better labour organization and overall results'.[42] By the mid-1960s Eddowes had extended his interest beyond maize and potatoes and was investigating the use of chemical weed control on sugar beet. These three crops remained the main area of investigation, although in the 1970s research began on the use of whole crop cereal silage.[63]

The piggery remained a separate unit from the Farm and throughout the Kenney years remained in the capable hands of Peter Luscombe. On his retirement in the summer of 1977 control of the pigs passed to the Farm Manager, Tony Evans-Fisher, who took over from Fordham at this time.[45] The earliest significant change in the piggery in the 1960s was the move to indoor farrowing. Alterations began in 1963 and by the end of that year thirty sows were housed intensively and thirty remained in the fields.[64] In 1967 the decision was taken to bring all in-pig sows onto concrete. In February 1972 farrowing was improved after completion of a crate house for sixteen sows at a cost of around £4,000. By this time the breeding herd had expanded from sixty to 120 breeding sows.[65] At the end of the Kenney Era the number was 145.

The arrival of the Post-Diploma course in pig technology in 1966 encouraged the expansion of the piggery in order to create a reasonably sized commercial unit. Extra fattening accommodation was a high priority and in 1966 a range of six all-enclosed feeding houses, each for eighty pigs, was erected.[6] The stock was increased to utilize the facilities. In April

1968 the total herd numbered 926. In 1973 the figure stood at 1,237. In 1972 parts of the older buildings were demolished to make way for a new weaner section, completed at a cost of around £9,500 in 1973.[65] The story of the Kenney years was very much one of piecemeal development with only alterations and small-scale work taking place, [59] although the need for an entirely new unit away from the existing site was recognized.[66] The growing problem of high feed prices and low pork prices ate into profits, limiting plans for major expenditure.

Post-diploma teaching in pig technology also emphasized the need for better experimental and investigation facilities. In 1966 it was agreed to build a new sixty-four pig experimental house. This enabled pigs to be fed in groups, allowing precise trials on different foods and additives to be accurately assessed.[19] Research began in July 1967 and was primarily undertaken as post-diploma

13.9 Feeding pigs in the Pig Experimental Unit: Shropshire Star & Journal.

projects.[67] Most of the experiments, as before, were to do with feeding trials and kept abreast of market developments. In 1976, for example, the use of growth promoters was being investigated. One of the most interesting investigations was into the use of fish silage in place of fish meal, which was becoming more expensive. The idea originated with Luscombe.[68] Luscombe's inventiveness was immortalized in a poem in the Cat-A-Mountain of 1969. It tells of the farrowing of one Cissy, who has been fed on 'Luscombe Bill's fearsome swill':

Outside, the neighbours all a titter
Laid odds upon poor Cissy's litter.
Alas, alas alack-a-day
Their estimates went all astray,
For Cissy with alarming frequence

Produced the following in this sequence:

Two knives, a fork, a mouldy date,
A tin of beans and half a plate;
Then at this junction - hardly nice,
Passed swiftly into Paradise.

The fish offal was pureed with formic acid and proved as good as the fish meal, apart from the risk of tainting the carcase.[67]

Last but not least, the NIPH in the 1960s was still a powerful force, but unbeknown to many the seeds of decline had already been sown. The serious problem in enrolment of poultry students not only weakened the NIPH's financial position, but also undermined its core role. Allied to this was the growing isolation of education from the Poultry Industry. At the beginning of the 1960s all were aware that in order to attract support a major reorganization of the plant was called for, but uncertain finances and the proposed move of the College to DES control left decisions on hold. However, good order was maintained and the status quo of the flock preserved. Those extensions and modifications which did take place were all directed towards intensification. By 1964 the bulk of the plant was intensive, even the rows of rearing pens with Buff Rock and Light Sussex hens, which had been such an attractive feature in the field in front of the Administrative Building, were removed.[64] Not only did these developments bring the NIPH closer in line with Industry, but higher productivity and better cost cutting with a reduction of labour offset declining profit margins. In 1961 the stock number was given as 12,891 and employees numbered seventeen, including some estate workers. This process continued. In 1973 the stock number was given as 20,090 and the number of employees as three full-time and two part-time. The experimental programme continued, but with funding problems it took the form of ad hoc studies rather than a set programme.[64]

The highlight of the 1960s was the opening of the Poultry Husbandry Experimental Unit in 1966. A separate research unit of national importance had been the ambition of Dr

13.10 Opening of the Poultry Husbandry Experimental Unit by the Right Hon. Fred Peart, Minister of Agriculture, 20 September 1966: Newport Advertiser.

Temperton as early as 1957. Now, in December 1964 the British Egg Marketing Board (BEMB) invited the NIPH to set up a self-contained research unit to examine practical husbandry problems of the commercial egg producer. The decision was a credit to Dr Temperton and to the standing of the Institute. Temperton would act as Director and would personally overview the research and a Senior Scientific Officer and research assistants would be employed. Dick Wells became the officer in charge.[8] A man of high reputation in the Poultry Industry, Wells was to be appointed Temperton's successor on the latter's retirement in 1974. The BEMB would give a capital sum of £84,000 and an annual grant of around £13,000. The College rented out three acres, close to the Rabbitry and Waterfowl Section, to the BEMB at a peppercorn rent.[69] The unit was to consist of a laboratory and staff offices, flanked by service buildings, two controlled environment houses for rearing and accommodation for laying stock, both battery and deep litter.[6] The Unit was officially opened by the Minister of Agriculture, the Right Hon. Fred Peart, on 20 September 1966.[31]

The Poultry Husbandry Experimental Unit never achieved a firm financial footing. Problems intensified when in 1969 the Government made the decision to wind up the BEMB. The BEMB ceased in April 1971 and the Unit passed to the College.[70] Interest on seventy-five percent of the Board's investment capital of £180,000 was given over to the administrative costs of the Unit. This amounted to around £13,000. However, in 1970 the Unit was already needing approximately £20,000 a year to keep running. Industry gave little support and in 1973 the allocation of all interest from the BEMB's investment capital did not improve the situation greatly.[71] Nonetheless, in 1976 the financial situation was felt to be sufficiently good for the BEMB Trustees to reach the decision to continue funding with a grant of £15,000, effective from 1978.[72]

With increasing financial worries for the NIPH as a whole, it was considered that a modernization programme of the plant was the only way to survive commercially. Obsolete and decrepit buildings dating from before the Second World War were still in use, but it was agreed to clear these away.[73] A five-year building programme was prepared in 1969 to extend the system of housing.[4] An investment of £40,000 was to be made, including grant money from the DES.[70] A heated deep pit battery unit for 3,000 laying birds and an intensive rearing unit for 5,000 table poultry were completed in 1970. By January 1973 a further laying battery unit, this time for 10,000 birds, was in use. However, the modernization programme was hampered by the fluctuating economic situation of the Poultry Industry with its associated repercussions for the trading account of the NIPH. A modest profit had been made in 1974, a high profit the year before and a large loss the year before that. Finances had to be juggled as wages and feed prices rose, egg prices rose [74] and then dropped and the needs of instruction clashed with those of commercial productivity with several uneconomic units having to be maintained.[70] With some stabilizing of the financial situation,[75] as the improvements began to take effect, and with the continuing support of the Governors, the building programme continued. In 1975 a second deep pit laying house was erected.

On 3 November 1976 the NIPH celebrated its Golden Jubilee by inviting George Dodd, Chairman of the Governors, to plant a maple tree outside the teaching block.[76] Three of those who had attended the opening ceremony were present that day: Frank Dudley, the Institute's first statistician, who had retired in 1971 after forty-five years of service,[77] Albert Rowley and Cyril Block, who had been Bursar from 1946 to 1963.[78] The fifty years had seen a chequered history, but despite the recent difficulties, there was a spirit of

13.11 Planting a tree in honour of the NIPH Golden Jubilee, 3 November 1976. From left; Dick Wells first, Principal Kenney second, Mr & Mrs Frank Dudley third and fourth, Albert Rowley fifth, Willard Moore sixth, C.W.G. Block seventh, Mr Roland Ward eighth, Mrs & Dr Harold Temperton ninth and tenth, Dennis Doody, Head Gardener, eleventh, George Dodd, Chairman of Govenors, planting the tree: Roger Price.

optimism. Kenney's fears that the NIPH would have to close in 1973 as poultry student numbers plummeted had not materialized and numbers, although still dangerously low, now showed signs of slow improvement. At the time of the Jubilee they numbered nineteen. Furthermore, the plant had been much improved. Three months after the celebrations, in February 1977, a report on the NIPH by Dick Wells showed an up-to-date facility.[59] The main commercial enterprise centred on two deep pit units and three smaller units. In all, battery birds amounted to 27,250. The majority of replacement pullets were reared on site in a double-storeyed deep litter unit.

Table poultry were confined to a specialist broiler house and annually 27,500 birds were produced in batches. Turkeys and ducks were ancillary concerns.

Reginald Kenney retired in 1977. After a very few years of retirement, he died in 1986 at the age of seventy-three.[79] Harper Adams on his departure was at a significant point in its history. During the Kenney years the College had seen a seventy-three percent growth in student numbers from 232 in 1962 to 402 in 1977 [80] and had absorbed dramatic changes in its regulation with the transfer of responsibility to the DES and in the nature of the very courses on offer. Even its appearance had altered greatly with expansion beyond the Caynton Road. The relentless cycle of challenge and change, although slowed down by the savage financial climate of the 1970s, was not to stop. Now, it was the turn of a new Principal and it was to the greatest benefit of the College that the man chosen thrived on challenge.

The Harper Adams Club[a]

The first official gathering of the Harper Adams Club, a meeting of around forty people, took place at the Grand Hotel in Birmingham on 30 November 1920. An earlier club had been started in November 1913, at a meeting in Birmingham which coincided with the Fat Stock Show. The First World War effectively ended the meetings. Club membership enabled ex-students to maintain the traditions of Harper Adams, but more importantly, in these early days, it allowed them to keep in touch with progress. They were invited, for example, to inspect experiments at the College and so obtain early and accurate knowledge of advances.

Members were to meet twice a year, with one meeting at Harper Adams and the autumn meeting to alternate between Birmingham and London. The first meeting of the Club at the College took place on 21 July 1921, when the College Jazz Band performed and there were recitations, sketches and songs. By the 1930s low attendance at meetings was giving cause for concern and in 1932 it was decided to hold future open days in association with the College Sports Day. The event proved a success with forty ex-students attending on 9 May 1933. Essentially, reunions were simply for enjoyment, but pleasure was mixed with fund raising for items such as library books and sports cups.

General Club activities were suspended during the Second World War, but started again soon afterwards with a meeting in Birmingham in 1946. A war memorial fund was set up, to which Club members contributed generously, eventually raising £116. The original intention was for this money to finance a new sports pavilion but there is no record of this being built. Instead the Club paid for the refurbishment of the College library, including a memorial plaque dedicated to those students who had died in the Second World War.

The Club came into its own during the Price years. The Principal and Mrs Price injected their own style into the Club. As one person attending a reunion remarked: *'They were*

13.12 First gathering of the Harper Adams Club, 21 July 1921.

amongst us all day - ever their charming and delightful selves'. By 1950 the summer reunion at the College was once again an event on its own, with cricket and tennis of an afternoon, followed by tea and the General Meeting and a dance. The next year a series of Easter refresher courses was initiated, with lectures and demonstrations over an entire weekend. In 1956 Harper Adams Day, usually held in June, was extended to a weekend in July and proved very popular indeed.

The Refresher Course, held in alternate years, initially remained, but in the late 1950s it merged with the Harper Adams Weekend and the one event was held in April. Gerard Poulter, student 1954-56, remembered the format: *'The first full weekend reunions had a conference style with formal lecture sessions interspersed with social events - farm walk, dinner dance and children's party. This was an opportunity not only to catch up with developments at College and in agriculture generally, but also the interaction between the different age groups was thoroughly enjoyed by all.'* There was a strong family atmosphere, due in no small part to Mrs Price's encouragement. From 1953 until 1965 autumn meetings continued to alternate between Birmingham and London, to coincide with the Smithfield Show or the Dairy Show. The meeting would be followed by a dinner and frequently a theatre visit and was usually attended by approximately forty members.

On his arrival in 1962 Principal Kenney proposed that the work of Secretary/ Treasurer should be continued by a member of the College staff and that office facilities should be made available to the Club. Kenney was a keen supporter of the Club, regularly attending meetings, promoting Club initiatives and encouraging members of staff to attend Club events. The introduction of a wine and cheese evening in June 1971 to recruit final year students was very successful and continued until 1988. The London/ Birmingham meetings ceased in 1965 and from 1966 an autumn dinner was held each year at a hotel in Newport, to coincide with the Rag Dance at College.

At the Club's autumn meeting in 1957, it was suggested that there should be a meeting place for old students at agricultural shows. By 1958 the Farmers Weekly had agreed to allow old students to use their stand as a meeting place at the Dairy Show, the Royal Show and Smithfield. In 1963 the first Harper Adams Club/College tent was provided at the Royal Show. On 1 July 1969 the pavilion was opened by the Earl of Bradford, Chairman of the Board of Governors. The cost of £2,200 was funded jointly by the Club, commerce and the College, although theoretically owned and staffed mainly by the Club. The pavilion was refurbished by the College in 1988 with the assistance of a donation of £200 from the Club and from then the College assumed full financial responsibility for its upkeep and staffing. It continues to be important, not only as a meeting place for past and present students, but also for marketing the College.

Club membership numbers were strong, but in the mid-1970s there was a tendency for year groups to form their own splinter associations such as the Glee Club, Sodbusters and others. As support dwindled, in 1979 the autumn dinner disappeared from the Club's annual programme of events. By the 1980s there were fears that the splinter groups were posing a threat to the Club itself. During 1981-88 the Club's social diary was dominated by the Club Weekend. It was generally a very successful event and included the Annual General Meeting and a buffet dance, and a farm walk on the Sunday morning. By the late 1980s the Club was in crisis. The career and family demands on Club officers meant that they were unable to give the time to organize the events which they would have liked or which were needed. Similarly, with the massive expansion which was taking place at the

13.13 Opening of the Harper Adams Pavilion at the Royal Agricultural Society showground, Stoneleigh, 1 July 1969.

College, it became almost impossible for staff to administer Club affairs in odd moments of spare time. The President Elect, Bob Ackroyd, student 1971-74, had a meeting with Principal Harris to discuss the situation. As a result, the Club rules were radically revised and in October 1991 a part-time co-ordinator, Julie Brook, was employed by the College to maintain a computer database of ex-student addresses and to assist the Club. It was at this point that the Club began its present revival. Under the revised membership system numbers soon increased to over 3,000. The appointment of the Club Co-ordinator was formalized as Alumni Officer within the Liaison and Marketing Unit.

The 1990s has seen the birth of other Harper Adams alumni associations, notably Harper Adams in Ireland. This finally came to fruition when the first meetings of Harper Adams in Ireland were held in April 1995 at Hillsborough and Omagh. Harper Adams in Ireland has been particularly active in student recruitment, publicity for the College and assisting with contacts in the Industry. It was not long before a Welsh branch of the Harper Adams Club came into being. Harper Cymry had its inaugural dinner at the College in autumn 1997. It has already started to make itself known with successful evening receptions for past, present and prospective students and other contacts at the Royal Welsh Show and the presentation of an annual award to a Welsh Harper Adams student on the first day of the Show.

Over the years the Club has made various valuable financial contributions to the College. However, the major contribution of the Club has been the immense support given to the College in a whole range of ways from employing students on pre-college practical years and providing sandwich placements, to advising and assisting with careers. They recommend their sons, daughters and other young acquaintances. Amongst Club members there is a great fondness for the College and concern for Harper people. The Harper Adams Club will play a full part in the College's Centenary celebrations with the launch of a Centenary Fund, its most ambitious fundraising plan so far, the naming of a new rose cultivar 'Harper Adams', to commemorate the occasion, and an Open Day for all former students and staff in July 2001.

Chapter Fourteen —

Diversification & Degrees 1977-94

Many of those involved with Harper Adams in the mid-1970s felt the College to be standing still. Beyond Harper Adams national developments in agricultural education were taking place, but the College seemed uncertain of its direction. As one of a small group of direct grant agricultural colleges in a time of considerable Government reorganization in Higher Education, it was necessary for Harper Adams to assert its presence. It was not simply a question of definition of role, but rather one of anticipating the best route to take and fine-tuning that route to make progress. It required confidence and firm leadership to ride the white waters of Government reports and papers. The fact that the man chosen to be the new Principal so clearly perceived 'the speed of the change in Higher Education' to be not simply 'daunting, but also exciting' was all important at this point in Harper Adams' history.[1] It was the essential quality to be added to his many existing attainments.

Tony Harris was well known to the College through his earlier Shropshire connections and his arrival was greeted enthusiastically by the staff. Resigning as Lecturer in Crop Production at the College in 1958, he became Vice Principal of the Shropshire Farm Institute at Walford and in 1967, following a Kellogg Fellowship at Purdue University in the United States he took up the post of Principal of Merrist Wood Agricultural College, Surrey.[2] As Principal of Harper Adams, and with such experience of college management behind him, he recognized the need for a 'hands on' approach to management: *'You do need to be very closely in touch with what is happening, whether it's students as individuals, or staff as individuals or the Farm or whatever it might be.'* [3] Tony Harris was to give firm direction to College strategy, but more importantly, he was a people's man. He combined 'strong ambition and a will to achieve with a genuine modesty and approachability'.[2] His Vice Principal, Wynne Jones, wrote: *'His ability to remember the names and backgrounds of students present and past is almost legendary. This ability derives from the fact that to him each individual*

14.1 Principal Tony Harris 1977-94: Shropshire Star & Journal.

student is important, he makes time to get to know them, his door is always open and he is accessible to students. [...] All who met him were left with the impression that you were the important person that he wished to speak to.' [2] The students were among the first to benefit from his eminently reasonable approach to problems, as he moderated outdated rules and brought in an age of greater liberality. Significantly, Tony Harris ensured that there was student representation on the Governing Body within months of his arrival. In 1988 students were accorded full Board membership.

Tony Harris also possessed that rare ability to allow people their head. As he so succinctly put it: *'I don't believe any one person takes credit. A principal can only create the environment for others to achieve.'* He worked with people, recognizing the skills of those in his team: *'He was open to ideas from other people and was glad to take them on if they would benefit the institution.'* [4] The downside was that, as a perfectionist, he expected his colleagues to live up to the high standards that he set himself. His commitment to Harper Adams twenty-four hours a day was a hard act to emulate. Ripples there inevitably were, but the confrontational atmosphere of the past years disappeared virtually overnight. As Peter Watson Jones, Chairman of Governors 1985-90, recalled: *'The interchange of people between the various offices and creating a team must be highlighted as a consolidating factor. [...] The College was unhappy in the Reg Kenney Era, but out of those problems grew a greater trust and growth.'*

14.2 Investigating soil forces on a cultivator tine using strain measuring equipment, c. 1970: Power Farming.

Described by the Earl of Selborne, an acknowledged leading figure in agriculture, as an inspiring leader and 'a highly effective promoter of the interests of the College', all recognized that Tony Harris's brand of dynamism could not easily be overlooked. Alan Costley, Head of Education Development and Quality Assurance, recalled the Principal was driving for progress right from the very beginning: *'When Tony Harris arrived, he was saying "Why haven't you done this? What are you going to do about that?" [He was] pushing in every direction.'* Management of the College was crucially brought under the spotlight. The Principal sought to move away from a traditional approach to running Harper Adams: *'My view of agricultural education is that it has always been highly competitive and there really is no place for a college unless it realizes it is in business as well as providing education.'* [3] In his eyes Harper Adams had to realize that it had a product to sell: *'[...] Our courses must be marketable - we need to develop them, promote them and sell them with confidence. Equally the customer (student) has to have confidence in them and the industry has to have confidence in the product of the course (student). In this way running a College and running a business have a lot in common.'* [5] The reaction of the real world to the product was paramount.[6] The needs of the market were to be noted and acted upon, to eradicate all the dangers of an educational institution becoming isolated from the Industry that it served *(see photos pages 263-65)*. Peter Watson Jones expanded the point: *'The jobs and careers have to be monitored by the Governors and the Principal, which we do. We say "How are these lads doing in Industry and Commerce" and if they say "They're no good. They don't need this and they don't do that. No more shall we have any of your lot. They're useless", then that course will close because you can't sell those students or they can't find good careers.'*

A good business was never to grow complacent. Tony Harris had every conviction that the College must take control of its future: *'We live in a dynamic environment. Whatever we plan or think internally we must be conscious of and react to outside stimuli.'* [5] In the case of Harper Adams it was essential to take on board the new ideas and developments in agricultural education. Fortunately, the new Principal was a master at understanding Government policy and predicting trends. Wynne Jones added: *'Allied to this foresight is his ability to get it right. These gifts enabled him to ensure that Harper Adams was always one step ahead of the field and in a position to take advantage of any initiatives or opportunities.'* [2] In 1978 his foresight led to the creation of the College's first strategic plan. It was the forerunner of the numerous rolling plans and college strategies of the 1980s. Today the Strategic Plan and the Annual Operating Plans form a pivotal role in College management. Strategic planning was rare in a period when many institutions concentrated on annual budgets alone: *'I felt there was a need to look ahead at the time. [The plan] arose from discussion with the DES and the recognition that Government funding was planned up to four years ahead. The only way to secure one's place within that funding was to inform them of what the College would wish to be doing for several years ahead. [...] It was a response to a recognition of how one had to play the game at that period of time.'* [3] In playing that 'game' Tony Harris was ably assisted by John Roberts, Head of Animal Science, who was heavily involved in the first rolling plan, and subsequently by Lieutenant-Colonel John Taylor, who had replaced Duggie Nixon as Bursar after seventeen years of service, in the autumn of 1979, and Derek Corlett, the Director of Finance.[7]

Reading the strategic plans of these years, a clear idea of the issues confronting Harper Adams emerges and at the same time there is confirmation of how closely the College was in step with Government policy, at best anticipating change and at the very worse rarely slow to take up the challenge. The key issue was funding. Although the Government sought expansion in Higher Education and encouraged substantial growth in student numbers, it inflicted arbitrary cuts in funding and made little provision for additional demands such as increased staff.[8] All this must be seen against a background of high inflation.[9] It was a time of immense frustration as clearly revealed by the Principal: *'A climate which is necessitating an increase in student numbers whilst at the same time severely restricting the provision of residential facilities. A climate which is demanding a reappraisal of the financing and organization of higher education institutions, major upheavals in the structure and content of courses, yet failing to allow the human resources for this to be undertaken.'* [10] In 1978 a Government report was published on the management of Higher Education in the maintained (public) sector.[11] The Report sought to rationalize the Local Education Authorities management and financial control of Higher Education and in February 1982 this led to the creation of the National Advisory Body for Public Sector Higher Education (NAB).[12]

In 1985 the NAB's authority was extended to include direct grant colleges, such as Harper Adams.[13] The College was one of three direct grant agricultural colleges of national standing, which had their capital and running costs financed by the DES. The other two were

14.3 Investigating soil forces on a cultivator tine using the soil tank.

Shuttleworth and Seale Hayne.[11] After 1985 Harper Adams still retained its direct grant status, but it was now an integral part of the national provision for Higher Education outside the university sector. With NAB came student-based funding and the need to perform a complex juggling act for financial survival. The budget for the first time would be based on the number of registered students and a whole new world of jargon was born, revolving around programme weighting, unit of resource and full-time equivalent students. This was in keeping with the Government's move to make Higher Education more cost-effective and to promote the generation of income. Already, a significant step had been taken in 1983 when the Government insisted that all catering and residential activities were to be self-funded for both capital and revenue requirements.[14] Tony Harris described the effect as 'devastating', but 'the outcome of that was it made us think very early on how to run the College totally as a business'.[3] Henceforth, the College facilities had to work to the full throughout the entire year, including vacations. The introduction of short training courses for Industry began in earnest together with lettings to other organizations who wished to use the College's facilities. By the close of the Harris years, clients using College facilities included not only those pursuing short training courses, but also people as diverse as the British Wheel of Yoga and the British Cave Research Association.

Soon, Harper Adams' experience of survival in the world of alternative funding was to pay dividends. Further centralization and rationalization in management and funding within Higher Education came about with the Education Reform Act of 1988 and the creation of the Polytechnics and Colleges Funding Council (PCFC) in 1989.[15] Henceforth Harper Adams ceased to be a direct grant college and like all other Higher Education institutions it became a corporation with a board of governors fully responsible for the mission and direction of the College. This was a significant increase in responsibility for the Governing Body and it necessitated changes in composition to draw in people from wider backgrounds than the academic disciplines on offer at the College. Thus, the necessary experience and expertise to run a multi-million pound business was provided. The college grants system was replaced by an annual contract between institutions acting as corporate bodies and the PCFC.[16] To encourage colleges to recruit more students the tuition fee paid from public funds was heavily weighted, a policy which was to prove very effective in increasing the number of students in Higher Education.

In 1991 the White Paper 'Higher Education; a New Framework' and the accompanying Further and Higher Education Bill of November brought in further rationalization with the amalgamation of the former Universities Funding Council (UFC) and the PCFC.
In 1992, as a consequence the Higher Education Funding Council for England (HEFCE) was formed. At the same time polytechnics were given the opportunity to adopt the title of university. As Tony Harris said: *'[Developments meant] that the College was in an even bigger league, standing with every university as well as the Polytechnics and all the colleges of Higher Education, a situation which was a major change and a bit frightening at first, but one in which we managed to hold our own and one in which the colleges tended to show that the smaller institutions can be efficient.'* Early in the life of HEFCE the Government realized that Higher Education was expanding beyond its planned projection rate (thirty percent) and by 1994-95 the rapid period of growth was being brought to an end by restriction in recruitment above an agreed ceiling for each institution.

The growing need for financial self-sufficiency and cost-effectiveness were strong spurs to the dramatic growth in student numbers at Harper Adams in the 1980s. In the late 1970s student

numbers had reached capacity, with the deferral list of 1978 running into 1981 and the introduction of an embargo on further allocation of vacancies for some courses. Recruitment was restricted by available facilities and academic staff numbers, which stood at twenty-eight in January 1978.[8]
In turn, these areas were unable to expand due to limitations in Government funding. Few considered it was the time for growth, but Tony Harris and his Vice Principal, Ray Mortimer, recognized that a larger institution could better survive the financial storms: *'One could clearly see even in the late 70s, early 80s, that we were going to be more and more dependent on what we earned. Therefore the need to grow and to spread overheads, became acute. There was recognition that the College had to get larger to be cost-effective, to be efficient.'* [3] The advantages would not stop there, the creation of sufficient critical mass to provide expertise for development and specialized functions would be essential for future survival.

Whilst survival for the College would depend on its size and vigour, it had to be viewed against a background in which the number of eighteen-year-olds in Great Britain was set to peak in 1982-83 and thereafter to fall drastically.[5] Between 1986 and 1995 there was a drop in the number of eighteen-year-olds by one third.[17] However, enlarged, the College could offer better facilities and a wider range of courses and in turn attract higher

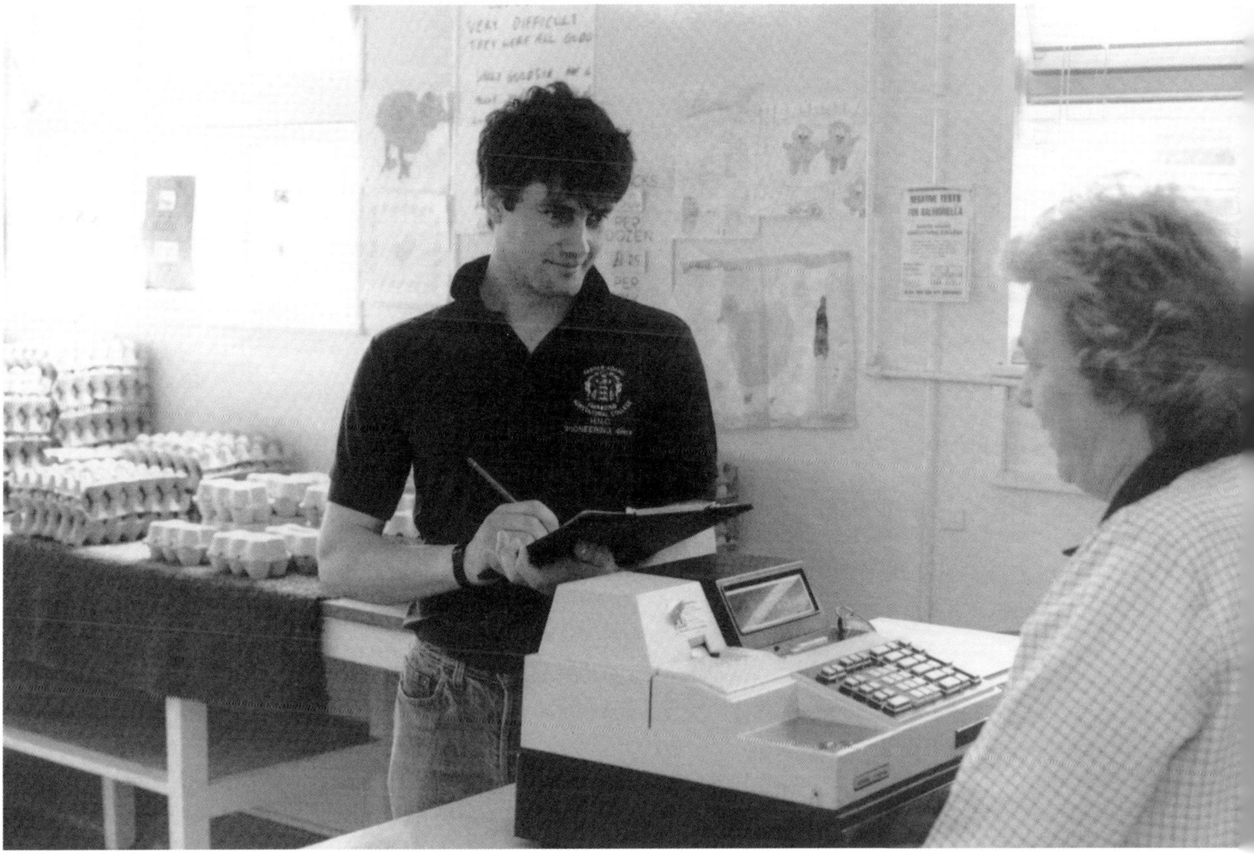

14.4 HNC Agri-food Marketing and Business Studies project reviewing the Harper Adams farm shop, 1989.

recruitment. Fellow institutions, recognizing the situation, took a different approach. There was 'enormous pressure' on the smaller colleges to amalgamate with each other or with other institutions.[18] In the late 1980s Harper Adams' direct grant agricultural college colleagues, Shuttleworth and Seale Hayne, merged with other institutions, namely Cranfield and the Polytechnic of the South West respectively.[19] Significantly, Harper Adams decided not to go down that path and it remains an independent corporation.

Harper Adams entered a period which witnessed the most dramatic expansion in its history. With a juggling of timetables, more efficient use of teaching facilities, the expedient introduction of self-service catering in 1979,[8] heavy reliance on external accommodation, adaptation of buildings for teaching use and a willing hard-working staff limitations were overcome and student numbers were encouraged to rise. For the session 1977-78 there were 402 students. In autumn 1981 the recruitment target of 604 was reached. Henceforth numbers were not envisaged to rise dramatically due to the declining number of school-leavers. However, it was not until the mid-1980s that recruitment difficulties began to make an impact. In 1981 there had been 392 applications for the HND Agriculture course, but by the summer of 1987 this figure had fallen to 211. Apart from the problem of a declining number of school-leavers, the difficulties in farming in the 1980s did little to encourage interest in agriculture as a career.

The effects of these difficulties in student recruitment were moderated at Harper Adams by a positive approach to marketing the College and by introducing a 'broadening portfolio of courses'.[15] By the session 1989-90 student numbers had increased significantly to 751 and there were once more concerns about the accommodation available. In 1990 recruitment across almost all courses was declared to be buoyant with a record enrolment of 882. By the session 1991-92 student numbers had passed beyond the thousand mark, a landmark in the history of the College and the College became 'the largest individual provider of Higher Education for the industries, professions and trades associated with rural land and rural communities'. Enrolments at Harper Adams accounted for forty percent of the national figure for HNDs and HNCs in agriculture and related subjects.[20] In 1994, when the Government's requirement for a standstill in Higher Education numbers began to take effect, the number of students at Harper Adams had reached 1,525.[21] There was also potential for considerable further growth in that only the first and second cohorts of students had embarked on recently introduced courses. It is of interest to note that by 1994 over a third of the students were women. The significant increase in the number of women students had arisen with the introduction of degree courses and greater diversification in course provision and the attendant excellent career opportunities.

The range and level of courses increased significantly during the Harris years *(see photo page 266)*. Courses on offer in 1978 were Pre-Entry, HND Agriculture, HND Agricultural Marketing and Business Administration, HND Agricultural Engineering (new in 1978 and the first HND course in agricultural engineering in the United Kingdom),[14] OND Poultry and the Post-Diploma Crop Protection course. Sixteen years later the picture was very different. There was the Foundation course, which was the renamed Pre-Entry course [22] and which now assisted entry to a range of HND courses, not simply HND Agriculture. There was also the Higher National Certificate (HNC) in Agricultural Marketing and Business Studies, which was introduced in 1988 as a two-year sandwich course for students with Further Education qualifications. It was the first HNC to be offered at Harper Adams.[23] As regards the HNDs on offer, the range covered

14.5 REALM class with Barrie Slatter, estimating the age of a tree.

Agriculture, Agricultural Marketing and Business Studies, Agricultural Engineering, Science (Technology of Food) with Business Studies (in association with Reaseheath College), Golf Course Management (in association with Reaseheath College) and Animal Care (in association with Rodbaston College). Also, degree teaching had arrived with BSc Honours in Agriculture, Agriculture with Animal Science, Agriculture with Land and Farm Management, Agriculture with Agricultural Marketing, Agri-Food Marketing and Business Studies, Rural Enterprise and Land Management, Rural Technology with Business Studies and Rural Environment Protection and BEng Hons in Agricultural Engineering. Finally, there was greater emphasis on post-diploma work, with the Post-Graduate Diploma in Crop Protection and a taught MSc, as well as MPhils and PhDs by research.[21]

Clearly, Harper Adams had recognized and acted upon the move towards 'the multi-purpose countryside', with its emphasis on changing patterns of economic activity in the form of new industries and enterprises.[24] Of all the new developments at Harper Adams the inclusion of subjects such as rural enterprise and land management and also rural environment protection was of the greatest significance. However, these advances were 'deliberate but not radical' and were firmly kept within areas related to agriculture, which was to remain at the core.[3] Despite the lean spell which agriculture was going through in the 1980s, Harper Adams was careful not 'to convey to the customer [the farmers and potential students] that there was a movement away from their prime interest'. Geography and the tradition of Harper Adams dictated continued strong links with the farming community. By 1992, although the number of production agriculture students was still increasing, they formed a decreasing percentage of the total student population.[6]

Similarly, Harper Adams did not ignore the European issue. It was during this period that attempts were first made to develop an European dimension to the courses. By 1988 there was the opportunity to study languages at the College [25] and soon after a language resources room was created in the library.[21] By the early 1990s languages could be studied as part of an HND or degree course. These moves coincided, in 1988, with Harper Adams' entry into various consortia with European colleges and universities, aimed at encouraging staff and student exchanges and the placement and supervision of sandwich students. The European Union provided funding for these developments through TEMPUS and ERASMUS programmes. Strong links were forged with institutions in France, Spain, the Netherlands, Hungary, Bulgaria, Greece and Turkey *(see photo page 150)*. In eastern Europe the expertise of Harper Adams staff was sought to help develop curricula as their education began to cater for a free enterprise economy.[23]

Developing close ties with fellow institutions was not limited to mainland Europe. Broadening the portfolio brought Harper Adams into closer co-operation with other institutions much closer to home, such as Reaseheath, Rodbaston and Pershore. It was in the College's best interests to maintain and extend links with other institutions to ensure access to a broader range of resources and experience.[26] To this day this policy is actively pursued. Such a policy had first appeared in the 1970s with the creation of the Post-Diploma courses.[27] By 1992, as a result of initiatives within the new regional structure proposed by the Further Education Funding Council,[28] the College was actively pursuing new partnerships with the land-based Further Education Sector in the region for delivery of Higher Education courses.[29] Harper Adams was to provide assistance for these courses taught mainly at and by other institutions and, above all, the College was to be responsible for the monitoring and validation of academic standards.[30] However, the major slowing down in public spending placed severe restrictions on Harper Adams' aspirations in this direction.[31]

College policy at this time, in line with Government thinking, not only sought to broaden the portfolio of courses, but also to widen access to Higher Education. The policy of broadening the path of entry to Higher Education beyond the uniform requirement for specific O- and A-Levels went back as far as the Robbins Report of 1963. As Tony Harris wrote in the Cat-A-Mountain for 1991: *'Without doubt one of the most rewarding and exciting features of Higher Education today is the opportunity it offers to young people to make good shortcomings in their school attainments and to mature entrants to upgrade qualifications and retrain.*
The barriers to entry which used to be so evident have disappeared to the benefit of the system and the individual.' Coincidentally, the compulsory requirement for a pre-college practical year was dropped at Harper Adams in the late 1980s, with the exception of those without any practical experience whatsoever. This decision was described as 'regrettable but necessary' to avoid losing students to other centres less demanding on practical entry requirements.[19]

The HND continued to be of prime importance at Harper Adams. In 1988, for example, the College had the largest enrolment of students

Assessing sheep health with Nigel Dodman.

Food stuff analysis with Paul Rose.

Preparing to download design data to the plotter, 1993.

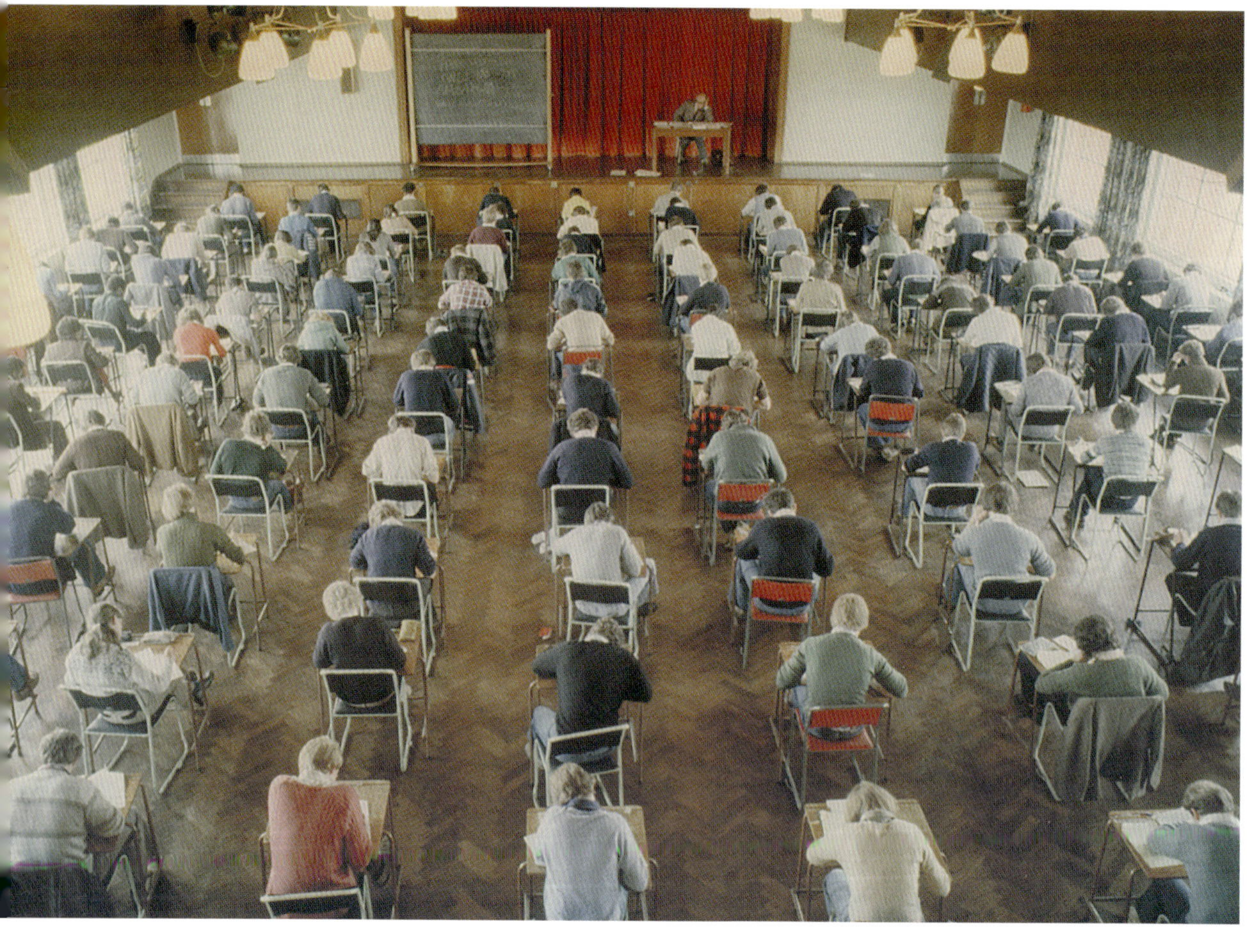

Examinations in Queen Mother Hall, 1980s.

First Graduation Day, 15 September 1988.
Front row from left: Peter Watson Jones, Chairman of Governors, The Earl of Selborne, Principal Harris.
Back row from left: Vice Principal Dr Wynne Jones, Richard Matson, Governor.

Harper Adams Open Day for prospective students and their parents, 1990s.

Marketing Harper Adams at the Royal Agricultural Show, July 1997.
From left: Richard Jopling, Head of Liaison and Marketing,
two prospective students, Ivy Loat, Registry staff.

Visit of the Prime Minister, the Rt Hon. Margaret Thatcher, 7 June 1989.
Mrs Thatcher visits the Soil Hall surrounded by the media.
Ian Gedye and Paul Hemmingway are seen behind Principal Harris.

for HND Agriculture in the country. In 1978 a new approach had been proposed. There was to be less formal teaching and a greater degree of student-centred and problem-based learning: *'What one teaches now is [...] how to seek information, how to use information, how to analyse it and reach a decision and solve problems.'* [3] Such skills were of greater importance for survival and success in 'a less predictable agricultural future'.[6] It was also at this time that the sandwich period was extended from nine to twelve months.[32]

Undoubtedly, the most important development at Harper Adams was the introduction of degree teaching. It was vital for Harper Adams to develop a culture of Higher Education and to move into opportunities for degree teaching and post-graduate work. The Robbins Report of 1963 had looked to a future where particular agricultural colleges might offer degrees and as early as 1966 the College had certain aspirations to introduce a three-year technological agricultural course leading to a degree awarded by the Council for National Academic Awards (CNAA).[33] However, such plans disintegrated in the face of bureaucracy. The idea was revived in the 1970s, but there was hesitancy on the part of Kenney.[4] When, in early 1977, ideas for a degree were eventually submitted to the CNAA, they were turned down on account of insufficient resources.[34] On his arrival Tony Harris, believing in the need for a vocational degree from his experience in the United States, drew on 'pent up energy' and grass-root ideas developed by members of academic staff, such as John Roberts,

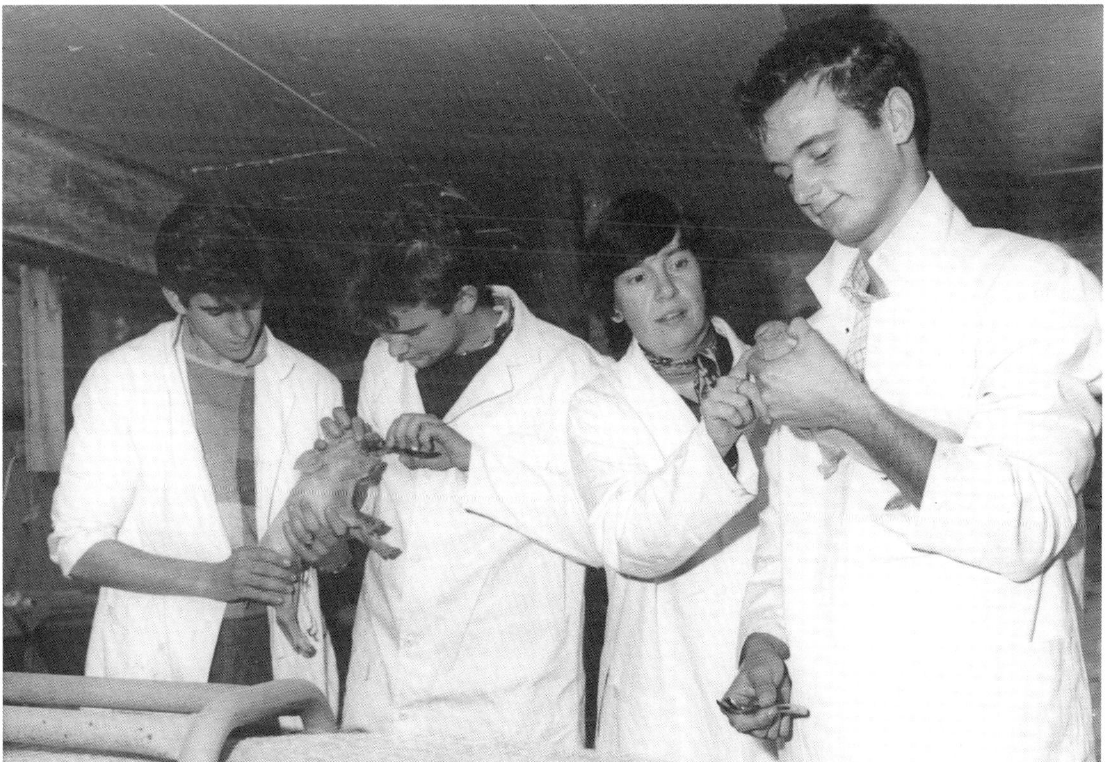

14.6 Practical class with Dr Margaret Parry.

Bruce Cox, Henry Hollinshead and Alan Costley.[3] With a college strategy in place and resources promised by the DES, Harper Adams was able to convince the CNAA of its ability to provide degree level education. Approval was gained in 1980 and a CNAA sandwich degree in agricultural technology was offered in 1981,[35] with the first students graduating in 1985. The Cat-A-Mountain of 1982 hailed the arrival of degree students thus: *'We now have an elitist, more intellectual gaggle of students, who, despite their superiority, hide it quite well and lower themselves to the level of the rest of us. In fact, the only thing that gives away the identity of a BSc are the glasses and executive briefcase!'*

At the instigation of the College, the degree course covered a four-year period and included an all-important sandwich year, which ensured *'a thoroughgoing agricultural understanding which would underpin the science and business principles'.*[3] The course ran in association with the Polytechnic Wolverhampton and students spent one day a week there.[36] The combination of 'expertise and facilities' gave 'strength and quality to the development and operation of the course'.[37] In 1983, at a CNAA Review, Harper Adams was given sole responsibility for delivering and maintaining the academic standards for courses leading to the Council's awards.[38] Approval was also given for the Post-Graduate Diploma in Crop Protection. From September 1987 the CNAA approved the course as an Honours degree and also gave the College delegated authority to make changes in courses validated by the Council. In the late 1980s and early 1990s a range of specializations was introduced to the Degree in Agriculture. Recruitment was buoyant for degrees, with Harper Adams by the late 1980s obtaining twenty-five percent of the intake in the whole university sector for the BSc. Hons in Agriculture.[23] In 1989 the CNAA Institutional Review 'expressed such confidence in the processes adopted [to ensure quality of academic provision]' that the College was given delegated authority for the validation of its courses, 'despite its relatively small size'.[23] On the dissolution of the CNAA in 1992, after lengthy discussion with students and careful consideration by the Academic Board, the Governors sought accreditation for the College's degrees from the Open University Validation Services (OUVS). The rigour adopted by the College under the guidance of CNAA and OUVS led to the Higher Education Quality Council at their visit in May 1994 recommending that the College should apply for its own degree awarding power for its taught courses - an objective which was achieved in 1996.

With the arrival of degree teaching came the return of the Awards Ceremony, last seen at Harper Adams in July 1913.[39] In the intervening years certificates had simply been sent through the post.[40] In the 1980s there was clear support from the students for such an occasion.[41] The first Harper Adams Graduation Day took place on 15 September 1988, with the Earl of Selborne as Honorary Guest *(see photo page 267)*. The attendence was 600. The revival of the Graduation Ceremony led to the creation of the first Harper Adams Honorary Fellowships in 1989. These were awarded to people who had performed outstanding service to the College or to the industries, trades and services served by the College.[42] To date Peter Luscombe (member of staff 1948-77), Peel Holroyd (student 1958-60 and Senior Food Technologist and Agriculturalist with Marks and Spencer), Robin Faccenda (student 1959-61 and Managing Director of Faccenda Holdings Ltd), Peter Watson Jones (Chairman of the Governors 1985-90) and Tony Harris have been appointed Fellows.

The diversity of courses on offer in the early 1990s was only possible with the introduction of modularisation in 1988. Indeed, Harper Adams was one of the first colleges to move into modularisation of its courses. As Alan Costley, appointed in 1988 Education Development

Manager in response to the ever growing demands of curriculum development, explained:

The fundamental principle of modular schemes is that they are based on packages of learning, defined by credit volume and level, which are assessed as soon as they are completed. Successful completion leads to the award of credit which can be accumulated towards an academic award - a certificate, diploma or degree. Modules are therefore the building blocks of awards which can be used in a variety of ways: we started by using them in prescribed courses with no choice, then in programmes with both core and optional subjects, and then - once a large portfolio of modules had been developed in this way - we devised new routes through it to match student needs. And it doesn't end there. Credit can be acquired at different times and in different places during working life according to circumstances and needs so that individuals can earn relevant qualifications through lifelong learning.

By 1990 the programme to convert all courses to a modular structure was complete. The introduction of modularisation changed the tone of the College: *'The students can't enjoy themselves for three years and then work like stink to get through the final exams and come away with an HND, because they have to work continuously to pass the modules studied in each semester. [...] The element of work is more now. [...] Now, it's a long process right the way through.'* [43] The speed and range of change must have sometimes seemed all too much for the lecturers. As Reg Anslow, Head of Crop Production, wrote on his retirement in 1991: *'[In past years] a student was either a Crop Prot, an Agric, an AMBA, a Poult or a Pre-Ent. Now, when a student asks for technical advice it takes half an hour to sort out which project this is, in which module of which semester of which course. I think that ear-tagging of students is the only answer. When both ears are full, with tags for each module passed, the student is freeze-branded with "HND" or whatever.'* [17] To accommodate the modular system it was necessary to move from a term-based organization of the curriculum to semesters. The College year was split into two semesters of sixteen weeks each for study purposes, with the sandwich changeover in mid-February.[15]

As well as suiting individual needs, modularisation assisted the process of widening access to Higher Education by means of continuing education (lifelong learning) and part-time studies. Traditionally part-time studies at Higher Education level had been considered an unsuitable option in an industry dominated by very small units scattered over a wide area, with only low numbers within reasonable travel distance of a centre of learning.[44] However, with the fast rate of technological developments and changing attitudes, Industry needed to enhance and update the skills and knowledge of its workers by means of 'in-career vocational study'. Indeed, in the Green Paper of 1985, 'The Development of Higher Education into the 1990s', it was stated that 'the provision of continuing education should be one of the principal parts of higher education's work'. With its considerable experience in the provision of short courses for Industry and modular provision, Harper Adams recognized that it had an important contribution to make to lifelong learning in the agricultural and rural sectors. Margaret John from the Animal Production and Science Department was appointed in 1993 to assist Alan Costley in the development of lifelong learning - a development which was greatly helped by the award of additional funds from the Department of Employment and later, in 1996, from HEFCE. The outcome was a system for accrediting lifelong learning for the Land-based Industry, whether in the form of short courses, distance learning or full-time study.

Harper Adams was involved in a steering committee, set up in 1981, to co-ordinate, promote and monitor lifelong learning. The Committee covered the West Midlands and promoted collaboration between the National Farmers Union, the Agricultural Development and Advisory Service, the Agricultural Training Board and various colleges.[45] On its behalf the College submitted a project, which received DES approval for PICKUP funding, the first agricultural project to be so chosen. As a result of Government initiative the Professional, Industrial and Commercial Updating (PICKUP) programme had been created to explore the potential for training to improve skills, techniques and knowledge mid-career.[46] Harper Adams was to be the base for the West Midlands. Starting in 1984, the project, under Project Director Colin Henderson, Head of Management and Marketing, set out to assess the potential for learning systems in agriculture, identifying the needs of and the ways of getting to clients and developing a curriculum and teaching methodology.[46] In March 1988, under the same project, a PICKUP officer was employed, broadening the range of short courses and conferences for Industry and ensuring their support.[47]

It was the PICKUP funding which gave a considerable boost to short-course work for Industry at Harper Adams. Short courses had been listed as an objective in the first strategic plan of 1978 and increasingly, it was recognized that planned progressive short courses did have a future worth investigating.[44] Such was the growth in short courses that in 1988 work began on creating a short-course centre in the Main Building. This was named the Temperton Room and was financed by the fund set up in memory of Dr Harold Temperton, who had died in 1984. Courses ranged from those on rural planning issues offered in association with the Country Landowners Association to courses for staff in the supply trade involved with agrochemicals (BASIS), animal medicines (AMTRA) and also with selling fertilisers. In the early 1990s short-course work was formally brought within its own unit, with the name the Professional Development Unit appearing in 1992. The Unit's task was to develop programmes to support the academic departments in the development and delivery of short courses to professional bodies and the land-based industries in general.[30] There was also a trend towards delivery of courses by the College away from site. In 1993 a move was initiated to establish centres around the country for the delivery of professional development courses by Harper Adams for the Rural Practices Division of the Royal Institute of Chartered Surveyors. A year later twenty-three venues from Aberdeen to Bodmin were involved.[21] The demand for short courses of all types was such that it became necessary to offer them throughout the year.

With all the many and varied demands on the College personnel the need for a staff development policy was recognized. This need was reinforced by the Government's heightened concern that expansion in Higher Education should not jeopardize the quality of teaching provision. Greater degrees of efficiency and professionalism were aspired to, as procedures for external scrutiny and internal monitoring and review were tightened throughout the 1980s.[48] Responsibility for the maintenance and enhancement of quality provision increasingly devolved to the College.[38] A staff development policy was agreed as early as the summer of 1978, with a senior member of staff taking specific responsibility for staff development and the formulation of an annual staff development plan emanating from consultation with members of staff on an individual basis.[49] Staff Development Days began in 1984, with workshops and seminars on matters such as assessment and curriculum development. In 1988 Alan Costley, in his

new role as Education Development Manager, became responsible for both curriculum and staff development.[38]
Thus, throughout the 1980s, members of staff were encouraged to develop their professional competence, to gain greater awareness of technical and educational developments and above all to be 'contemporary in terms of what they are teaching'.[4] They were actively encouraged to become involved in research and consultancy, to take secondment to Industry, to attend conferences and short courses, to participate in staff exchanges, with notable exchanges in the mid-1980s involving New Zealand and Australia, and so the list could continue. The number of staff was growing too. In 1990 academic full-time staff numbered fifty, with fifteen new staff being appointed within the last two years and eight of these in new posts.[50]
By the end of the 1980s a more formal structure for assessing staff development needs had developed across the whole College with the introduction of an annual individual staff performance review and development process.

Intricately linked to the issue of staff development was the creation in the late 1970s of an organizational structure which could foster growth and maximize staff involvement. A matrix structure was chosen. As the Principal explained: *'We had Heads of Department on one axis. We had Course Managers on another axis and all the support services on a third axis. [...] This gave three sets of people that had to interact with each other and that meant they drove each other and I believe that provided a tremendous stimulus to achieve the growth that was required.'* The Heads of Department were staff and subject orientated. The Course Managers were responsible for all aspects of course management and operated across the departments. They were assisted by year managers where courses were of more than a year's duration. The Vice Principal as Director of Studies co-ordinated the implementation of the three arms of the matrix and the implementation of academic policy as agreed by the Academic Board.[51]

With the best will in the world, the expansion in the range of courses and all the developments and improvements on offer would have had little significance without interest from the customer, the student *(see photo page 268)*. To ensure this, the essential tools were publicity and marketing. Before the end of Tony Harris' first year at the College, the Open Day for parents was revived and a sherry reception was held for all who had helped the College over the year. In 1980 there was a special open day for the people of Telford, organized in association with Wrekin District Council. The attendence was 1,300, with twenty-seven trailer tours being organized. Themed farm open days and conferences were promoted, such as the joint BOCMS and Harper Adams Sheep Conference, which in 1981 received coverage on radio and television as well as in the press. Once again the College promoted itself at agricultural shows, whether at the Shropshire and West Midlands, where in 1982 it had a stand for the first time in many years, or at the Royal Smithfield, where in 1987 the College's first appearance was rewarded with unqualified success.
By 1989 Harper Adams had an increasingly high profile at an ever growing number of shows and events across the country, from Muck [89] to British Growers Look Ahead.
On a more direct recruitment drive, in 1988 the Academic Registrar, Richard Jopling, was appointed Course Marketing Manager with responsibility for establishing and maintaining links with schools and colleges.[52] That year a national sixth form conference was launched at the College.
In 1990 the scope of Richard Jopling's work was widened *(see photo page 269)* and he

received the title Schools and Colleges Liaison Officer.[53]

In 1988, to promote press coverage of the College, a public relations firm was engaged. Naturally, the best way of achieving good press coverage was to host a visit by a distinguished individual. Among the big names to visit Harper Adams were, for example, in 1992 John Gummer, Minister of Agriculture, who was presented with a pair of boxer shorts emblazoned with the Rag motif. On 7 June 1989 the Prime Minister, Margaret Thatcher made a surprise visit to the College *(see photo page 270)*. The unforgettable event, surrounded by intense security, was recalled by the Principal: *'She literally took the place by storm. I've rarely seen the equal. She was so vibrant from the minute she stepped out of the car. We'd faxed her a brief which we'd had to prepare and she must have read that totally and absorbed it, because everyone she met, she knew and she could ask the right questions. She was so at ease with students. [...] Her questioning was absolutely to the point. [...] It was just one of those outstanding occasions that left everyone breathless.'* A lady of equal fame if less political candour heightened Harper Adams' fame nationally when she enrolled as an HND Agriculture student in 1988. An estimated seven million fans tuned in to BBC Radio 4's 'The Archers' in December 1988 to hear the wedding of David Archer to one of Shropshire's most famous students, Ruth Pritchard, played by actress Felicity Finch. Altogether 1,000 car stickers declaring 'Ruth Archer does it at Harper' were produced.[54]

In 1982 Harper Adams' weather station also played its part in publicizing the College. The weather station at Harper Adams started keeping records in 1903 and in 1924 equipment was increased and the College was recognized as an official crop weather station under the Ministry of Agriculture.[55] It has maintained a continuous record longer than any other station in Shropshire. For many years following the Second World War, recordings were taken by Bill Burrell, the NIAB Senior Regional Trials Officer. Of the fascinating data he compiled, perhaps the most significant was on the night of 9-10 January 1982, when the temperature dropped to minus 26.1C. This was the lowest temperature ever recorded in England, only 0.5C above the United Kingdom record in a notorious frost pocket at Braemar, Scotland. This 'achievement' ensured the College a place in the Guinness Book of records.

Eight years later the College was again making the national newspapers. Harper Adams found itself in the national spotlight as police descended on the College in the wake of the arrest of Kevin O'Donnell on 22 May 1990, a Northern Irish student studying poultry husbandry at the College. O'Donnell had been arrested in London as a suspected IRA terrorist, having been caught in possession of firearms after his car had been stopped by the police. He was charged with possessing firearms with intent to endanger life. Police activity had been heightened in the wake of the bombing of the Tern Hill Barracks in February 1989 and the London bombings a year later. Convicted of the lesser offence of possession of firearms, O'Donnell was deported to Northern Ireland, where, in 1992, he was shot dead in a gun battle with undercover troops.[56]

Such publicity, although not sought and clearly disruptive of College life, helped to keep Harper Adams before the public eye. Behind the scenes, events of a less dramatic nature were nonetheless impacting on the College. The everyday pressures arising from the numerous changes in agricultural education were always in evidence. The move away from public financing towards cost-effectiveness and self-funding was not easy to accomplish, especially in the face of Government policy to expand opportunities in Higher Education. Add to this the whole issue of attracting students amid such factors as the recession in farming, the changing profile of agriculture and the declining numbers of school-leavers, and the challenge facing the College, although by no means unique, was daunting. Strategic thinking saved the day. By introducing policies such as a broader range and level of courses, wider access and encouragement of lifelong learning, short-course work and utilization of facilities through lettings, Harper Adams was able to sway with the winds of change. The College not only expanded, it had become 'street wise', confident of asserting its role in the new order.

The Steps to Degree Awarding Status 1980-96

1980 CNAA approval of Harper Adams as a centre to deliver degrees in co-operation with other institutions

1983 CNAA approval of Harper Adams as a centre with sole responsibility for delivering and maintaining academic standards of courses validated by the Council

1987 CNAA approval of first honours degree and the granting of delegated authority to make changes in courses validated by the Council

1989 CNAA institutional review and granting of delegated authority for the validation of courses

1992 Accreditation by Open University Validation Services and delegated authority for approving appointment of external examiners

1994 Higher Education Quality Council (HEQC) Quality Audit recommending the College to apply for degree awarding powers

1996 Privy Council approval of degree awarding powers

14.7 Graduation Day.

Chapter Fifteen —

Enterprise, Enhancement and Enjoyment 1977-94

All aspects of the College life seemed to be in a state of momentum. The following verse, which was penned by a degree student, Philip Colison, and appeared in the Cat-A-Mountain of 1991, humorously highlighted the fact that change at the College was far reaching. Could life ever be the same now that sheep were being milked?

The Times They are A'Changing!
It used to be in terms you know,
but now we're in semesters.
They used to milk cows you know,
but now it's Border Leicesters.

They used to serve good beer you know,
but now it's not so good.
They used to buy more land you know,
but now all they buy is wood.

They used to be quite keen you know,
but now they go on strike.
They used to use a fork you know,
but now they use a spike.

There used to be a poultry course,
but now it is to cease.
They used to own their tractors,
but now they're all on lease.

There used to be a few girls you know,
but now there's lots of chicks.
We all used to be farmer boys,
but now some of us are FRICS.

Cows used to have their own you know,
but now they have each others.
And all the little calves you see,
are sisters if not brothers.

Harper's always been the best you know,
but all these things have altered.
Yet the more you see, the more you know,
Harper Adams can't be faulted.

For much of the history of the College, sheep were of little significance: *'There were very few facilities for experimentation or observation with sheep. They were a commercial flock and as such made a profit and that was all there was to it.*

They were useful indeed for the demonstration of a normal flock on a normal farm.' [1] Under the Principalship of Kenney sheep ceased to be kept, when in the early 1960s toxoplasmosis made the flock an uneconomic proposition.[2] In 1977, with a renewed interest in sheep arising from newly imported breeds, genetic improvements and increased profitability, the decision was reached to reintroduce sheep as a 'useful demonstration unit for students'.[3] A frequent lambing flock of fifty Finn-Dorset ewes was acquired and Dr Mike Tempest, Head of the Animal Production Department and a specialist in sheep production, took charge.[2] The farm buildings at Ancellor were expanded and modified to accommodate the research work *(see photo page 287)*. The aim was to investigate the economics and management of prolific ewes on a frequent lambing system, every eight months, compared with annual lambing. The work involved the application of outcomes from fundamental research in a commercial situation and attracted very wide interest in the farming community. This was the start of an intense period of development work with sheep which later looked at using different terminal rams for finished lamb production with both Finn-Dorset and Lleyn ewes. In 1988 Charles Minter, Senior Lecturer in Sheep and Beef Production, started investigations into the milking potential of sheep, an area in which little research had previously been undertaken.

Sheep were not the only area of research at Harper Adams. Indeed the growth in entrepreneurial spirit and competitiveness in the 1980s did much to encourage research across the entire College.[4] Furthermore, research had always been seen as an important element in support of the teaching programme and, with entry into degree teaching, this fact was reinforced by the CNAA's requirement that the College should have a research programme. There were many benefits. Research returned Harper Adams to the national stage and the accompanying publicity attracted students and staff. At the same time research cemented important links with farmers and the Industry as a whole, something very much encouraged by the Government.[5] Tony Harris was very conscious to avoid 'academic drift' and the move away from practicalities and application.[6] Traditionally research at Harper Adams had been applied and this was to continue. However, despite all the clear advantages to research, it had either to be profit-making or at the very least to be self-funding. There was intense competition to obtain a share of the limited Government and commercial funding that was available. From the early 1980s first Fred Harper and then Wynne Jones, in their role as Vice Principal and the latter also in his role as Director of Research, had a vital part to play in obtaining funds and taking the research programme forward.

The 1980s saw the creation of the Silcock and Temperton Fellowships, which provided significant support for the research programme. The family of the late Richard Silcock sought to endow a fellowship at Harper Adams associated with animal production and it was proposed that a trust should be established for the purpose.[7] The Trust would be administered by BOCM Silcock Ltd, represented by the Silcock Family and the Governors.[8] The Trust, amounting to approximately £55,000, made its first award in September 1985.[9] In 1989 the Temperton Fellowship was officially launched, based on a trust fund established in memory of Dr Harold Temperton *(see photo page 288)*. Contributions had come from Industry and from the sale of the Diamond Jubilee History of the NIPH. The sum of £25,000 was earmarked for the encouragement of research at the College by personnel involved in the British Poultry Industry.[10] In 1984 the NIPH had been awarded a Teaching Company Scheme by the Science and Engineering Research Council, the first use of the scheme in agriculture.[11] The Scheme ran in conjunction with the Polytechnic Wolverhampton and

15.1 Presentation of the Silcock Fellowship deeds, 1983. From left: Peter Duckworth, stepson of Richard Silcock, Peter Watson Jones, Chairman of the Governors, and Michael Dowdall, Chairman of BOCM Silcock Ltd.

C. E. S. Potter Ltd and gave the opportunity to take up research to someone already in employment, in this particular case the opportunity to investigate ventilation systems in poultry houses.[12]

Research was a real growth area at the College and, increasingly, this research was undertaken by post-graduate students. The first post-graduate student at Harper Adams to be awarded a Ph.D. was Mike Gooding in 1989. By 1993 those involved in postgraduate research at the College numbered fifteen. It was Dr Paul Davies and Peter Kettlewell of the Crop Production and Science Department who first highlighted the potential of postgraduate student research.[13] In 1981 they obtained a Ciba Geigy/SERC post-graduate scholarship to study the effect of propiconizole on the flag leaf of cereals. Their success in attracting additional funding continued and the research activity was unstoppable. By the late 1980s and early 1990s their success was shared by colleagues. Dr David Parry's investigation of the effects of climate change on crop disease was of national importance. In 1991 the decision was reached that a Crop and Environment Research Centre (CERC) should be established to raise the profile of crop research still further under the direction of Dr Patrick Fair, Head of the Crop Production and Science Department.[14]

As regards animal production, Harper Adams' tradition of research into pigs continued. For example, in 1985 a Ministry of Agriculture post-graduate agricultural studentship covering a two-year period was awarded to allow investigation of the interaction between building design and heat balance in the growing pig. Hitherto such awards had been the preserve of universities. Pig feeding trials continued, with this work being notably the preserve of Dr Paul Rose of the Animal Production and Science Department.[13] The tradition of research into animal production was extended, not only to include sheep and pigs but also beef, with various feeding trials for intensive beef production, including the recently introduced Belgian Blue cattle. Research work with cattle and sheep was facilitated in the early 1990s with the building of a series of yards specifically for research purposes. Similarly, research within the Agricultural Engineering Department took off in 1989. That year Peter Bloxham, Senior Lecturer in Agricultural Engineering, and Dick Wells, in collaboration with Eurotech International, won a contract with the EEC to finance a project on industrial co-operation between Israel and the EEC in the field of poultry equipment. Peter Bloxham also received a wide range of funding to investigate all aspects of farm waste. The Centre for Farm Waste Management was created with NAB funding.[4] In 1992, in association with Parry and Kettlewell, Peter Bloxham obtained £145,000 from the Farm Electric Centre to investigate the aeration of slurry and its value for crop and grass production.

In 1992 for the first time the College participated in the University Funding Council Research Assessment Exercise 'with gratifying results'. The College obtained the rating of two in a ratings list which ran from one up to five. Institutions granted a rating of two and above qualified for research funding from HEFCE. Although the grant was relatively small, the recognition of Harper Adams as a research centre in such august company provided a spur to staff to increase the rating at the next research assessment exercise in 1996.

Closely aligned to the growth of research and the provision of short courses for Industry was the development of an increasing volume of consultancy work. In 1986 the Centre for Agri-Food Marketing Studies (CAMS) was set up by Dr Paul Custance, Bill Howard and Stephen Parsons, of the Management and Marketing Department. It was to undertake marketing and business research and consultancy on a contract basis for organizations in the Agri-Food Sector. The significant work undertaken by CAMS for companies and organizations across the country broadened over the years to include studies involving the rural economy.

Research and consultancy links with Industry were again encouraged by the setting up of the Enterprise Unit. The Unit was established in October 1992, with part funding from the Department of Trade and Industry to the tune of £118,728 over four years. It was the first time that such an award had been made to agriculture.[15] Its role was to facilitate and co-ordinate a diverse range of attractively priced research and consultancy, linking the College with industry, commerce and the professions.[16] Its creation was seen as a great boost to the entrepreneurial activities of the College. Richard Bruce was appointed to win contracts and co-ordinate the work. He drew on the expertise within the academic staff, who quickly realized the benefits of this entrepreneurial activity and by the termination of the grant in 1996 were able to assimilate the work within the departments.

Research continued to have a high profile at the NIPH, but the story of the NIPH as a whole was one of steady decline. Despite sterling efforts to revive it, little could be done when the Poultry Industry was saddled with a bad press and battling against recession. Primarily, there was the problem of recruitment of students *(see photo page 289)*. Both taster courses for school children and a series of careers days to alert careers personnel to the prospects on offer failed to improve matters. Recruits for the session 1990-91 were just eleven despite a target figure of twenty-four.[17] With the availability of a Further Education diploma course at the Lincolnshire Agricultural College, the NIPH opted to discontinue teaching the Diploma from June 1992. The College decided to concentrate on poultry education at Higher Education level,[18] with poultry husbandry being taught as optional modules within the HND and degree courses.[19] Poultry education still has a high profile and the title NIPH has been retained to demonstrate the College's continuous commitment to the Poultry Industry.[19] From September 1991 the staff of the Poultry Department were transferred to the Animal Production and Science Department.[20] It was fitting that in 1986, before such events overtook the NIPH, the Diamond Jubilee was celebrated and a book of its history was launched.

Poultry research was centred on the Poultry Husbandry Experimental Unit, which in the mid-1980s was renamed the Harper Adams Poultry Research Unit. Here too the story was one of difficulties. In 1980 Dr Chris Belyavin, student 1969-71, was appointed Senior Scientific Officer to succeed Dick Wells who had earlier moved to take charge of the NIPH.[21] Together with Wells varied work was undertaken, such as important research on controlled feeding which was becoming standard practice in the Industry.[22] In 1983 experimental facilities were expanded, with the opening of a new deep pit battery unit of 4,800 birds, financed by commercial companies and poultry organisations. This was the first major development since the Unit's opening.[21] However, although research brought reputation and status to the Harper Adams Poultry Research Unit, funding was an ever increasing problem. Since its establishment the Unit had never achieved financial self-sufficiency and in the 1980s low egg prices and poor support from an industry in recession did little to help. In 1987 the choice was clear, either the Unit had to restructure and operate within limited means or it had to close. Restructuring did take place, but the inevitable decision was reached in favour of closure on 30 September 1991.[23] The facilities were transferred to the Animal and Crop Departments for use in their growing research programmes.

The commercial side of the NIPH fared considerably better than education and research. The late 1970s and 1980s saw a continuation of the modernization programme. Between 1977 and 1981 the hatchery was improved,[21] although the hatching and breeding side remained on a small demonstration scale.[7] In 1981 the poultry meat processing plant was similarly modernized with the introduction of a fully automatic slaughter and plucking line, in accordance with modern practices and hygiene regulations.[7] There was expansion in turkey production [21] and in early 1986 a third layer unit for 10,000 birds was completed.[24] Egg production remained the central enterprise at the NIPH and in the late 1970s graded eggs arrived with the establishment of the NIPH as a registered packing station. It was a financially sound move, allowing sales to local shops. Free range production was reintroduced in 1983 [21] and by 1991 the flock had expanded to a unit for 4,000 birds *(see photo page 290)*.

In 1994 construction of a 60,000 bird unit with packing and chilling facilities got underway with a leasing arrangement of £475,669 *(see photo page 291)*.[25] This was part of a major streamlining at a time when the financial viability of the plant was a 'cause for great concern'. The educational requirement for small operational units in a time of growing food costs and static, even declining prices was a disadvantage to say the least. Thus, commercial broiler production was stopped. As with the Poultry Research Unit the more substantial buildings at the NIPH which were no longer required, for example the mill, were put to alternative use to increase research facilities and as teaching, tutorial and student rooms. This was of great benefit to an expanding and financially restricted College.[16]

The main College site was experiencing equally dramatic changes. However, the essential character of Harper Adams was never jeopardized.[6] Indeed, the traditional appearance of the College was recognized for its marketing potential: *'If one looks at the pavilion, the cricket pitch and the College all together, it is a superb setting and conveys the feel of a quality residential establishment.'* [6] The decision in the late 1970s to move car parking from the front of the Main Building to the original site of the Principal's garden was a controversial one, but one which greatly restored the integrity of the building. Considerable effort was also made to mellow the hard lines of the Ruabon brick and to improve the amenity value of the grounds by the creation of shrubberies and the judicious planting of specimen trees.[6] Perhaps, the saddest loss to Harper Adams' heritage was the felling of the avenue of limes at Ancellor which succumbed to the drought of 1975-76. Replanting did take place as a part of a comprehensive tree planting scheme across the College.[2] Out of respect for the College's rich tradition and in a bid to carry that into the future, the decision was made in 1987 to form the College Archives, under the care of Dr Heather Williams.[26] The collection of items ranging from photographs and student notes to administrative papers and taped memories provides a full picture of College life over one hundred years. An ongoing collection policy ensures that today's College is not overlooked.

It was the overall policy of growth which stimulated improvements at Harper Adams. More students enabled a dramatic improvement in facilities, which in turn promoted student recruitment.[6] The primary need was for a new science facility. The old laboratories were outdated and inadequate for the work now expected of them, notably at degree level.[27] On 16 May 1985 Her Royal Highness, the Princess Margaret opened the new Science Building, funded by the DES at a cost of £677,000 *(see photos pages 292-96)*. As one student declared: *'Of all the recent building developments this must be the most important in securing a good future for the College.'* [28] There were seven teaching laboratories, two preparation rooms and a series of smaller rooms:[21] *'[The building] was light. It was clean and students responded to it. They felt that they had got a new facility that was worth having. [...] We had a large capital sum for equipment. Each student in the laboratory would have a brand new £500-£600 microscope and that was the sort of thing which I think they felt raised their status in some way.'* [13]

The Main Building was redesigned, with new offices, teaching rooms and a relocated staff common room. Better facilities were made available for those students who lived off campus with the addition of shower and changing rooms. About 300 non-resident students had been sharing accommodation facilities provided for two hundred and twenty residents.[29] As student numbers grew, the 1990s saw further refurbishment and change of usage in the face of the growing need for staff and student accommodation. Part of Adams Hostel and the North West Rooms were converted to offices.

The library was recognized as occupying the central position in the teaching, learning and research processes,[30] but it had been designed for only 220 residential students. With the growth in student numbers in the early 1980s there was an urgent need to increase the staffing, the stock of periodicals and books and the study area. A second full-time librarian was appointed in 1981 to the post of Assistant Librarian [7] and clerical support staff were also employed. In 1982 Derek Taylor was appointed as Head of Library and Learning Resources with responsibilities covering audio-visual services and computing as well as the library. To create more space, staff offices along the top corridor and the tutorial rooms in the lower corridor were absorbed into the library and learning resources complex.[31] However, this could only be a temporary expedient. The change in teaching method to student-centred learning, the need for every student to be computer literate and the growth in the number of degree and postgraduate students necessitated a major extension of the facilities available.

Action was taken in the summer vacation of 1992 by building a major extension to the library *(see photo page 297)* and converting Lecture Theatre B to a computer centre containing four rooms of computers to which students had unrestricted access, at a cost of £348,000. Computing facilities were also provided for students on the NIPH site. In 1993 plans were introduced to install a college-wide computer network equipping all staff with computers - a plan which came to fruition in 1995.[32]

The College had good reason to be proud of its excellent laboratory and learning resource facilities. They were the envy of agricultural colleges and university departments. The arson attack on the laboratory and library during the night of 7 December 1992 was understandably heartbreaking *(see photo page 297)*. The prompt discovery of the fire by a student and the response of the Fire Service minimized the structural damage to the buildings, but the cost of refurbishment and the replacement of books and equipment amounted to £250,000. About 40,000 books and journals and over one kilometre of shelving had to be cleaned. Thankfully, of the 1,599 volumes which were damaged or destroyed, the majority were able to be replaced.[33]

The perpetrators were not caught, but the incendiary devices used were identical to those being used by the Animal Liberation Front who were involved in similar attacks across the country at this time.

It was not only in the area of learning resources that impressive improvements were taking place. An extension to the machinery block was started in November 1980 at a cost of £174,000. It more than doubled the size of the Engineering Department and Ian Gedye, Head of the Engineering Department, declared the new premises to be 'one of the best machinery teaching facilities in the country'.[7] On 25 February 1987 the Soil Hall was officially opened by Jeremy Coleclough, President of the Agricultural Engineers Association. The creation of a soil hall, at a cost of £75,000, was 'a first for Harper Adams in agricultural education in the UK'. The large covered soil working area of 1800m^2 was for all-weather use. It increased the effectiveness of the instruction in soil-engaging implements and enabled an expansion in the range of test and development work undertaken by the engineering students.[26]

The 1980s saw a great need for additional hostel accommodation, but for most of the period there was little prospect of major expenditure with the requirement from 1983 that student accommodation had to be self-funded.[34] It was not until September 1991 that Gloucester Hall, the first development, with its ensuite facilities for eighty, was first occupied.[35] The Hall, its total cost amounting to £934,000, was officially opened by the Duke of Gloucester on 8 June 1992 *(see photo page 298)*. On its completion there was already the need for another hostel [36] and in July 1994 there was the official opening of Harris Hall, with ensuite facilities for one hundred. Harris Hall cost £820,000. As Ian Robson, Head of Student Services, declared: *'Ten years ago, if you'd told a student that they'd have an ensuite bedroom with their own shower, wash-basin and things like that in their*

15.2 The Soil Hall opened in 1987.

room, I don't think they would have believed you.' Throughout the 1980s improvements in existing hostel accommodation took place. From the mid-1980s College houses, originally purchased or built for use by employees, were designated hostel annexes.[9] In 1990 this policy was extended and the College began to purchase houses in the surrounding area for student use as an economic and flexible option.[37] In 1994 the number of students accommodated on campus was just over 500.[38]

As well as accommodation, from 1983 catering had to be self-funded. In 1978 catering had become self-service, with waitress service still maintained for the staff.[39] With the growing pressure of student numbers self-service allowed the College to serve 500 people without extending the dining room. There was, however, the need for enlargement of the kitchens and this was done at a cost of around £70,000.[40] In 1992 the decrease in the demand for meals at weekends from non-resident students was recognized and the main catering service was restricted to weekdays. Over the weekend and for limited periods throughout the week a cafeteria facility was provided. This was situated in the former bar in the Student Union Building.[41] In 1982 laundry facilities were modernized to accommodate the growing student population.

Recreational facilities are important for full-time students, particularly in a rural situation such as Harper Adams *(see photo page 298)*. Many of the younger prospective students are influenced in their choice of college by

Carwyn Ellis, to right, and student recording sheep in Ancellor Yard.

Inauguration of the Temperton Fellowship, at the Farmers Club, 1989.
From left: Principal Harris, Peter Watson Jones, Chairman of the Governors, The Earl of Selborne, The Baroness Trumpington, Minister of State in Agriculture, Fisheries and Food, and Peel Holroyd.

Poultry husbandry students on practical work with turkeys.

Dick Wells, Head of the NIPH, and students with reintroduced free range poultry.

60,000 Bird Unit at the NIPH, c. 1994.

The Science Building opened in 1985.

Visit of HRH The Princess Margaret on the occasion of the opening of the Science Building, 16 May 1985. The Princess on arrival being introduced by Sir Gwilym Williams, Chairman of the Governors, to Lady Williams.

HRH The Princess Margaret talking with students.

HRH The Princess Margaret talking with Alan Costley, Senior Lecturer in Entomology.

HRH The Princess Margaret meeting staff and students.

Fire damage to the library as a result of the arson attack of 7 December 1992.

Library, 2000.

Gloucester Hall (right), completed in 1991, Ward Hall (centre) with Harris Hall, opened in 1994, under construction.

Live band in the main bar, c. 1993.

The disco area in the refurbished student bar, known as 'Uncle Ken's', c. 1983.

Lounge bar in the Student Union Building, 2000.

The dairy cubicle building, 2000.

New tree planting in Caynton Gorse, 1993.

Ladies hockey team, 2000.

First XV rugby match, 1993.

First XI football team, 2000.

Swimming event.

the social life and therefore the College had to ensure that the recreational facilities kept pace with the growing student numbers. The facilities for 200 resident students in the SU Building and Queen Mother Hall had become overcrowded. The bar, always such an important rendezvous for students, was refurbished and named 'Uncle Ken's' in honour of Warden Ken Mathias [42] *(see photo page 299)*. It was then extended to provide a disco area, but it was obvious that more space was needed, not only for the student population but also for visiting students and guests at social and sporting functions when numbers often exceeded a thousand. The only way forward was to build a new facility adjacent to the SU Building, with a large bar which could also accommodate large social events and an up-market lounge bar where students and visitors could relax in comfort *(see photo page 299)*. The new facility, costing £486,000, was completed in 1991 and became the hub of student life.[23] The development also freed Queen Mother Hall to be refurbished and used for external conferences and events to generate income.

Harper Adams has always had a fine tradition on the sports field, a tradition which continues to attract students to the College. Two squash courts and a clay pigeon shooting range were added to the facilities in the early 1980s, but the pressure on the sports pitches continued. The purchase of the neighbouring farm in 1991, at a cost of £90,000, provided the solution with the construction of two further rugby pitches and an additional soccer pitch. Good sporting and social facilities continue to be a major factor in developing and maintaining the excellent spirit which exists in the student body.

The creation of new facilities and the improvement and redeployment of existing facilities is a continuous process in a progressive and expanding college, but the development which took place at Harper Adams over the period 1991-94 was exceptional. With student numbers more than doubling between 1989 (751) and 1994 (1,525), the Governors undertook a major capital investment programme amounting to well over £3 million to provide the necessary teaching, learning, research, commercial and residential facilities. Starting with minimal financial reserves the programme was not without its risks. Bank loans were arranged to cover major projects, but buoyant student recruitment, income generated from entrepreneurial activity, and very strict financial control of revenue expenditure saw the College through this expansion. The bold approach adopted by the Governors was well and truly vindicated and provided a sound foundation for further expansion.

With the arrival of Tony Harris there were no major changes in policy concerning the Farm. The Farm continued to operate as a commercial unit, but at the same time it fulfilled the role of 'an essential teaching and learning resource, a base for research and development and a centre of interest for those involved in agriculture and the general public'.[6]
The practical character of the HND remained and students of the 1980s were very much encouraged to take part in activities on the Farm, with those who wished joining rotas for milking, lambing etc.[43] As Angela Billington, nee Holding, student 1982-86, recalled:
'It wasn't the theory the whole time. You did your theory and it was always backed up by practical work. Whether it was doing things with sheep, or the dairy, or tractors or whatever, or work in the fields.' The Principal acted as Farm Director with overall responsibility for the Farm, especially the integration of teaching and research into the farming system. However, the daily running of the Farm was once again delegated to the Farm Manager, Tony Evans-Fisher.[6] In answer to the needs of the education programme the farming system was diverse and intensive as was appropriate to the area. Tony Harris above all sought 'a good example of a commercial mixed farm'.[6]

Towards the end of the 1970s farm profitability nationally was starting to decline, a trend which was followed, apart from minor fluctuations, until the mid-1990s. Then a temporary resurgence occurred before the severe collapse in 1997 and ensuing years. To combat this situation farms increased in size and farming systems became more specialized with accompanying growth in individual enterprises. The ever increasing capacity of farm machinery led to the development of machinery rings, the growth of contract farming and much wider use of contractor services. Therefore, the challenge at Harper Adams was how to maintain profitability with a diverse farming system to meet the needs of the education programme. One of the major limitations was that the budget for the Farm had to be submitted each year to the DES and the Farm was required to make an annual financial contribution to the College Revenue Account.[31] The difficulty in getting capital expenditure for the Farm from the DES was considerable as it had to be considered alongside all the other demands for capital from other direct grant colleges for educational purposes.[6] However, after prolonged negotiation the DES agreed in the early 1980s that the Farm could be regarded as an entirely commercial operation, with the opportunity to apply to the Ministry of Agriculture for capital grants and to borrow money from normal credit sources.[44] At last Harper Adams was free to farm, a significant move which allowed major development to take place.

In keeping with commercial farming it was decided that there should be three substantial enterprises, namely dairy, pigs and poultry, from which sufficient profit could be gleaned to maintain a diversity of smaller enterprises.[6] The acquisition of additional land was also a high priority. During the early 1980s the dairy unit was updated and enlarged. This included replacement of the old, dark, cramped parlour with a new 8/16 herringbone parlour,[12] redevelopment of the cubicle area, installation of a new calf-rearing unit and yards, installation of new silos and improvement in the slurry handling system.[45] By 1985 it was clear that the dairy unit was running below capacity and extra milk quota was sought to make it more viable as a commercial concern and more creditable as a teaching platform. The need for expansion and upgrading in farming was unrelenting and in 1993 a further upgrading of the parlour took place *(see photo page 300)*. The eastern end of the cattle building was demolished and a feeding passage and ninety-six cubicles were built. By the close of the Harris years the dairy herd stood at 180 Holstein Friesian cows.[33]

The pig unit was also modernized to improve output and labour economy. On Tony Harris' arrival the piggery was in a state of deterioration and improvement was essential to maintain progress in a competitive market.[46] Between 1981 and 1984 the pig unit was redeveloped and enlarged.[47] Among several improvements was the erection of new weaner and finishing houses, with an environmental control system. All the pigs from the existing herd could be accommodated.[42] Thereafter only minor improvements occurred. By the close of the Harris years the pig unit comprised 185 Large White and Landrace crossed breeding sows with progeny taken through to bacon weight. In the light of the intense specialization which had occurred in the Industry, dicussions were already taking place on the future of the pig unit at Harper Adams.

There was a conscious policy to extend the acreage of the Farm, notably to acquire arable land.[34] The latter was especially important to maintain a realistic teaching and research programme and to ensure commercial viability.[27] The limited size of the Farm, 367 acres in 1979,[48] was a problem.[31] Land had to be acquired, whether by purchase, tenancy or management agreement.[16] In 1979 six acres of land was purchased [47] and in 1986 the tenancy was agreed on Brassington's Farm. Five years later the forty acres were bought. In 1989 twenty acres of land were similarly acquired and a year later a further forty-five acres at Caynton House Farm [18] became College property.[47] In 1991 the College took up the tenancy on Willowmoor Farm, forty-eight acres on the slopes of the Wrekin, in order to extend the education and research activity of the Animal Production and Science Department.[35] A year later the tenancy of Shray Hill Farm, 169 acres, was obtained to take the area being farmed to over 700 acres.[41]

Harper Adams was very keen 'to broaden our whole interpretation of land use on farms', especially with the introduction of environmentally sensitive practises.[12] As the policy statement says: *'We consider the wildlife habitats that are present on the estate to be an extremely important asset and we are fully committed to a policy of preservation, protection and enhancement.'* [48] A conservation plan was developed, which included the management of existing woodlands and shelter belts to demonstrate a range of techniques which could enhance the amenity, wildlife and recreational value of the Farm.[16] As early as 1979 a grant from the Countryside Commission allowed the clearance of over mature trees and brought about a policy of replanting and fencing, in all covering about a third of the College woodland. Throughout the 1980s existing hedge-rows were improved and new ones were planted.[34] In 1987 the overgrown stagnant pool in Marlpit Leasow was cleared to create a freshwater pool to maximize its wildlife and landscape value, again financed from grant aid. In 1984 the College had reintroduced the use of field names, taken from the 1841 Edgmond Tithe Map, to replace the numbers and letters previously used. The acquisition in 1988 of Caynton Gorse, an area of woodland, was a valuable resource to the teaching of land management courses.[49] Parts of Caynton Gorse and the later purchase, Newpool Plantation, obtained in 1989, were cleared and replanted in stages with grant aid and part of Newpool Plantation was restored to a small lake *(see photo page 300)*. In 1994 a small arboretum was established in front of the old Poultry Research Unit with the support of the County's Countryside Officer.

And what of the Harper Adams students of the 1980s? One who knew them only too well, could even mischievously put them into categories:

> *If you walked a few students past me now, I could still tell you which ones are the agrics, which ones are the AMBAs, which ones are the food marketing, which ones are the engineers, just by their manners and the way they are. There are still typical agrics, no disrespect to farmers, but they still look like agricultural students who haven't cleaned their shoes for three months, probably haven't ironed their clothes for a while, that sort of thing.*
> *You get the AMBA people, the girls particularly, who are always smart, always well dressed. Alright they may be wearing agricultural clothes, but they are totally different.*
> *Then you get the engineers, who are always walking around in overalls, with their hands up to their elbows full of oil and grime.*[38]

Student life may have changed in some of the detail. Expectations as to facilities, for example, were naturally considerably higher, although even in the late 1980s students accepted a room with a wash-basin as standard accommodation.[38] However, overall, student life remained just as much a time to enjoy as ever it was. The student of the 1980s may have used computers and owned a ghetto-blaster or perhaps, even a television, but as with his or her predecessors, good fun could still be had drinking coffee and talking. As Ed Pridgeon, student 1986-92, said: *'I am wondering what we did, [...] apart from sitting in each other's rooms and drinking various things. I had a kettle and my friend had a TV. So, between us we weren't too badly off.'* He added: *'It's amazing how many people you can fit into a little room.'*

Angela Billington remembered doors to the rooms of Ward Hall always being open, inviting people to step in 'to have a coffee, have a chat, sit on the bed'. Everyone was 'very, very friendly'. Angela had been impressed by the College's pleasant atmosphere from the start: *'I was put off University. It was big and in the middle of a city and I had never lived in a city. Harper Adams was very rural and it was a lovely area.'* She was one of a growing number of female students at the College. In 1980 Harper Adams students elected the first female President of the Student Union, Fiona Fitzgerald. By this time there was greater mixing of the sexes, although halls had segregated floors or blocks.[38] Furthermore, rules regarding visiting hours were relaxed.[50] The new halls of residence allowed integration with the arrival of ensuite facilities.

Above all, the family atmosphere was as evident at Harper Adams in the 1980s as ever it had been, with friendships formed across the courses and strong staff-student interaction in the social dimension.[13] Regular features of the year were the student and staff revues *(see photo page 312)*. The anti-establishment feelings of the 1960s belonged very much to the past.[51] Indeed, from the session 1992-93 final-year students took over the responsibility of wardening the halls from members of staff and technicians. The feeling of community was described by Ian Robson, Head of Student Services, as 'a bonding', which despite his experience of many other places, he had never witnessed anywhere else to such an extent. His predecessor, Ken Mathias, summed it up thus: *'Harper Adams students looked after their own and there was no question about that.'* Ed Pridgeon saw himself as part of 'an extended family', with someone always there to help you out. There was a timeless charm to the ritual of Coffee Time in the mid-1980s. Everyone gathered: *'Coffee Time at eleven o'clock in the Common Room was always a real get together time. Eleven o'clock you went into a canteen and had your plastic orange mug of coffee and you always used to have left-over cold toast from breakfast or there used to be bits of bacon, stuff like that [..] and you used to go up to the Common Room.'* There was also the opportunity to purchase the much-relished Harper Adams 'Toastie'. In the Coffee Bar, throughout the day, 'loads and loads of bread and great big bins of cheese from the wholesalers' would be transformed into toasted sandwiches, which sold "like nothing on earth".' [50]

Sport very much dominated student leisure hours and the range of non-sporting clubs in the 1980s was limited, despite attempts to revive interest in such matters.[52] Quite a scam was for the students to set up bogus clubs in Freshers Week and to enrol members, spending their subscriptions in the bar.[53] Entertainment was by no means limited. For example, the tradition of student plays, mostly bedroom farces, continued until 1990. The College year kept to a calendar of events. For all, the introduction to Harper Adams began with Freshers Week, regarded by Ian Robson as the ideal way to bring new students into the family fold. Existing students regarded it as an opportunity not to be missed, for central to events was the Harper Adams tradition of the initiation ceremony. For the victims it was something to be endured in the knowledge that their turn would come to inflict even more weird and wonderful activities on the new arrivals in the following years. Integration of a more relaxed kind came as a result of Ken Mathias' very successful Pie and Peas Nights. These had the objective of giving new students the chance to mix throughout the courses. Ken Mathias warmly recalled in the later years 'doing about 600 pints of home brew'. Rag similarly continued, although in a much-reduced form,[43] with students now taking part in Newport Carnival and raising money by selling the Rag magazines and organizing various events in Rag Week in the Summer Term and throughout the year *(see photo page 311)*. Events for the 1985 Rag included a toga party, a raft race on the Newport Canal and a pancake evening.[28]

All students needed formal dress for the numerous dinners which were held. Dances, discos and bands were regular attractions *(see photo page 312)*. The Eurythmics visited in the early 1980s and swam nude in the swimming pool.[51] Other guest bands included Katrina and the Waves, Doctor and the Medics and Madness. Although by this time the majority of students possessed a car, albeit very often old and in need of constant tinkering, weekend attractions were well attended and there was no room to move in Queen Mother Hall.[43] The End of Session or Summer Ball was, as it always had been, the climax of the year. As before, much care and attention was lavished on the decorations and all was done amid great secrecy to heighten the impact of the night.[38] One year a castle was created, with gas guns firing balls at people and an entrance through the mouth of a huge dragon breathing dry ice. It was always a challenge to see who could survive through to breakfast.[43] With the development of the new student facilities and the vastly increased number of students in the 1990s the End of Session Ball extended into the sports field with marquees for dancing and the introduction of a full-scale fair for additional entertainment. Today around 1,500 people attend at around fifty pounds a ticket *(see photos pages 313-14)*.[38]

Social drinking remained at the heart of many a student's life, whether on campus or in one of the local hostelries. Angela Billington has fond memories:

> *On a Friday night the place to go was the Three Horseshoes at Sambrook. On a Friday night everyone went there. We called it the Three Wellies. It wasn't called the Horseshoes. The landlord was John Whitehead, who was a brilliant chap and he loved having students and both rooms were always packed. The Lion at Edgmond was another place where we all used to go on a Friday lunchtime, because that was the place where you could get sausage, egg and chips for about £1.50. Or sometimes it used to be the Wheatsheaf at Cheswardine and there were certain places that students were always renowned to be at on a certain night of the week.*

Forever tied to this lighter side of student life must be the pranks. Ken Mathias had the challenge of dealing with students who used the underground ducts to travel around the campus and flaunt authority. They were successfully flushed out 'like rabbits' one night after a concerted effort by staff. On another occasion the big white feet of a yetti appeared 'going up the Queen Mother Hall step by step'.[50] A prank which has become something of a tradition is performed on the night of the Rugby Club Dinner:

> *It started off that they used to be dropped by the Principal's House and they would strip first-years naked and make them run past the Principal's House into College. You can imagine now that tradition, when you've got probably a hundred first-years playing rugby for the first time. They go to the Dinner and the person who is organizing the run thinks they're at Tibberton crossroads, whereas they're in fact at Crudgington crossroads. The students are dropped off stark naked and have to run all the way to College on the coldest night of the year. We managed to get in the Daily Telegraph.*[38]

In 1991 Tony Harris received an OBE in recognition of his services to agricultural education. As regards Harper Adams, he had steered the College through some of the most challenging times in its history. Competitiveness, cost-effectiveness and entrepreneurial spirit were now the essential tools to shape any kind of a future in education and these he had readily taken on board from the very first. On Tony Harris' departure in 1994, the College was in a healthy position, very much fitter to face the next round of Government thinking. It stood firmly on the national stage and was moving confidently towards degree awarding status, possessed of the efficiency and strength to meet its future, accepting change as a challenge never a threat.

Sport at Harper Adams

To a large number of students over the years sport has played an important part in the life of the College. Documents show that soccer, rugby, cricket, cross country running, athletics, tennis, netball, hockey, shooting, table-tennis, squash, Gaelic football, lacrosse, swimming, basketball and golf have all had representative teams from the College at some time or other *(see photos pages 301-02).*

Some are more recent such as Gaelic football (given our number of Irish students) and squash (with the relatively new squash courts), whilst others such as cross country running, table-tennis and athletics ceased some 15 years go.

The remainder have seen regular College teams compete in competitions over the years against local colleges in the early days (one presumes that transport was a problem) and more recently against colleges and universities all over the United Kingdom and, in the case of rugby, this included mainland Europe.

Soccer, rugby and hockey were the winter favourites with cricket during the short summer term being the most common.

Soccer or, in the days of the 40s and 50s, football, provided strong teams competing in local (Midland) leagues. A win in the Midlands cup final in 1957 shows the strength of soccer at the time. Although small in numbers, Harper were able to hold their own against much larger colleges and universities but outside the College this never seemed to be recognized.

Regular fixtures against agricultural institutions such as the Royal, Seale Hayne, Writtle, Wye, Shuttleworth and others provided the highlights of 'block fixtures' at all sports with the main competition being that of 'Harper' against the 'Royal'. To most students at each college this was billed as the 'Agricultural Varsity Match' and still has pride of place with today's teams.

During the 60s, 70s and 80s a group called the United Kingdom and Ireland Agricultural Students Association (UKIASA) was formed to provide an umbrella association for agricultural colleges. Harper Adams played a major role in both hosting and organising these events which took on the whole spectrum of sporting activities. More recently, since 1992, Harper were accepted into what was then the Universities Athletic Union (UAU) which has now changed its name to British Universities Sports Association (BUSA) involving every university and college of higher education in England, Scotland and Wales. This organisation covers all sports (42 in total) and provides inter-University and national competition for University teams and individuals. During the eight years Harper have entered teams and individuals in most sports and have achieved a high level of success.

There have been notable achievements in a variety of sports. In clay pigeon shooting there have been winners of the National Team competition for 3 years and the College has provided the individual winner for the same years. In rugby, in the National Premier Division, Harper Adams has reached the quarter finals of the National Tournament on two occasions and the semi-final of the European Cup. Individual success has been achieved by Matt Farmer in the Downhill Mountain Bike, winner in thre successive years, and ten male students and one female student have represented their country at student level. Numerous individuals have reached the final stages of national tournaments in golf, canoeing, wind surfing, modern pentathlon, rifle shooting and athletics.

Not bad for a small college!!

Ian Robson

Chapter Sixteen —

Harper Adams University College 1994-2000

The new Principal took up post in September 1994. The choice of a New Zealander, working as Principal of Marcus Oldham Farm Management College in Victoria, Australia, was an innovative one, indicative of the wider horizons which the College now sought.[1] Graham McConnell possessed both charm and enthusiasm and brought with him an air of informality. His term of office lasted but two years and in the best interests of both parties involved, he resigned in July 1996. This difficult period is summarized by the then Chairman of the Governors, Richard Matson, on behalf of the Governing Body:

With Tony Harris' impending retirement in 1994, the Governors were well aware that to find a successor, who would continue the progressive path that the College had been on, would be a difficult task. We were concerned that someone with successful management experience was more important than a successful academic record. We advertised worldwide - and of the shortlisted nine candidates, three were from overseas. Mr Graham McConnell, who had for seventeen years managed the Marcus Oldham Farm Management College in Victoria, Australia, was appointed (see photo page 315). Marcus Oldham was not a publicly funded college but one financed by fees, with considerable help on capital projects from an ongoing public appeal which Graham McConnell and his wife, Janet, had been given much credit for its success.

Following Graham McConnell's appointment to succeed Tony Harris certain difficulties arose with his style of management and after discussions with Graham McConnell he agreed to step down in July 1996. The Governors immediately sought a successor. Dr Wynne Jones had been Vice Principal for eight years.

His abilities were well known to the Governors and he was appointed Principal with effect from 1 August 1996 (see photo page 315).

During 1995/96, the College had undergone the exhaustive investigation by HEFCE into its competence which led to the Privy Council giving it the right to award its own taught degrees. Many people contributed to this process but as Vice Principal and Director of Academic Affairs and Research, Wynne Jones' role had been vital. He brought together and motivated the staff at a difficult time, resulting in one of the most significant landmarks in the College's history.

It was to be the first of a number of assessments of the College over the next two years that established Harper Adams as the leading faculty of agricultural education in this country. Dr Jones has provided a continuity in the management of the College, from Tony Harris' period of growth, and improvement through to the end of the century, when its outstanding quality of education became recognized and acknowledged not only by the industry it serves but also by the board of assessment of standards in Higher Education.

Along with his excellent track record at the College, Wynne Jones possessed a strong academic background. He had been awarded a PhD in applied animal genetics at the University of Reading. Prior to his arrival at the College in 1988 he had been Head of the Department of Animal Production at the Welsh Agricultural College, Aberystwyth, now the Welsh Institute of Rural Studies and part of the University of Wales. He was a very popular choice and the College settled down to the task in hand - survival in the face of rigours of the Government's drive for increased efficiency *(see photo page 316).*

Student numbers rose modestly in the mid-1990s, despite the Government's moratorium on growth.[2] For the session 1994-95, when constraints began to be felt, the total number of students stood at 1,525. Five years later the number was 1,665, including forty-two associate students. The skills of the staff, learnt of long years of necessity, ensured maximum recruitment. Always there was an astute appreciation of the opportunities on offer rather than acceptance of the restrictions within the regulations laid down. The sandwich component and newly introduced courses fortunately allowed Harper Adams a degree of flexibility. Even more important, however, was the growth in research and taught post-graduate students and in part-time associate students, who were not limited by the moratorium and who were aiming at upgrading their qualifications in an increasingly competitive world. The numbers game was nonetheless a risky one. Funding Council policy was to reduce the grant by the level of two students for every one student who was recruited over the limit. As Wynne Jones declared: *'It was an exercise in fine judgement to recruit to a level which allowed for a proportion of students to drop out.'* [3] The Annual Review for 1998 proudly declared: *'Harper Adams enjoyed a substantial lead for undergraduate enrolments as compared with any other institution in our sector.'* The most popular course remained agriculture, with the College 'taking the lion's share of the national pool of degree and HND Agriculture students'.[4] As Principal Jones stated: *'The number of students entering agriculture is declining and has been declining for a number of years and we're under no illusion that Harper has been taking an increasingly greater slice of a shrinking cake and that we have to be very clear all the time in our strategy that as Industry concentrates we are in the position to be the best [...].'*

Nevertheless, despite working to levels of maximum recruitment, there was the problem that public funding per student was shrinking in real terms. Government policy, as illustrated clearly by the Dearing Committee of Inquiry into Higher Education of 1997, continued the trend towards public accountability and financial efficiency. The Dearing Report was

Round Bale Push to the Royal Show, Rag 1986.

Student Revue, 1992.

Live band in Queen Mother Hall, 1980s.

End of Session Ball, the dinner, 1995.

End of Session Ball, the morning after, 1998.

Principal Wynne Jones 1996 ongoing.

Principal Graham McConnell 1994-96: Roger Price.

Harper Adams University College, 10 October 1993.

The new piggery, 2000.

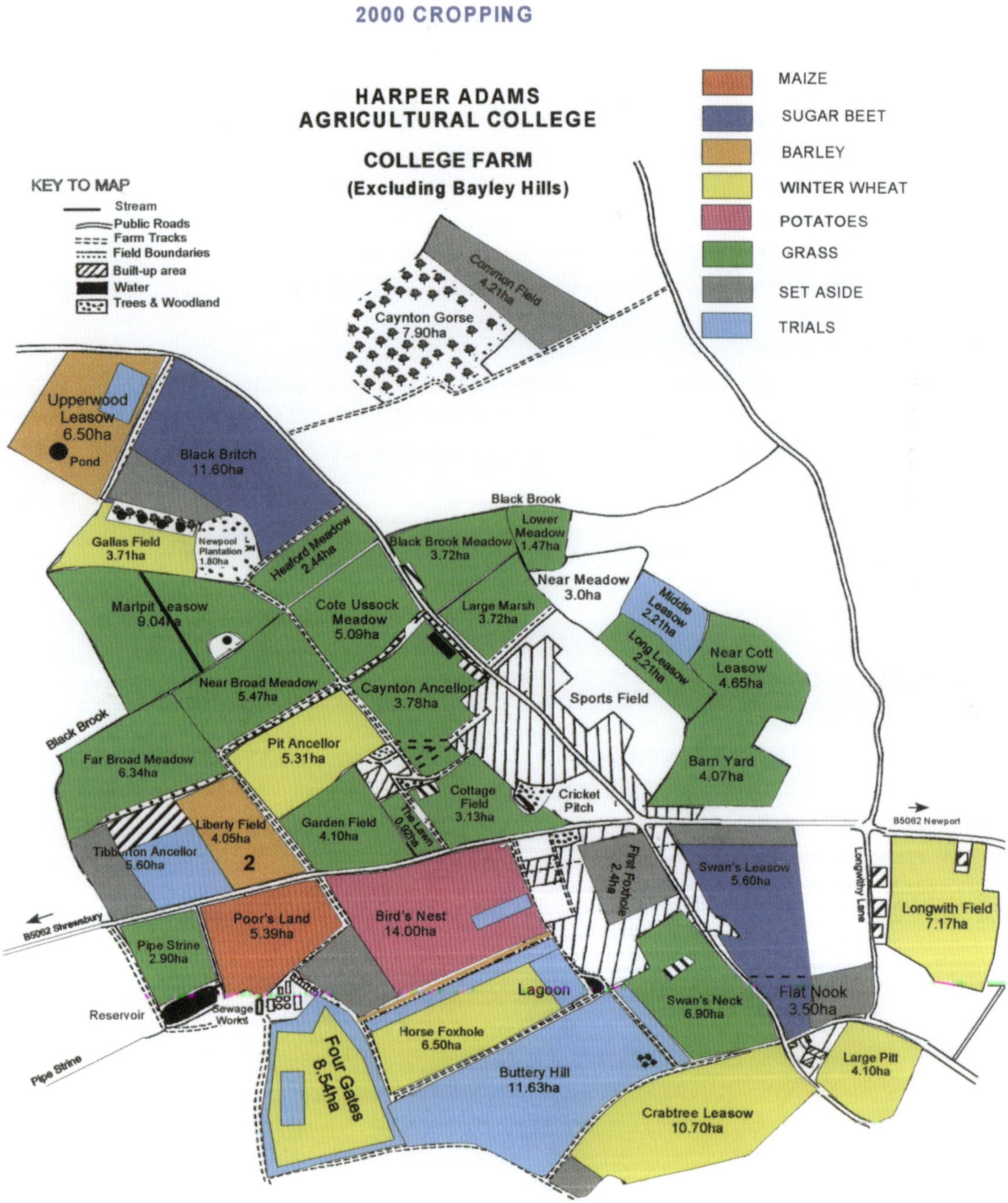

Plan of the College Farm, 2000.

the first major review of Higher Education since the 1960s.[5] Thus, over the period 1996-2000 there was the prospect of continuously reducing expenditure, but fortunately the College, due to the quality of teaching and research resources, won recognition in 1998 as a specialist provider in a high cost sector of education (agriculture) and was awarded a ten percent premium on its funding. This offset, at least temporarily, the potential damaging effects of a continual reduction in its base funding.

However, capital developments, which for the most part had to be self-funded, were dependent on the generation of income. Hopes centred on an extension for the Learning Resources Centre and a further hostel. The first stage of the building work to unite the library, the computing centre and the audio-visual unit was completed in 1998. Laboratory facilities were also refurbished and an additional hall was built for the Engineering Department at a cost of £63,000.[4] There was also a further extension of the farm-based livestock research buildings. This included the replacement of the piggery, at a cost of a third of a million pounds *(see photo page 317)*. The new piggery was built to a standard to conform with Home Office Licence requirements and to promulgate improved welfare. Thirty-five hectares of farmland at Bailey Hills, adjacent to Edgmond village, was purchased in 1996. Student Union facilities were enhanced, with the creation of new offices and a board room and improvement of the main student bar.[6] The Student Union was to become a more professionally managed body.[7] In 1998, for the first time, the post of President was made a sabbatical one due to the pressure of work involved. Hannah Butler became the first student to take up office in this capacity.[4] With a full-time President, assisted by a support officer, the Union was able to strengthen its financial position and increase the investment it was making for the benefit of future students.

Harper Adams in the mid-1990s moved away from a period of rapid growth, but this is not to say that the College stood still. The policy of anticipating trends, both in education and farming, was actively pursued in the face of the deepening recession in British agriculture in the late 1990s. Principal Wynne Jones likened the College's strategy to the game of rugby: *'A good player will watch when somebody is kicking the ball up in the air and move immediately in anticipation of where the ball is going to land, rather than try and run after the ball once it has landed.'* New ideas were tried, new links forged and the perimeters of experience were ever stretched. Consequently, the College has evolved along with the Industry it serves.

At the root of policy through the 1990s was the principle of diversification in line with the changing nature of the Industry. Agriculture courses became more focussed along the food supply chain,[5] and in 2000 the new Degree course in Food and Consumer Studies was introduced. The dramatic decline in farm incomes in the late 1990s led to the introduction of measures encouraging farmers to enhance the rural environment and to diversify the farm economy. Many farmers sought new ways of supplementing the farm income by introducing non-farming enterprises to the farm or by employment off the farm. Recognizing these momentous changes would impact on the training provided and the fact that the modular system was in need of updating, the College undertook a major review of its provision in 1998-99 under the banner of 'Modules for the Millennium'. The aim was to make the provision more coherent in terms of student choice and progression for both full-time and associate students. Also with the increasing tendency of students to pursue Higher Education locally, some of the modules and the courses became more generic in nature, especially in the areas of engineering and business studies. The new modular system was validated early in 2000 and provides the opportunity for an ever widening portfolio of courses. With the need for and interest in rural development,[8] in 1999 the College introduced a Countryside Development Unit under the leadership of Steve Parsons. The Unit involves itself with research, participation in rural projects and organizing appropriate conferences. Throughout these developments the College has ensured the core subject remains agriculture.

It was foreseen that increased competitiveness within the Industry would lead to demand for higher level skills. New ventures included a 'raft of masters programmes' in agricultural marketing, food marketing and quality management, crop protection, animal health and management and organic production.[9] Degree teaching continued to receive growing emphasis. By late 1997 the proportion of degree students had reached sixty-six percent of the total student population. In September 1995 the Degree in International Agriculture was offered for the first time, described as 'an exciting prospect for the expansion of the course programme and developing further links with Europe'. The degree was a joint venture with Larenstein International Agricultural College in the Netherlands, with a substantial part of the course revolving around student exchanges between the two and a placement period outside the home country.[6] Issues over quality assurance and the lack of interest of United Kingdom students in the course are placing its future in doubt, but the experience gained will be of value for future joint international ventures. The College in the 1990s increasingly sees its role in partnership with other institutions, most notably on a regional basis with those in Further Education. Such partnerships allow the College to tap into Government encouragement of Higher Education, particularly at sub-degree level, in Further Education colleges and thus to 'increase student numbers without heavy capital expenditure'.[9] Tangible outcomes have been the signing of a strategic alliance with Reaseheath College on the joint development of courses and the introduction of a degree in animal health run in association with Rodbaston College.

Harper Adams' commitment to degree teaching was rewarded in September 1996 with the granting of taught degree awarding status.[10] The College's early decision to remain independent meant that this was a necessary step in the competitive world.[5] The Dearing Report of 1997 endorsed the establishment of the category of university college for those institutions which possessed degree awarding status as well as those who hitherto had been federal colleges of a larger university. This was a significant milestone for Harper Adams, which began using the title from 1998. The title gave Harper Adams 'a very clear position in the market place'.[5] As the Annual Review of 1998 declared: *'The inclusion of the word "University" is more than a mere formality, it is an acknowledgement of the excellence of teaching, research and overall quality of Higher Education at this centre.'* The next target for Harper Adams is to obtain research degree awarding powers.

The excellence on offer at Harper Adams was perfectly illustrated in 1998, when the Teaching Quality Assessment of the Higher Education Quality Assurance Agency, established in 1997, awarded the College a score of twenty-three out of a maximum of twenty-four points. Throughout the 1990s the Government was continuing to place greater emphasis on quality and standards in Higher Education.[11] The College's positive features were outlined. These included a curriculum which offered a number of starting points at different levels and provided a number of pathways to meet individual needs. Other highly commended points were the high quality teaching which actively involved students and the excellent resources.[4] In 1998 the staff structure was reorganized in a bid for even greater efficiency in the wake of the substantial changes which had taken place since the late 1980s:

Functions were disaggregated and a layer of management, the Vice Principal's post, was removed. Greater flexibility was achieved by creating schools comprising a number of groups. Within the wider school structure, it was easier to respond to new demands. New subject areas which could not be tackled within the smaller department structure, could be introduced. A greater level of budget delegation to schools and to groups, in the schools, became possible so that staff could own their budget. Any entrepreneurial income accruing from departmental activity could return to the generating department.[5]

Thus the School of Management was created under Professor Brian Revell and the School of Agriculture under Professor Andrew Cobb. At the same time the Principal received the title of Professor.[10]

In a very practical way the continuing quality of the educational provision can be illustrated by the fact that Harper Adams produces students who are highly suited to the market place. The employment record of Harper Adams students within six months of leaving is nearly thirty percent higher than the national average for all universities and colleges in the Higher Education sector.[4] As Ian Robson, Head of Student Services, stated: *'A Harper Adams student is employable. It's the sandwich element - the maturity, the work they do on sandwich, the work they do here and how they mature and the groups of friends they've got around them. Harper people employ Harper people, because they know they're good workers.'*

As the Agriculture Industry faced increased competitiveness, declining profits and financial stringency it was inevitable that redundancies would follow. Those remaining in employment invariably needed a broader and updated range of skills, especially with the increased pace of technological development.[8]

The answer lay in lifelong learning, which was very much supported in the Dearing Report. This was an especially important growth area following the Government's limitation on any expansion in full-time Higher Education. In 1996 Harper Adams secured an annual sum of over £200,000, to be paid over a four-year period and to be used to enhance participation in lifelong learning.[5] This was the culmination of a two-year research programme, funded by the Department of Employment and Education, into the special needs of land-based industries.[10] Lifelong learning at its most flexible combines short courses, part-time study and work-based learning, all linked to a credit accumulation system. Thus, side by side with the encouragement of lifelong learning goes the continuing policy of developing 'a broad range of short courses for a broad range of companies and institutions'.[10] In 1998-99 forty-seven short courses were run, attended by 698 delegates and involving 4,256 delegate days of training. This generated a fee income of £238,827. The complexities of timetabling staff to undertake such a flexible programme together with the need for specialists to develop the content of short courses has led to a system where courses are managed and delivered by staff in the Schools. They are supported by the Professional and Career Development Services staff which includes the Programme Manager, the Conference and Short Course Service and the Careers Adviser.

Essential to the establishment of a Higher Education culture at Harper Adams is research. As before, one object of research is to underpin teaching. On occasion it can lead to the creation of new modules in existing courses and new undergraduate programmes.[6] However, research has always to be self-financing and ideally should make a profit. Neither does Harper Adams waver from its policy of making research and advice available to the rural community. Research remains essentially applied. Similarly, research is an ideal way of forging sound links with Industry. As Wynne Jones explained: *'[Such a policy] enables the College to act as a conduit for reconciling the science "push" with the industry "pull"* '.[6] From 1992 research funding had been linked to an external review process. In the mid-1990s the College sought to improve its research rating in order to obtain greater funding and in 1996 the College was awarded a rating of 3b, a significant advance on the 1992 rating.

The Crop Production and Science Department continues its good record of successfully attracting funded research and there are many beneficial links with commercial companies. In 1996, within the Department six dedicated researchers were supported by fourteen post-graduates. In 1997 a specialist building for the Crop and Environment Research Centre (CERC) was opened at a cost of £350,000. Its task was to undertake research in crop development and biotechnology, allowing complex techniques like DNA 'fingerprinting', which allows the identification of potato nematodes, to be carried out under one roof.[10] It does much to boost the College's profile in the area of published research.

Other important research is undertaken by the Animal Production and Science Department. For its part, it possesses the facilities of the new Animal Science Research Centre. Research is primarily of a nutritional nature, such as the study of the fat metabolism of sheep, rumen degradability and synchrony studies. Poultry work is similarly nutritionally orientated, but there are also investigational studies of perchery design and leg weakness in broilers.[4]

Harper Adams frequently seeks partners in its bid to broaden research opportunities. A promising new venture is the provision of dairy research and consultancy in the form of the Dairy Research Consultancy (DRC Ltd). This consortium, composed of the Scottish Agricultural College, the University of Reading, Myerscough College and Harper Adams, was created in 1995.[6] It aims to provide a consultancy, information and training service utilizing the research and technical resources of the four institutions.[12] The Milk Development Council, milk purchasing companies and feed manufacturers are all customers. The Crop Agronomy Research and Information Group (CARIG) is a similar joint venture, composed of Levington Agriculture, Crop Technology Ltd and Harper Adams. CARIG began in 1996.[6] Harper Adams is in the heart of the country's fastest growing potato region and the group was set up to provide information and training services. CARIG is now a major centre for potato research, undertaking work for companies and the British Potato Council.[10] It grows the British Potato Council reference crops for the West Midlands, monitoring the growth and development of varieties.

As the 1990s drew to a close, Higher Education institutions were being increasingly encouraged by the Government to become more involved with industry, commerce and communities. Harper Adams was already experienced in this field. The College's considerable links with Industry through its research, consultancy and training programmes and the high quality of its educational provision resulted in 1999 in the award of £550,000 over four years to develop further its already substantial technology transfer programme. This was under HEROBIC (Higher Education Reach-Out to Business and Community Fund). A development which will fulfil Harper Adams' mission and bring benefit to the diverse industries, professions and communities associated with the countryside.

The College Farm 2000

The College currently owns 278 ha of land, of which 207 ha are farmed, the remainder being buildings (16 ha) and woodland (11 ha). A further 76 ha of farmland is rented. The core location of the farm remains consistent with that of 100 years ago *(see photo page 318)*. The College Farm operates a range of enterprises covering dairy, poultry, arable (potatoes, sugar beet, cereals, forage maize, grassland and feed mill).

The pig, beef and sheep operation of the College Farm are managed from the Animal Production and Science Group whilst the Crop Production and Science Group operate an extensive programme of crop experiments and trials covering some 50 ha of land based at the College or on adjacent local farms or nationally.

The National Institute of Agricultural Botany and other organisations are located at Harper Adams and operate farm trials at the College.

The dairy is located adjacent to the feed mill and considerable investment has been made in sophisticated milking parlour facilities, housing, manure handling and forage storage. Individual feeding stations for the recording of feed-intake data support the research work. Forage is based on grass, maize, whole crop cereals and grass silage. All the cattle are fed by a complete feeding wagon where essentially the forage and concentrates are mixed into one ration.

The pig unit was completely refurbished in 1999 *(see photo page 317)* . The primary purpose of the unit is to provide for the teaching, research and development functions of the College. The scale of enterprise required for commercial viability could not be accommodated within other constraints.

Poultry, both commercial and experimental, feature prominently in the farm-based activities. Commercial activities are restricted to egg production. Experimental work continues with eggs, meat (both chicken and turkey), nutrition and housing strategies. The demands of intensification require the College to be very strict and rigid on hygiene control and therefore bio-security measures need to be adopted when students are entering the animal units.

The feed mill provides food for all the poultry on the farm whilst a very small more sophisticated feed mill provides rations for the experimental activities of the enterprises.

Most of the farm tractor work is undertaken by farm staff except for silage making and corn harvesting where the specialist machinery is expensive and the size and scale of the College enterprise make the hire of specialist contractors more attractive financially.

Stock numbers include 200 dairy cows, 85,000 laying hens, 120 sows, 100 beef cattle, 220 sheep.

Breeds

The breed of dairy cows is the Holstein whilst the majority of the beef cattle are continental cross dairy breeds. The ewe flock tends to be based on cross-bred ewes crossed either with a traditional Suffolk breed or a continental breed namely Charollais or Texel. Both the pig and poultry enterprises make use of hybrid breeding stock acquired from specialist breeding companies.

Environment and Conservation of Bio-diversity
The farm is located in a nitrate vulnerable zone and therefore careful auditing and rationing of organic manures needs to take place. The College estate adopts the concept of integrated crop management and this has currently been extended to integrated farm management to ensure coherent standards of husbandry and rotation. Superimposed on this are participation in the Countryside Stewardship and Arable Stewardship Schemes aimed at enhancing the bio-diversity of both flora and fauna of the farm and showing students how the imperatives of modern farming are reconcilable with those of husbanding the environment. This important element is embraced in the curriculum for all students.

The College Farm provides for commercial, teaching and research functions. Latterly, increasingly, it is a focus for farmers to see the latest technology embraced within coherent systems to enhance efficiency and profitable farming as part of the developing role of knowledge transfer. Computing and information technology application feature prominently in equipment control, management and marketing decisions taken on the farm.

Professor Wynne Jones

Conclusion: Harper Adams through a Hundred Years

History can very often become simple nostalgia. This is not to dismiss those who read this book with the aim of rekindling memories of their happy days at Harper. The College is the sum of its people and a very important part of this book has been the memories of those people. Indeed, the book would have failed in one of its most important aims if it did not encapsulate the spirit of the College. Harper Adams inspires a very special kind of loyalty. Its distinctive character has spawned fond memories for its students for a hundred years and is set to go on doing so. Behind the human story, however, is an equally absorbing one, that of a college fine-tuning to the demands made upon it, whether they arose from changes in farming, the rural scene, student attitude, education expectations or simply the funding environment. Just as an organism dies if it does not respond to nature, then a living institution must respond to the challenge of change.

The history of one hundred years at Harper Adams is not surprisingly intricately linked to the two themes of agriculture and education. The Twentieth Century included many difficult years for agriculture, with its fluctuating fortunes, dramatic changes in everything from Government policy to farming techniques and by the end of the period, a drastic reappraisal of the very role of farming in the countryside. Throughout, the College kept in close touch with the Industry that it served and ensured it catered to the developing demands of the market. Similarly, as succeeding governments have extended Higher Education from the few to the many, the College has responded. It increased student numbers and embraced new methods of teaching. It adopted advances in technology and met demands for changing Higher Education qualifications by providing a curriculum, varied in subject matter, range and flexibility. To manage these frequent changes, the structural organization of the College has been constantly adapted and amended. In recent times the Government has made Higher Education institutions highly accountable for the public funding they receive, at the same time as encouraging institutions to seek increased income from private and commercial sources. Responsiveness to the fluctuations of financial reality has always been at the heart of College policy. Today the College moves to the tune of the strategic plan, a document which epitomises the College's philosophy of meeting the challenge of change. However, the simple fact of recognizing how the College is placed and deciding where it needs to go is far from new.

So, as the College passes the landmark of its Centenary, its size and scale, its objectives, the character of its courses and the very nature of its students can seem quite alien to the boys public school character of its very early days and indeed they are. However, one theme runs throughout these hundred years, namely the ability to respond to and even to anticipate change. A list of the developments and achievements so evident today could give little impression of the substantial strength which has anchored the College through some of the most changeable years known to agriculture and to education. Harper Adams' strength is born of its heritage and it is this fact which lies at the heart of all the Centenary celebrations.

Harper Adams Chairmen of the Board of Governors

1901-06	Sir Thomas Fletcher Boughey, Bart., JP.
1907-16	Colonel Sambrooke Thomas Higgins Burne, BA, JP.
1916-26	William Heath Lander, JP.
1926-41	Colonel Harold Platt Sykes.
1942-56	Thomas C. Ward.
1956-73	The Rt Hon. The Earl of Bradford, TD, JP, DL.
1973-77	George H. Dodd
1977-85	Sir Gwilym T. Williams, CBE.
1985-90	Peter Watson Jones, BA Ag Cantab, FRAgS.
1990-94	The Viscount Boyne, KCVO.
1994-99	Richard T. Matson, FRAgS.
1999 ongoing	Anthony F. Burgess, FCA MBIM.

Harper Adams Student Union Presidents

1907	J. C. Brown
1922	C. E. Pearson
1922	V. Liversage
1923	G. W. Spencer
1924	E. F. Martin
1925	F. C. Bobby
1925	D. J. B. Jervis
1926	W. B. Hutchinson
1927	N. F. Dixon
1928	C. Lacey
1929	R. M. Burke
1930	W. H. Parker
1931	R. P. Rice
1932	W. M. Swindells
1933	J. A. Collier
1934	R. A. Dyke
1935	A. J. Higginson
1936	E. W. Wood/C. O. Wright
1937	G. W. Hart/P. L. Poole
1938	G. R. Adams
1939	P. G. Delingpole
1940	H. J. M. Brown
1941	J. Cotterill
1942	J. R. Mason
1943	S. D. Virtue
1944	T. Jones
1945	D. Southall
1946	D. J. Oatley
1947	Ll. Griffith-Jones
1948	D. Greenhalgh
1949	K. Base
1950	M. H. Clay
1951	D. I. W. Davies
1952	C. J. Hayes
1953	W. R. H. Woods
1954	D. E. Roberts
1955	J. A. C. Weir
1956	J. S. Wood
1957	G. N. Parker
1958	F. S. Oakeshott
1959	T. McCluggage
1960	D. R. Withinshaw
1961	P. V. Hamilton-Hunter
1962	T. R. Mould
1963	R. J. Esslemont
1964	J. B. Mathias
1965	K. Henton
1966	T. F. Maxwell
1967	W. J. Wilder
1968	A. C. Borrett
1969	J. Martin
1970	D. C. Ralli
1971	C. M. C. Rowland
1972	R. D. Williams
1973	P. Heydon
1974	B. Florendine
1975	S. Burt
1976	M. Pinard
1977	C. Briscoe
1978	W. S. McArthur
1979	D. A. Pascall
1980	C. Harris
1981	Fiona Fitzgerald
1982	B. McKenzie
1983	C. Evans
1984	Liz Maxted
1985	J. Sanderson
1986	M. P. Miles
1987	J. R. Galloway
1988	D. V. Lee
1989	J. B. Pearce
1990	I. W. Burdass
1991	D. J. McNeil
1992	S. J. H. Wragg
1993	D. E. Jones; Rachel Griffiths
1994	R. T. Lloyd
1995	P. C. Chaplin
1996	H. J. V. Broom
1997	M. C. P. Ellwood
1998	Hannah Butler
1999	Rachel Marsland
2000	I. Napper

Harper Adams Club Presidents

1920-27	The Viscount Lewisham
1927-29	R. A. Dalley
1929-31	A. Roebuck
1931-33	F. C. Bobby
1933-36	G. K. Parkes
1936-38	C. E. Pearson
1938	L. Poulter
1939	R. H. Axworthy
1940-45	R. T. Wiglesworth
1946-48	R. H. Axworthy
1948	L. Poulter
1949	R. A. Dalley
1950	W. H. Parker
1951	P. H. Jeavons
1952-53	F. R. Mason
1954	Ll. Griffith-Jones
1955	J. A. Alexander
1956	J. E. Harrison
1957	J. L. Smedley
1958	Mrs H. M. Warner
1959-60	C. J. Hayes
1961	J. S. Parker
1962	D. L. Jones
1963	N. V. Elcock
1964	J. S. Tavernor
1965	I. B. Howie
1966	D. W. Smith
1967	G. J. Pickering
1968	C. L. Johnson
1969	I. P. Guthrie
1970	A. K. Heathcote
1971	A. Bostock
1972	G. E. Fawdry
1973	J. D. Poulter
1974	P. D. Hebblethwaite
1975	G. Solari
1976	M. S. Keightley
1977	P. W. Singleton
1978	K. Stevens
1979	D. G. Poulter
1980	P. V. Hamilton-Hunter
1981	E. G. Clent
1982-83	R. P. Simkin
1984-85	J. R. Twigge
1986-87	C. Briscoe
1988-89	R. Tomley
1989-90	L. Morris
1991-92	R. J. Ackroyd
1993-94	B. S. F. Gunn
1995-97	Mrs R. J. Barsley
1998-99	E. A. Pridgeon
2000-01	Mrs R. J. Barsley/E. A. Pridgeon

List of Illustrations

Chapter 12

Chapter 13

Colour Plates

References

Abbreviations

FGP Finance & General Purposes Committee

HAUC Harper Adams University College

KMA Kilvert Minor Adams papers, Private Collection

LH Liddle and Heane papers, SRRC

NA Newport Advertiser

PRO Public Record Office, Kew

SC Shrewsbury Chronicle

SCA Shropshire Chamber of Agriculture

SCC Shropshire County Council

SRRC Shropshire Records and Research Centre

General Bibliography

Edith H. Whetham, The Agrarian History of England and Wales 1914-39 (Cambridge: Cambridge University Press,1978).

Mike Soper, Years of Change ([n.p.]: Farming Press, 1995).
B. A. Holderness, British Agriculture since 1945 (Manchester: Manchester University Press, 1985).

T. C. Morgan, 'Development of Agricultural Machinery over the Century', Century of Progress 1875-1975, Shrophire and West Midlands Agricultural Society Centenary Brochure ([n.p.]: Shrophire and West Midlands Agricultural Society, 1975).

C. S. Orwin and E. H. Whetham, History of British Agriculture 1846-1914 (Newton Abbot: David and Charles, 1971).

Stewart Richards, Wye College and its World, A Centenary History (Wye: Wye College, 1994).

Edith H. Whetham, 'The Mechanisation of British Farming 1910-45', Journal of Agricultural Economics, 21(1970), 317-31.

Brian Hill, Britain's Agricultural Industry (London: Heinemann Educational Books, 1975).

General Material in the HAUC Archives

Governors' Minutes

Principal's Reports

Chapter 1

1. HAUC, Newscuttings 1901-04, SC, 27 September 1901.
2. Ibid. *[Staffordshire Advertiser ?]*, September 1901.
3. Ibid., *NA*, 28 September 1901.
4. *Report of the Departmental Committee appointed by the Board of Agriculture and Fisheries to inquire into and report upon the Subject of Agricultural Education in England and Wales, Minutes of Evidence*, Cd 4207 (London: HMSO, 1908), pp. 204-208.
5. LH, Papers High Court of Justice Probate Divorce and Admiralty Division (Probate), in the Goods of Thomas Harper Adams Deceased, Particulars of Estates and Valuation of Messrs Pooler and Hammonds, 1101/104.
6. LH, Papers High Court of Justice (Probate), Brief for the Defendant, 1101/104.
7. SRRC, Ralph Purton, Shropshire Pedigrees, Adams, 6001/4360. The author is grateful to David Smith for his help with research on the Adams Family.
8. HAUC, Volume Adams Legal Material, 100/4.
9. SRRC, *NA*, 18 June 1892.
10. LH, Marriage Settlement John Adams and Sarah Harper 13 November 1804, 2451/2.
11. SRRC, H. D. G. Foxall, Field Map of Tibberton.
12. KMA, Articles for Building a House at the Ansellor Farm, John Carline and John Adams 30 December 1806.
13. SRRC, Ancellor Estate Papers, Indenture of Lease and Release, Robert Pigott to William Adams the Younger 9-10 August 1779, 214/8.
14. KMA, Invoice John Carline to John Adams 14 February 1809.
15. KMA, Schedule of Anceller House.
16. LH, 1101/104-107 and KMA.
17. Ancellor Estate Papers, Indenture Robert Pigott and others 13 May 1695, 214/1.
18. Margaret Gelling, *The West Midlands in the Early Middle Ages* (Leicester: Leicester University Press, 1992), pp. 48-52.
19. SRRC, Edgmond Baptisms, MF 215.

20. Ancellor Estate Papers, Settlement by John Adams Esquire of Freehold, Copyhold and Personal Estates in favour of his three Sons 1 February 1832, 214/26.
21. KMA, Document to settle the Question of the Contribution of £100 Annuity to Mrs Adams.
22. LH, Notes taken on the Examination of Mrs Sarah Adams on Commission, 1101/106. The author is grateful to George Baugh for his help in transcription.
23. LH, Abstract of Will of the Late Thomas Harper 13 May 1829, 1101/106.
24. HAUC, Correspondence Tom Gollins 1995-96, 100/10.
25. Dutton, 'Moreton Mill Methodist Church', *The Wesley Historical Society (Shropshire Branch)*, 5 (December 1999).
26. LH, Evidence of Edwin Hazledine Harper, Professor of Music, Haverstock Hill, Middx 1101/105.
27. LH, Marriage Settlement Thomas Harper Adams and Sarah Dickin 26 May 1841, 2451/8.
28. St Peter's Churchyard, Edgmond, Adams Memorial.
29. LH, Papers High Court of Justice (Probate), Case for Council to advise and prepare Statement of Defence, 1101/104.
30. LH, Death Certificate of Thomas Harper Adams 7 June 1892, 2451/8.
31. HAUC, Copy of the Will of Thomas Harper Adams, 100/1.
32. LH, Papers High Court of Justice (Chancery), Office Copy Affidavit of Mrs Minor in Support of Summons, 1101/104.
33. HAUC, Interview with Vera Bradley, 102/110.
34. LH, Boughey v. Minor Proofs of Witness, 1101/104.
35. LH, Notes taken at the Examination of Mrs Sarah Adams on Commission at Edgmond House, on Saturday the 10th day of December 1892, 1101/104.
36. LH, Papers High Court of Justice (Probate), Copy of Dr Brereton's Notes, 1101/107.
37. The author is grateful to Dr C. H. Lisk, Linden Hall Surgery, Newport, for the conjecture.
38. LH, Papers High Court of Justice (Probate), Thomas Harper Adams Proofs of Witness, Dr Thorpe, 1101/107.
39. LH, Thomas Harper Adams Deceased, Boughey v. Minor, Copy of General Correspondence referred to in Plaintiff's Affidavit of Documents, Letter from Kenyon-Slaney 12 December 1890, 1101/104.
40. LH, Copies of the Wills of Thomas Harper Adams 31 December 1890 and 20 March 1891, 1101/104.
41. The author is grateful to Mrs Irfana Jones, Gwynne & Sons, Wellington, for legal information.
42. SCA, Minutes, 25 January 1879.
43. Michael Argles, *South Kensington to Robbins, an Account of English Technical and Scientific Education since 1851* (London: Longmans, 1964).
44. SCA, Reports, [...] & Proceedings of the Central Chamber, Council Meeting of the Central Chamber, March 1878.
45. Andrew Humphries, *Seeds Of Change: A Hundred Years Contribution to the Rural Economy, Society and the Environment* (Newton Rigg: Newton Rigg College, 1996).
46. SCA, Reports & Proceedings, Report 31 December 1879.
47. SCA, Minutes, 25 February & 3 March 1888.
48. *Report of the Departmental Committee appointed by the Board of Agriculture and Fisheries to inquire into and report upon the Subject of Agricultural Education in England and Wales*, Cd 4206 (London: HMSO, 1908), pp.3-6.
49. Paul Sharp and John Dunford, *The Education System in England and Wales* ([n.p.]: Longman, 1990).
50. Clive Beale and Geoff Owen, *Writtle College: the First Hundred Years 1893-1993* (Chelmsford: Writtle College,1993). Paul Brassey, 'Agricultural Science and Education', *The Agrarian History of England and Wales 1850-1914*, ed. E. J. T. Collins (Cambridge: Cambridge University Press, [in print]).
51. *Shropshire Chamber of Agriculture 1866-1986*, ed. by E. R. Dinnis ([n.p.]: SCA, [1986?]), pp. 3-4.
52. W. J. Slack, *Origins of the National Farmers' Union Shropshire County Branch* (Shrewsbury: NFU Shropshire County Branch,1950).
53. SRRC, SC, 12 December 1890.
54. Ibid., SC, 15 August 1890.
55. *Myerscough College, a Lancashire Centenary* (Myerscough: Myerscough College, 1993).
56. SRRC, SC, 7 November 1890.
57. Ibid., SC, 19 December 1890.
58. LH, Examination of Miss Davey, 1101/105.
59. Advertiser Offices, Newport, *NA*, 24 January 1891. The author is grateful to Mrs Judy Stephens and the Newspaper Research Group of the Newport & District History Society for help.
60. Ibid., *NA*, 28 February 1891.

61. HAUC, Newscuttings 1963-77, *NA*, 18 February 1966.

62. Advertiser Offices, Newport, *NA*, 6 June 1891.

63. Ibid., *NA*, 13 June 1891.

64. Ibid., *NA*, 18 July 1891.

Box 1

a. *A New English Dictionary on Historical Principles*, ed. by James A. H. Murray (Oxford: Clarendon Press, 1901). Georgina F. Jackson, Shropshire Word-Book (London: Trubner,1879). The author is grateful to Freddie Custance and Russell Evans, SRRC, for their help.

Chapter 2

1. LH, Death Certificate of Thomas Harper Adams 7 June 1892, 2451/8.

2. Copy of the Will of Thomas Harper Adams, 100/1.

3. HAUC, Attendance Book of Foundation Meetings 1896-1963, 101/5.

4. *Shropshire: A Short History with Genealogies and Current Biographies*, ed.by John Grant (London: London & Provincial Publishing, [n.d.]).

5. *The History of the Newport and District Agricultural Society: The First Hundred Years 1889-1989*, ed. by David R. B. Adams ([Newport]: Newport and District Agricultural Society [1989]).

6. *County Biographies: Shropshire*, ed. by John Dale (Birmingham: Hammond, 1904).

7. Shropshire, *Mate's County Series* (Bournemouth: Mate, 1906).

8. Alfred Higgins Burne, *A Memoir of Sambrooke Thomas Higgins Burne* ([n.p.: Burne, 1954]).

9. Advertiser Offices, Newport, *NA*, 27 January 1900.

10. Ibid., *NA*, 18 June 1892.

11. LH, Valuation Pooler and Hammonds, 1101/104.

12. LH, Grant of Probate of Thomas Harper Adams 24 February 1893, 2451/8.

13. The author is grateful to Mrs Irfana Jones, Gwynne & Sons, Wellington, for legal information.

14. HAUC, The Harper Adams Foundation Scheme 29 June 1896, 1/1A.

15. *Report upon Agricultural Education*, Cd 4207, pp. 204-208.

16. B. R. Mitchell, *British Historical Statistics* (Cambridge: Cambridge University Press, 1988).

17. *Report upon Agricultural Education*, Cd 4207, pp. 208-214.

18. PRO, Papers Ministry Agriculture and Fisheries 1891-1907, Letter C. R. Liddle to the Secretary of the Board of Agriculture 20 July 1893, MAF 33 13 A1018/1901.

19. HAUC, Plan and Particulars of Edgmond House, to be sold at Auction 10 July 1894, 100/2.

20. HAUC, Misc. File Hedworth Foulkes, Board of Agriculture Matters, Information provided for the Board June 1907.

21. Paul Brassey.

22. *Report upon Agricultural Education*, Cd 4206, pp. 3-6. No reference is made in the text to agricultural education in Scotland or Ireland.

23. Roger Sayce, *The History of the Royal Agricultural College, Cirencester* ([Stroud]: Alan Sutton, 1992).

24. Andrew Humphries.

25. G. E. Jones and B. K. Tattersfield, 'John Wrightson and the Downton College of Agriculture', *Agricultural Progress*, 55 (1980), 69-77.

26. Paul Harris, *The Silent Fields: A Hundred Years of Agriculture at Reading* (Reading: Reading University, 1993).

27. A. D. Hall, 'The Development of Agricultural Education in England and Wales', *Journal of the Royal Agricultural Society of England*, 83 (1922), 15-34.

28. *Myerscough College.*

29. Clive Beale and Geoff Owen.

30. *Cheshire Notes and Queries*, 3 (1898), 53-55.

31. Tony Flint, John Mitchell and Jerry Roberts, *Centenary Celebrations at the Faculty of Agricultural and Food Sciences Sutton Bonington Campus* (Nottingham: University of Nottingham [1995?]).

32. *Sparsholt College Hampshire. Seventy-Fifth Anniversity. Special Edition Magazine* ([Sparsholt: Sparsholt College, 1989]).

33. Uckfield and District Preservation Society, *Village into Town: The Growth of Uckfield 1788-1988* (Uckfield: Uckfield and District Preservation Society [1988?]).

34. A History of the County of Stafford, ed. M. W. Greenslade and J. G. Jenkins, *The Victoria History of the Counties of England*, 2 (Oxford: Oxford University Press for the Institute of Historical Research, 1967).

35. SRRC, SC, 8 November 1895.

36. Ibid., SC, 9 November 1900.

37. SRRC, T. M. D. Jones, 'Development of Education Provision in Shropshire 1870-1914' (Unpublished Master's thesis, University of Keele, 1969), MF 90.

38. D. C. Cox, *Shropshire County Council: A Centenary History* (Shrewsbury, SCC, 1989).
39. Newscuttings 1904-10, *NA*, 21 July 1906.
40. Newscuttings 1901-04, *NA*, 28 September 1901.
41. Ibid., *NA*, 27 April 1901.
42. The author is grateful to Mr Graham Talbot, Newport, for this information.
43. Graham Talbot, *They said I should write a Book* ([Newport]: Graham Talbot, 1997).
44. Newscuttings 1901-04, *Edgmond Parish Magazine*, May 1901.
45. HAUC, Misc. File Hedworth Foulkes, Early Correspondence, Letter R. P. Liddle to Hedworth Foulkes 28 December 1900.
46. *The Building News*, 15 October 1897.
47. HAUC, Memorandum on the History and Work of the Harper Adams Agricultural College, Newport, Salop, 7 March 1933, 101/7.
48. Newscuttings 1904-10, *Staffordshire Weekly Sentinel*, 8 May 1909.

Box 2

a. Newscuttings 1901-04, *Wellington Journal*, 23 February 1901.

Chapter 3

1. HAUC, 'HAAC in 1904', Cat-A-Mountain, 1950.
2. 'A Decade', *Ibid.*, 1925.
3. Newscuttings 1901-04, *SC*, 27 September 1901.
4. HAUC, Prospectus [1902?], 6/5.
5. Newscuttings 1901-04, *NA*, 27 April 1901.
6. *The Building News*, 15 October 1897.
7. Interview with Dorothy Neate, 102/100.
8. HAUC, Principal's Report, 1936.
9. HAUC, FGP, 13 June 1947; Governors' Minutes, February 1949.
10. HAUC, Student File 224.
11. Newscuttings 1901-04, *NA*, 4 October 1902.
12. Interview with Albert Rowley.
13. Interview with Mellor Boyns, 102/104.
14. 'A New Start', *Cat-A-Mountain*, 1988.
15. 'Fifty Years Ago - I was there', *Ibid.*, 1977.
16. 'A Retrospect 1922-44', *Ibid.*, 1944.
17. HAUC, Finance Committee, 14 November 1928, 104/18.
18. 'Obituary', *Cat-A-Mountain*, 1953.
19. Newscuttings 1901-04, *NA*, 28 September 1901.
20. HAUC, Board of Agriculture Returns, 1924.
21. Principal's Interim Report, July 1937.
22. *Report upon Agricultural Education*, Cd 4207, pp. 204-208.
23. Newscuttings 1904-10, *Wellington Journal*, 11 May 1907.
24. *Report upon Agricultural Education*, Cd 4207, pp. 208-214.
25. Newscuttings 1904-10, *Staffordshire Weekly Sentinel*, 8 May 1909.
26. Ibid., 21 August 1909.
27. Newscuttings 1901-04, *Standard*, 24 September 1901.
28. Ibid., *Country Gentleman*, 4 July 1903.
29. Ibid., 15 August 1903.
30. Newscuttings 1901-04 *[Staffordshire Advertiser?]*, September 1901.
31. Memoirs of Bernard Gregg, *Cat-A-Mountain*, 1987.
32. Prospectus [1902?], 6/4.
33. Paul Harris, Stewart Richards and C. Tyler, 'The History of the Agricultural Education Association 1894-1914', *Agricultural Progress*, 48 (1973), 1-9.
34. Newscuttings 1901-04, *Wellington Journal*, 23 February 1901.
35. Newscuttings 1904-10, *Wellington Journal*, 16 July 1908.
36. Prospectus 1908, 6/8.
37. Newscuttings 1904-10, *NA*, 22 May 1909.
38. Ibid., 5 August 1905.
39. Paul Brassey.
40. SRRC, *SC*, 9 November 1900.
41. Newscuttings 1904-10, *NA*, 2 June 1906.
42. Newscuttings 1901-04, *NA*, 3 August 1901.
43. 'Parliamentary Publications, Board of Agriculture and Fisheries, Grants for Agricultural Education and Research', *Journal of the Board of Agriculture* [?] (1903-4), 409.
44. Newscuttings 1901-04, *Stafford Chronicle*, 28 September 1901.
45. T. M. D. Jones.
46. Newscuttings 1904-10, *NA*, 13 March 1909.
47. Newscuttings 1901-04, *The Field*, 25 May 1901.
48. Newscuttings 1901-14, Nature Study Course Promotional Leaflet, p.5.
49. Newscuttings 1904-10, *NA*, 22 August 1908.
50. Newscuttings 1912, two unidentified articles on nature study at the College.

51. *Report upon Agricultural Education*, Cd 4207, p.178.
52. *Journal of the Board of Agriculture* (1903-4), 410.
53. Paul Harris, Stewart Richards and C. Tyler.
54. HAUC, *Guide to Field Experiments*, 1909.
55. Newscuttings 1901-14, *Guide to Field Experiments*, 1904, p.49.
56. *Guide to Field Experiments*, 1908.
57. Ibid., 1910.
58. Ibid., 1912.
59. HAUC, Report of an Inspection of Harper Adams Agricultural College, Newport, Salop, held by the Board of Agriculture and Fisheries in June 1913, 11/1.
60. Newscuttings 1901-04, *NA*, 18 July 1903.
61. Ibid., 8 August 1903.
62. Newscuttings 1912, *Staffordshire Advertiser*, 3 August 1912.
63. *Guide to Field Experiments*, 1905.
64. Newscuttings 1904-10, *NA*, 30 July 1904.
65. Newscuttings 1901-04, *NA*, 10 May 1902.

Chapter 4

1. SRRC, *SC*, 26 June 1914.
2. Ibid., 3 July 1914.
3. Report of an Inspection by the Board of Agriculture in June 1913, 11/1.
4. HAUC, Board of Agriculture Return, 1912.
5. Newscuttings 1911, *NA*, 29 July 1911.
6. The author is indebted to Mr Sheppy, Librarian for the Royal Agricultural Society of England at Stoneleigh.
7. Board of Agriculture and Fisheries, *Fifth Report of the Rural Education Conference 1912, Courses in Agricultural Colleges*, Cd 6151 (London: HMSO, 1912).
8. National Agricultural Examination Board, Report on the Results of the Fifteenth Examination for the NDA, *Journal of the Royal Agricultural Society of England*, 75 (1914), 243.
9. Newscuttings 1901-04, *NA*, 27 April 1901.
10. HAUC, Charles Crowther, Memorandum on the Post-war Needs of Harper Adams Agricultural College - Reconditioning, Modernization and Extension, 4 July 1944.
11. HAUC, Memoirs of Ruth Capron, 102/13B.
12. HAUC, Misc. File Laboratory Equipment etc., Sliding Partition 1913.
13. Charles Crowther, 'In the Beginning, a Series of Articles dealing with the Development of Educational and Research Institutes - Harper Adams Agricultural College', *Agricultural Progress*, 27, part 1 (1952).
14. Prospectus 1908, 6/8.
15. Interview with John Wickstead.
16. *Report upon Agricultural Education*, Cd 4206, Summary.
17. SRRC, *Harper Adams Agricultural College, Newport, Salop* ([Newport: Harper Adams College, 1907?]).
18. HAUC, The Harper Adams Foundation Scheme, Alterations, 14 May 1915, 1/1B; Alfred Higgins Burne; Report of an Inspection by the Board of Agriculture in June 1913, 11/1.
19. Prospectus [1902?], 6/5.
20. Newscuttings 1911, unidentified article.
21. 'The College Dairy', *Cat-A-Mountain*, 1953.
22. 'Fifty Years On', *Ibid.*, 1952.
23. HAUC, Misc. File Income from Trust Money etc., Soiling Farm.
24. Newscuttings 1911, *Staffordshire Weekly Sentinel*, 5 August 1911.
25. HAUC, Stanier Correspondence, 102/5.
26. HAUC, Farm Guide, 1911.
27. Ibid., 1912.
28. HAUC, J. C. Brown, *Soiling of Milch Cows Experiment 1918* ([Newport]: Harper Adams Agricultural College, 1918), 14/4.
29. 'Farm Notes', *Cat-A-Mountain*, May 1919.
30. HAUC, Finance Committee, 10 October 1928, 104/16.
31. HAUC, Misc. File Advisory Matters, Report of a Conference on Agricultural Education convened by the Right Hon. Walter Runciman, MP, President of the Board of Agriculture and Fisheries, at Stafford 23 July 1912.
32. Paul Brassey.
33. *Report of the Committee on Post-War Agricultural Education in England and Wales*, Cd 6433 (London: HMSO, 1943).
34. HAUC, Board of Agriculture and Fisheries, *Agricultural Education in Connection with Farm Schools and Farm Institutes* ([London]; [HMSO], March 1913).
35. Newscuttings 1912, *Staffordshire Advertiser*, 3 August 1912.
36. Ibid., SC, 29 November 1912; *Border Counties Advertiser*, 27 November 1912.
37. Newscuttings 1913, *Staffordshire Weekly Sentinel*, 12 July 1913.

38. Ibid., *The Times*, 21 February 1913.
39. 'Winnowings', *Cat-A-Mountain*, 1951.
40. *Report upon Agricultural Education*, Cd 4207, pp. 222-230.
41. M. E. Telford, P. H. Holroyd and R. G. Wells, *History of the National Institute of Poultry Husbandry* (Newport: Harper Adams Agricultural College, 1986).
42. 'Thirty Years On', *Cat-A-Mountain*, 1931.
43. Newscuttings 1917, *Staffordshire Weekly Sentinel*, 18 August 1917.
44. HAUC, *Guide to Field Experiments*, 1913.
45. Ibid., 1912.
46. Newscuttings 1917, *Border Counties Advertiser*, 15 August 1917.
47. Ibid., *NA*, 18 August 1917.

Box 4

a. Prospectus [1903?], 6/6.

Chapter 5

1. HAUC, Newscuttings 1916, *SC*, 4 February 1916, 11 February 1916.
2. *Report upon Agricultural Education*, Cd 4207, pp. 208-214.
3. Carol Twinch, *Women on the Land* (Cambridge: Lutterworth Press, 1990).
4. Newscuttings 1915, *NA*, 30 January 1915.
5. Ibid., *The Times*, 10 May 1915; *NA*, 17 August 1915.
6. Ibid. 1915 & 1916, *Lady's Pictorial*, 5 June 1915; *NA*, 5 August 1916.
7. Ibid. 1915, unidentified article.
8. Ibid., *NA*, 8 May 1915.
9. *Journal of the Board of Agriculture*, 25 (April 1918-March 1919).
10. The author is grateful to Peter Watson Jones for bringing his father's Diary to her attention.
11. SRRC, Shropshire War Agricultural Committee, Labour Committee, 26 November 1918.
12. Newscuttings 1916, *NA*, 3 June 1916.
13. Ibid. 1915, *NA*, 24 April 1915.
14. 'On the Making of a Fringe Biologist', *Cat-A-Mountain*, 1976; Principal's Report, 1919.
15. Newscuttings 1918, *NA*, 16 February 1918.
16. Ibid. 1915, *Science Monitor*, 7 December 1915.
17. Ibid. 1916, *NA*, 12 August 1916.
18. Ibid. 1914, *Staffordshire Advertiser*, 19 September 1914.
19. Ibid. 1917, *NA*, 1 September 1917.
20. Ibid. 1918, *NA*, 18 May 1918.
21. Ibid. 1917, *NA*, 22 September 1917.
22. *Guide to Field Experiments*, 1914.
23. Misc. File Advisory Matters, Provincial Advisory Council for Higher Agricultural Education, Minutes of the Meeting at Shire Hall Shrewsbury 25 November 1913.
24. Newscuttings 1917, *Staffordshire Advertiser*, 11 August 1917.
25. 'Roll of Honour', *Cat-A-Mountain*, July 1919, Spring 1920.
26. 'Harper Adams Club', *Ibid.*, 1971.
27. 'SUC', *Ibid.*, Autumn 1921.
28. 'Memories of Douglas Scotland' and 'Library', *Ibid.*, Spring 1922.
29. 'SUC', *Ibid.*, Spring 1921.

Box 5

a. *Cat-A-Mountain*, 1921,1922,1939.

Chapter 6

1. 'A Contrast', *Cat-A-Mountain*, July 1919.
2. Memoirs of John Capron, 102/13C.
3. Board of Agriculture Return, 1922.
4. Ibid., 1920.
5. 'Principal's Address', *Cat-A-Mountain*, Autumn 1919.
6. SRRC, SCC Minutes 1921-22, Agricultural Committee 7 May 1921, C46.5.
7. 'Editor's Chair', *Cat-A-Mountain*, May 1919.
8. Student File 31. The author is indebted to Julie Brook for her excellent research involving the student files.
9. Thomas E. Whittle, *A Triumph of Science. A Seventy Year History of the UK Poultry Industry* ([n.p.]: Poultry World [n.d.]).
10. HAUC, National Poultry Council Diploma Examination Board, National Diploma in Poultry Husbandry, Examiners Awards 1922-1935, 200/2.
11. Prospectus 1921, 6/11.
12. Advertiser Offices, Newport, *NA*, 13 August 1921.
13. Ibid., 8 December 1923.
14. Board of Agriculture Return, 1923.
15. SRRC, Harper Adams Agricultural College Letter Book, 3670/1/31.
16. Board of Agriculture Return, 1919.
17. PRO, Papers Ministry Agriculture and Fisheries 1891-1907, *Report of the Divisional Inspection 28 May 1921*, MAF 33 13 TE929.
18. The author is indebted to Tavia Maclean, Newport & District History Society, for the information.

19. Charles Crowther, Post-war Needs of Harper Adams Agricultural College, 4 July 1944.
20. 'Harper Adams Club', *Cat-A-Mountain*, 1952.
21. 'Charivaria', *Ibid.*, Autumn 1920.
22. PRO, Papers Ministry Agriculture, Certificate by A. M. Hickman n.d., MAF 33 13 TE929.
23. HAUC, Minutes of the Students Union, 2 July 1920, 123/11.
24. Memorandum 1933, 101/7.
25. PRO, Papers Ministry Agriculture, Note from the Ministry to Liddle 28 April 1921, MAF 33 13 TE929.
26. Minutes of the Students Union, 7 July 1920, 123/11.
27. PRO, Papers Ministry Agriculture, Letter from Liddle 5 January 1923, MAF 33 13 TE929.
28. Ibid., Letter from the Lords Commissioners of His Majesty's Treasury 18 June 1923.
29. Student File 148.
30. Ibid. 197.
31. Minutes of the Students Union, 7 March 1922.
32. 'College Notes' and 'Editorial', *Cat-A-Mountain*, Summer 1921.
33. 'College Notes', *Ibid.*, Autumn 1921.
34. Student Files 26, 137.
35. 'College Notes', *Cat-A-Mountain*, Spring 1922.
36. Charles Crowther, *Agricultural Progress*, 27, part 1 (1952).
37. Student File 204.
38. Interview with Tony Gale.
39. 'College Times', *Cat-A-Mountain*, 1947.
40. Misc. File Hedworth Foulkes, Superannuation.
41. SRRC, *SC*, 3 February 1922.
42. Student File 260.
43. Ibid. 153.
44. Ibid. 149.
45. Minutes of the Students Union, 6 February 1922.
46. Ibid. 24 March 1922.
47. 'Obituary', *Cat-A-Mountain*, 1938.
48. 'A Tribute', *Ibid.*, Summer 1922.
49. Student File 179.
50. Ibid. 167.
51. Ibid. 311.

Box 6

a. *Guide to Field Experiments*, 1919
b. *Cat-a-Mountain*, 1986.
c. Advertiser Offices, Newport, *NA*, 6 August 1921.
d. The Author is grateful to Reg Anslow for the information.

Chapter 7

1. 'Ave Atque Vale' and 'A Retrospective', *Cat-A-Mountain*, 1944.
2. 'A New Start', Cat-A-Mountain, 1988.
3. HAUC, Obituaries D. H. Robinson, 101/33-34.
4. Interview with John Wickstead.
5. *Cat-A-Mountain*, Winter 1923.
6. 'College Notes', *Ibid.*, Winter 1922.
7. Obituary W. Morley Davies, Ibid., 1981.
8. Finance Committee, 5 November 1924, 104/4.
9. Interview with John Large.
10. 'Thomas Oakley Wilson', *Cat-A-Mountain*, 1980.
11. Memoirs of and Interview with Martin Cross, 102/103.
12. Interview with Mellor Boyns, 102/104.
13. Ibid. Irene Doody, 102/109.
14. 'Howes Appreciation', *Cat-A-Mountain*, 1951.
15. *Who Was Who*, 6 (1961-70), (London: Adam & Charles Black,1972).
16. 'Charles Crowther', *Cat-A-Mountain*, 1990.
17. Interview with Albert Rowley, Additional Comments of Basil Bayne.
18. *Cat-A-Mountain*, Summer 1923.
19. Interview with Willard Moore,102/112.
20. 'Where were You in '32?', *Cat-A-Mountain*, 1985.
21. Interview with Tony Gale.
22. FGP, 10 September 1930, 104/30.
23. Ibid., 12 November 1930, 104/31.
24. HAUC, Charles Crowther, 'Science and the Agricultural Crisis', *British Association for the Advancement of Science*, 1923.
25. Advertiser Offices, Newport, *NA*, 11 August 1923.
26. Ministry of Agriculture Returns, 1925.
27. Conversation with Dorothy Gale, nee Saunders.
28. Obituary W. B. Thompson, *Cat-A-Mountain*, 1964.
29. *Farm Guide*, 1928.
30. FGP, 23 July 1937.
31. Finance Committee, 9 March 1932.
32. SRRC, SCC Minutes Book, Education, Horticultural & Rural Industries Sub Committee 10 April 1923, Letter from Charles Crowther to the Council 4 April 1923, SC15/2A1/1; Advertiser Offices, Newport, NA, 21 April 1923.

33. Memorandum 1933, 101/7.
34. SRRC, SCC Reports 1936-37, FGP Committee 7 November 1936, C46.5.
35. FGP, 9 February 1938.
36. Prospectus 1928, 6/13B.
37. Ibid. 1937, 6/26.
38. 'Fifty Years Ago', *Cat-A-Mountain*, 1975.
39. Interview with Margaret Moore, nee Brewer.
40. Minutes of the Students Union, 1 June 1922.
41. Interview with Albert Rowley.
42. 'Obituary', *Cat-A-Mountain*, 1956.
43. HAUC, Memoirs of Leonora Cottrell, nee Grice, 102/107.
44. Student File 167.
45. '51 Years in Retrospective', *Cat-A-Mountain*, 1977.
46. Interview with Connie Hedges and John Wickstead.
47. 'Harper as I know it', *Cat-A-Mountain*, 1962.
48. Charles Crowther, Post-war Needs of Harper Adams Agricultural College, 4 July 1944.
49. Finance Committee, 10 September 1936.
50. Newscuttings 1963-77, *NA*, 7 April 1967.
51. 'In Memoriam', *Cat-A-Mountain*, 1936.
52. Barbara Woodhouse, *Just Barbara* (London: Michael Joseph & Rainbird, 1981).
53. Board of Agriculture Returns, 1923.
54. A. D. Hall.
55. 'Editorial', *Cat-A-Mountain*, 1934.
56. Prospectus 1930, 6/17.
57. Interview with Tony Chart.
58. 'Memories of Bill Parker', *Cat-A-Mountain*, 1986.
59. Prospectus 1945, 6/30A.
60. Ibid. 1927, 6/13A.
61. 'Poultry Institute Club', *Cat-A-Mountain*, 1932.
62. Interview with Pat Davies.
63. 'SUC', *Cat-A-Mountain*, 1939.
64. *Ibid.*, Winter 1923.
65. *Ibid.*, 1935.
66. *Ibid.*, 1929.
67. 'Entertainments', *Ibid.*, 1933.
68. 'Entertainments', *Ibid.*, 1930.

Chapter 8

1. Memorandum 1933, 101/7.
2. Advertiser Offices, Newport, *NA*, 28 December 1922.
3. Colin J. Holmes, 'Science and the Farmer', *Agricultural History Review*, 36 (1988), 77-86.
4. Advertiser Offices, Newport, *NA*, 11 October 1924.
5. 'A Retrospective', *Cat-A-Mountain*, 1944.
6. Interview with John Wickstead.
7. Charles Crowther, Post-war Needs of Harper Adams Agricultural College, 4 July 1944.
8. Board of Agriculture Returns, 1923.
9. 'Fifty Years Ago', *Cat-A-Mountain*, 1975.
10. 'Obituary', *Cat-A-Mountain*, 1970.
11. 'A New Start', *Cat-A-Mountain*, 1988.
12. Board of Agriculture Returns, 1924.
13. 'Obituary', *Cat-A-Mountain*, 1989.
14. Ministry of Agriculture Returns, 1920.
15. Finance Committee, 5 November 1924.
16. Ibid., 14 November 1928.
17. Ibid., 1 May, 12 June, 23 October 1929.
18. Ibid., 11 February 1931.
19. Interview with Pat Davies.
20. Finance Committee, 13 May 1937.
21. Prospectus 1928.
22. 'Obituary', *Cat-A-Mountain*, 1981.
23. 'Farm Notes', *Cat-A-Mountain*, 1928.
24. Interview Mellor Boyns.
25. E. Crompton and D. A. Osmond, 'The Soils of the Wem District of Shropshire', *Memoirs of the Soil Survey of Great Britain, Ministry of Agriculture and Fisheries* (London: HMSO, 1954).
26. The author is grateful to Paul Beckwith for the information.
27. *Annual Catalogues of British Government Publications* (1951-60).
28. Interview with Frank Ward.
29. A History of Shropshire, ed. by G. C. Baugh, *The Victoria History of the Counties of England*, 4 (Oxford: Oxford University Press for the Institute of Historical Research, 1989).
30. Interview with Willard Moore.
31. FGP, 9 February 1938.
32. Prospectus December 1945.
33. Interview with Frederick (Jim) Wickstead.
34. A. D. Hall.
35. SCC Reports, Letter from Charles Crowther 4 April 1923, SC15/2A1/1.
36. Mary Boutflour, *Bobby Boutflour* (London: Crosby, Lockwood & Son, 1965).

37. 'No-Roots Boutflour', *Cat-A-Mountain*, 1978.
38. Prospectus 1927.
39. *Farm Guide*, 1929.
40. 'Query?', *Cat-A-Mountain*, 1927.
41. Prospectus 1932-33.
42. *Farm Guide*, 1932.
43. 'Harper Adams Day', *Cat-A-Mountain*, 1933.
44. 'College Times', *Cat-A-Mountain*, 1939.
45. 'The Work of Harper Adams College Pig Feeding Experimental Station 1926-31', *Journal of the Royal Agricultural Society of England*, 92 (1932); 'Pig Feeding Facts: A Summary of the Work of the Harper Adams College Pig Feeding Experimental Station 1926-38', *Journal of the Bath & West and Southern Counties Society*, 13 (1939).
46. Interview with Tony Gale.
47. HAUC, *NA*, 6 November 1926.
48. '51 Years in Retrospect', *Cat-A-Mountain*, 1977.
49. Interview with Albert Rowley, Additional Comments of Basil Bayne.
50. 'Impressions of the Duke of York's Visit', *Cat-A-Mountain*, 1927.
51. 'NIPH -Eggs On', *Ibid.*, 1976.
52. 'A Golden Jubilee', *Ibid.*, 1977; Newscuttings 1949-62, *The Shropshire Magazine*, December 1950.
53. 'Modern Science and Agriculture, Recent Progress in Poultry Husbandry, The Organization of Education and Research', *Agricultural Progress*, 8 (1931), 7-33.
54. Advertiser Offices, Newport, *NA*, 8 December 1923.
55. HAUC, General Purposes Committee, 2 January 1925.
56. Interview with Willaim Neate, 102/101.
57. Ibid. Margaret Moore, nee Brewer.
58. Finance Committee, 5 November 1924; Advertiser Offices, Newport, *NA*, 8 November 1924.
59. 'The New Poultry Director', *Cat-A Mountain*, 1928.
60. Interview with Albert Rowley.
61. 'Reminiscences', *Cat-A-Mountain*, 1932.
62. 'Opening of the NIPH by HRH the Duke of York', *Cat-A-Mountain*, 1927.
63. Interview with Cecil Warner.
64. 'NPI Notes', *Cat-A-Mountain*, 1939.
65. 'College Times', *Ibid.*, 1933.
66. 'Notes from the NIPH', *Ibid.*, 1938.
67. Temperton Scrapbook, p. 300. The author is grateful to Godfrey and Ron Temperton for the loan of their father's Scrapbook.
68. Thomas E. Whittle.
69. 'College Times', *Cat-A-Mountain*, 1938.
70. HAUC, Memorandum NPI Director, 104/55.
71. Temperton Scrapbook, p. 403.
72. Telford, Holroyd and Wells.
73. 'National Institute of Poultry Husbandry', *Cat-A-Mountain*, 1929.
74. Charles Crowther, 'In the Beginning'.
75. **HAUC**, Harper Adams Club, The Eighth Annual Autumn Meeting and Luncheon of the Harper Adams Club, 3 December 1929.
76. Prospectus 1934.
77. National Poultry Council Diploma Examination Board, National Diploma in Poultry Husbandry, Regulations and Syllabus of the Examinations as revised 25 September 1931, 200/3.
78. Finance Committee, 3 August 1939.

Box 8

a. Charles Crowther, 'In the Beginning'.
b. 'College Times', *Cat-A-Mountain*, 1925.
c. Interview with William Neate.
d. Tony Giles, See You at Oxford, A Celebration of Fifty Oxford Farming Conferences Over Sixty Years (Reading: Oxford Farming Conference & University of Reading,1995).

Chapter 9

1. Interview with Dorothy Neate.
2. Ibid. Margaret Moore, nee Brewer.
3. Newport & District History Society, *NA*, 3 October 1997.
4. Interview with Willard Moore.
5. Ibid. Tony Gale and Dora Gale, nee Saunders.
6. Ibid. Anthony Chart and Robert Tupper.
7. Ibid. John Large and Paula Large, nee Johnson.
8. *Cat-A-Mountain*, 1944.
9. *Ibid.*,1945.
10. Student File 1324.
11. *Cat-A-Mountain*, 1942.
12. *Ibid.*, 1940.
13. Interview and Memoirs Martin Cross.
14. *Cat-A-Mountain*, 1943.
15. 'Looking Back', Ibid., 1982.
16. 'On the Making of a Fringe Biologist', *Ibid.*, 1976.
17. 'Harper Adams Club', *Ibid.*, 1952.

18. PRO, Records Information, Records of the National Farm Surveys of England and Wales, 1940-1943 (London: HMSO, 1993).
19. Memoirs of Leonora Cottrell, nee Grice.
20. Telford, Holroyd and Wells.
21. '51 Years in Retrospect', *Cat-A-Mountain*, 1977.
22. 'Principal's Report', *Ibid.*, 1953; Newscuttings 1949-62, *Poultry World*, 24 September 1953.
23. Charles Crowther, Post-war Needs of Harper Adams Agricultural College, 4 July 1944.
24. Finance Committee, 14 April 1938.
25. 'Reminiscing', *Cat-A-Mountain*, 1977.
26. HAUC, Freehold Properties in the Parish of Edgmond, 31 January 1958.
27. Prospectus 1934.
28. Finance Committee, 6 August 1924.
29. *Farm Guide*, 1930.
30. HAUC, *Sport and Country*, 24 November 1944.
31. *Who Was Who*, 6 (1961-70), (London: Adam & Charles Black, 1972); Newscuttings 1960s, unidentified article, 19 June 1965.
32. HAUC, Obituary Archibald McVicar, 101/30.
33. *Cat-A-Mountain*, 1946.
34. Interview with Peter Luscombe.

Box 9

a. The author is grateful for the information provided by Karl Cieciora and Bernard Lowry.
b. Principal's Report, February 1942.
c. *The Times*, 17 May 1940.
d. *Ibid.*, 23 July 1940.
e. SRRC, Information on the Eleventh Battalion Home Guard, 3216/13.
f. 'B4 Platoon', *Cat-A-Mountain*, 1943; Interview with John Wickstead.
g. 'Editorial', *Cat-A-Mountain*, 1942.
h. Obituary Professor Mark Carpenter, *Ibid.*, 1979.
i. Interviews with Tony Gale and John Large.
j. A. W. Churchill, *From 'Stand-to' to 'Stand-down'* (Hereford: Jakemans Ltd, 1946).

Chapter 10

1. Interview with Irene Doody, 102/109.
2. Ibid. Tom Fordham, 102/108.
3. Ibid. Maurice Lancaster.
4. Ibid. Bob Bradley.
5. *Cat-A-Mountain*, 1960.
6. Newscuttings 1960s, *Express & Star*, 27 June 1964.
7. Ibid., *NA*, 10 June 1950.
8. *Cat-A-Mountain*, 1962.
9. Interview with Willard Moore.
10. Ibid. Pat Davies.
11. Ibid. Jim Wickstead.
12. G. C. Baugh.
13. FGP, 11 July 1947.
14. Prospectus 1961.
15. Michael Argles.
16. Newscuttings 1949-62, *Journal of the Ministry of Agriculture*, 56 (1949), No. 4.
17. *Cat-A-Mountain*, 1950.
18. Interview with Basil Bayne.
19. Ibid. David Roberts.
20. HAUC, Ministry of Agriculture's Comments on the Recommendations of the Report of the Committee on the Agricultural Colleges 1959, 10 February 1960.
21. HAUC, Letter from the Ministry of Agriculture to W. T. Price, 14 April 1955.
22. Ibid. Ministry of Agriculture to T. C. Ward, 30 April 1954.
23. FGP, 6 May 1955.
24. Interview with R. Derek Turner.
25. Ibid. Ian Gedye.
26. Ibid. Steven Colman (Steve Kelman).
27. Prospectus December 1945.
28. Telford, Holroyd and Wells.
29. Newscuttings 1960s, *NA*, 14 July 1961.
30. Ibid., *Birmingham Post*, 14 January 1955.
31. *Cat-A-Mountain*, 1959.
32. 'Twenty-Nine Years *Ibid.*, 1977.
33. Interview with Peter Luscombe.
34. Ibid. Maurice Eddowes.
35. Ibid. Anthony Warner.
36. 'Reminiscing', *Cat-A-Mountain*, 1977.
37. HAUC, Letter from W. T. Price to Ministry of Agriculture, 11 June 1955.
38. HAUC, Extracts from the Report of the Committee on the Agricultural Colleges 1954.
39. FGP, 19 May 1947.
40. Prospectus 1949.
41. Student File 2406.

42. HAUC, Extracts from the Report of the Committee on the Agricultural Colleges 1959.
43. *Cat-A-Mountain*, 1953.
44. Interview with Anthony Chart and Robert Tupper.
45. Newscuttings 1949-62, *Wolverhampton Chronicle*, 13 June 1952.
46. *Cat-A-Mountain*, 1952.
47. FGP, 23 March 1956.
48. HAUC, Letter from R. P. Liddle to the Board of Education, 25 August 1947.
49. FGP, 13 September 1946.
50. HAUC, Letter from the Ministry of Agriculture to W. T. Price, 10 February 1948.
51. Ibid. from R. P. Liddle to the Ministry of Education, 28 December 1951.
52. Ibid. from the Ministry of Education to R. P. Liddle, 18 March 1952.
53. Newscuttings 1949-62, *NA*, 26 November 1954.
54. Ibid., 15 June 1951.
55. Newscuttings 1949-62, *Wellington Journal*, 21 January 1950.
56. FGP, 24 June 1949.
57. *Cat-A-Mountain*, 1951.
58. FGP, 14 January 1949.
59. HAUC, Letter from W. T. Price to the Ministry of Agriculture, 3 July 1952.
60. FGP, 11 May 1951.
61. HAUC, Letter from the Ministry of Agriculture to W. T. Price, 26 March 1954.
62. Ibid., 4 November 1953.
63. HAUC, Report of the Meeting of Chief Education Officers, 13 April 1949.
64. FGP, 11 February 1949.
65. Ibid., 10 February 1950.
66. HAUC, Staff Meeting, 13 September 1963.
67. HAUC, Minister of Education, The Harper Adams Foundation Scheme, Alterations, 8 January 1951, 1/4A.
68. FGP, 9 August 1957.
69. Newscuttings 1949-62, *NA*, 18 September 1953.

Box 10

a. Interview with Irene Doody.
b. Ibid. Tom Fordham.
c. Newscuttings 1949-62, Express & Star, 20 July 1955.
d. Cat-A-Mountain, 1994.

Chapter 11

1. Newscuttings 1949-62, *NA*, 14 July 1961.
2. Thomas E. Whittle.
3. Interview with Maurice Lancaster.
4. *Cat-A-Mountain*, 1962.
5. Interview with Willard Moore.
6. *Cat-A-Mountain*, 1950.
7. HAUC, Letter from the Ministry of Agriculture to W. T. Price concerning the Proposed Transfer of the NIPH to the College, 8 June 1949.
8. FGP, 11 November 1949.
9. Ibid., 10 August 1951.
10. Temperton Scrapbook, p. 403.
11. Interview with Anthony Warner.
12. Ibid. Basil Bayne.
13. Ibid. Jim Wickstead.
14. Temperton Scrapbook, p. 382.
15. *Cat-a-Mountain*, 1948.
16. Newscuttings 1949-62, *Sport & Country*, 3 May 1950.
17. *Cat-A-Mountain*, 1951.
18. Newscuttings 1949-62, Sport & Country, 16 April 1952.
19. *Cat-A-Mountain*, 1952.
20. Newscuttings 1949-62, *Poultry World*, 24 September 1953.
21. *Cat-A-Mountain*, 1956.
22. Newscuttings 1949-62, *Sport & Country*, 14 April 1954.
23. David Charles, 'A Brief History of the UK Poultry Industry', *The World's Poultry Science Association United Kingdom Branch Fiftieth Anniversary 1946-1996* ([no details]).
24. '51 Years in Retrospect', *Cat-A-Mountain*, 1977.
25. *Cat-A-Mountain*, 1957.
26. Newscuttings 1949-62, *Sport & Country*, 17 April 1957.
27. *Cat-A-Mountain*, 1958.
28. *Ibid.*, 1959.
29. *Ibid.*, 1960.
30. Prospectus 1927, 6/13A.
31. 'Farm', *Cat-A-Mountain*, 1949.
32. Newscuttings 1949-62, *Daily Express*, 5 August 1949.
33. Temperton Scrapbook, p. 388.
34. Newscuttings 1949-62, *Sport & Country*, 28 September 1955, 13 April 1955.
35. *Cat-A-Mountain*, 1954.

36. Interview with Albert Rowley.
37. Temperton Scrapbook, p. 300.
38. Ibid., p. 302.
39. Newscuttings 1949-62, *Wolverhampton Chronicle*, 13 June 1952.
40. Telford, Holroyd and Wells.
41. Newscuttings 1949-62, *NA*, 26 October 1962.
42. Prospectus 1961, 6/35.
43. *Cat-A-Mountain*, 1961.
44. Interview with David Roberts.
45. FGP, 12 August 1955.
46. 'The Rabbitry', *Cat-A-Mountain*, 1965.
47. *Cat-A-Mountain*, 1964.
48. Interview with Tony Harris.
49. 'Twenty-Nine Years', *Cat-A-Mountain*, 1977.
50. Interview with Irene Doody.
51. Newscuttings 1949-62, *Farmer & Stock Breeder*, 22-23 September 1953.
52. Ibid., *Sport & Country*, 7 July 1954.
53. Prospectus 1957, 6/34A.
54. Interview with Peter Luscombe.
55. Newscuttings 1949-62, *The Shropshire Magazine*, September 1950.
56. HAUC, Information Sheet on the Farm, Pig Husbandry Experimental Station and Courses, n.d., 101/9.
57. 'HA Pig Husbandry Station', *Cat-A-Mountain*, 1955.
58. Interview with Tom Fordham.
59. HAUC, Notes by K. W. D. Campbell for the Attention of the Board of Governors, January 1947; Confidential Statement of K. W. D. Campbell to the Governors, 14 August 1947, 104/126A.
60. *Farm Guide*, 1951.
61. 'Reminiscing', *Cat-A-Mountain*, 1977.
62. G. C. Baugh.
63. *Journal of the Ministry of Agriculture*, 56, (1949), No. 4.
64. Interview with Robin Harris.
65. Newscuttings 1955-69, unidentified newspaper, December 1955.
66. Newscuttings 1949-62, *Express & Star*, 13 December 1955.
67. Interview with Maurice Eddowes.
68. Newscuttings 1949-62, *NA*, 6 April 1962.
69. *Cat-A-Mountain*, 1964.
70. Interview with R. Derek Turner.
71. Ibid. Bob Bradley.
72. 'Principal's Report', *Cat-A-Mountain*, 1953.
73. Newscuttings 1949-62, *NA*, 10 October 1952.
74. 'Obituaries', *Cat-A-Mountain*, 1982.
75. Interview with Ian Gedye.
76. Memoirs of Brian Lloyd-James.
77. Retirements, *Cat-A-Mountain*, 1970.
78. Memoirs of Don Findlay.
79. 'The Troffe', *Cat-A-Mountain*, 1971.
80. 'Editorial', *Cat-A-Mountain*, 1947.
81. Interview with Shirley Barber.

Box 11

a. Memoirs of Gerard Poulter.
b. 'Tommy Bates', *Cat-A-Mountain*, 1956.
c. Newscuttings 1949-62, *Express & Star*, 4 November 1959.

Chapter 12

1. 'Obituary', *Cat-A-Mountain*, 1987.
2. *Who's Who 1976* (London: A & C Black, 1976).
3. Interview with Maurice Eddowes.
4. Ibid. Christine Parnham, nee Justice, 102/114.
5. Ibid. Trevor Parnham,102/113.
6. Ibid. Mellor Boyns.
7. Ibid. Alan Costley.
8. Ibid. Ian Gedye.
9. Ibid. Bob Bradley.
10. Ibid. Jim Wickstead.
11. HAUC, Staff Meeting, 17 September 1962.
12. Interview with Maurice Lancaster.
13. Ibid. Peter Watson Jones.
14. HAUC, Schedule of Accommodation [1965-66?].
15. Newscuttings 1955-69, *Caterer & Hotelkeeper*, 5 September 1963.
16. Newscuttings 1963-77, *Birmingham Post*, 30 April 1963.
17. Staff Meeting, 21 December 1962.
18. FGP, 31 January 1964; Staff Meeting, 9 July 1963.
19. Staff Meeting, 20 December 1963.
20. FGP, 14 January 1963.
21. 'Principal's Notes', *Cat-A-Mountain*, 1967.
22. HAUC, A Review of Harper Adams to be submitted to the International Conference September 1975, 101/17.
23. HAUC, Extracts from the Report of the Committee on the Agricultural Colleges, 1959.

24. HAUC, Principal's Report on the College Finances [1963?].
25. HAUC, Meeting at the DES on 30 September 1964; Principal's Report on the Financial Problems facing the College with Special Reference to the Bank Overdraft [1964?].
26. HAUC, Principal's Report on Agricultural Education with Proposals for Future Course Structures and Estimated Student Numbers at Harper Adams Agricultural College [1972-73?].
27. HAUC, Principal's Report on Development Plans, 25 October 1965.
28. HAUC, Report on the Administrative Running of Harper Adams College, 9 December 1969.
29. FGP, 29 January 1965.
30. HAUC, Grant received from the DES, July 1975.
31. HAUC, Report of the Meeting held at the DES to discuss Capital Projects, 23 March 1966.
32. Staff Meeting, 4 March 1964.
33. HAUC, An Outline of Present & Proposed Future Developments at the College [Autumn 1962?].
34. Staff Meeting, 15 June 1965.
35. FGP, 21 January 1977.
36. Staff Meeting, 21 December 1966.
37. *Cat-A-Mountain*, 1977.
38. 'SU President's Letter', *Cat-A-Mountain*, 1968.
39. Staff Meetings, 20 January 1971, 10 September 1971; FGP, 23 July 1971.
40. Interview with John Roberts.
41. HAUC, Brief Progress Report and Future Developments [July 1968?].
42. Staff Meeting, 5 January 1970; FGP, 16 October 1970.
43. FGP, 23 April 1971.
44. Interview with Ros Barsley, nee Forrester.
45. 'SU President's Letter', *Cat-A-Mountain*, 1972.
46. *Cat-A-Mountain*, 1970.
47. FGP, 13 July 1973.
48. Ibid., 23 January 1976.
49. HAUC, Replies to the Points raised in the Committee's Letter, 3 July 1964.
50. Staff Meeting, 16 September 1966.
51. HAUC, Recommendations of the Finance & General Purposes Committee concerned with the Recommendations of the Report of the Committee on Further Education for Agriculture provided by the LEAs, 24 February 1959.
52. Michael Argles.
53. *Higher Education Report of the Committee appointed by the Prime Minister under the Chairmanship of Lord Robbins 1961-63*, Cd 2154 (London: HMSO [1963?]).
54. *DES Report of the Advisory Committee on Agricultural Education 1966* (London: HMSO, 1966).
55. HAUC, New Course Structure [March 1968?].
56. Staff Meeting, 21 September 1970.
57. HAUC, Report on New Courses [April 1969?].
58. Staff Meeting, 18 December 1968.
59. 'Principal's Notes', *Cat-A-Mountain*, 1973.
60. Staff Meeting, 28 June 1973.
61. Newscuttings 1963-77, *Shropshire Star*, 13 August 1976.
62. Staff Meeting, 10 September 1971.
63. *Cat-A-Mountain*, 1975.
64. Staff Meeting, 3 July 1969.
65. J. S. Hall, 'Developments in Formal Agricultural Education: a Selective Review', *Agricultural Progress*, 47 (1972).
66. HAUC, Principal's Interim Report on the Structure and Future Teaching Methods proposed for the Existing Two Year Diploma Course in Agriculture [January 1966?].
67. Staff Meeting, 15 July 1964.
68. Interview with Basil Bayne.
69. Staff Meeting, 20 October 1971.
70. Interview with Peter Luscombe.
71. Prospectus [1973?], 6/37.
72. 'AMBA', *Cat-A-Mountain*, 1971.
73. Staff Meeting, 15 September 1972.
74. FGP, 26 April 1974.
75. Academic Board, 29 November 1978.
76. 'Obituary', *Cat-A-Mountain*, 1982.
77. Staff Meeting, 13 May 1970.
78. Newscuttings 1969-76, *Farmers Weekly*, 23 July 1976.
79. Ibid. 1963-77, *Shropshire Star*, 25 May 1976, *Newport Advertiser*, 28 May 1976.
80. 'Overseas Club', *Cat-A-Mountain*, 1966.
81. HAUC, Report of the Committee set up to consider Future Courses of Education at the College, 24 October 1966.
82. Telford, Holroyd and Wells.
83. FGP, 14 July 1967.
84. Academic Board, 3 June 1971.
85. FGP, 17 October 1975.

86. Academic Board, 15 March 1978.
87. Staff Meeting, 26 March 1971.
88. Thomas E. Whittle.
89. Academic Board, 2 December 1971.
90. Newscuttings 1969-76, *Farmers Weekly*, 18 April 1975.
91. 'Principal's Notes', *Cat-A-Mountain*, 1976.
92. Staff Meeting, 13 May 1970.
93. Academic Board, 27 November 1970.
94. Joint Staff Student Committee, 13/11/72, 104/325.
95. Ibid. [1971?], 104/310.

Chapter 13

1. 'Harper Adams 1966-1968', *Cat-A-Mountain*, 1995.
2. Interview with Trevor Parnham.
3. Newscuttings 1955-69, *Newport Advertiser*, January 1963.
4. *Cat-A-Mountain*, 1970.
5. Interview with Irene Doody.
6. *Cat-A-Mountain*, 1966.
7. Newscuttings 1955-69, unidentified article, p. 64.
8. *Cat-A-Mountain*, 1967.
9. 'Those were the Days', *Cat-A-Mountain*, 1995.
10. Newscuttings 1955-69, unidentified articles, pp. 72-74.
11. Staff Meeting, 5 May 1971.
12. *Cat-A-Mountain*, 1969.
13. *Cat-A-Mountains*, 1962-1977.
14. Interview with Ian Gedye.
15. Ibid. Christine Parnham, nee Justice.
16. 'Editorial', *Cat-A-Mountain*, 1962.
17. Interview with Anthony Warner.
18. 'Motor Club', *Cat-A-Mountain*, 1961.
19. Interview with Peter Luscombe.
20. Staff Meeting, 16 September 1966.
21. Interview with Ros Barsley, nee Forrester.
22. FGP, 23 January 1970.
23. *Cat-A-Mountains*, 1960,1965,1966.
24. Staff Meeting, 17 September 1962.
25. 'SU President's Letter', *Cat-A-Mountain*, 1971.
26. Staff Meeting, 2 April 1965.
27. Ibid., 8 July 1965.
28. Ibid., 15 July 1964.
29. Ibid., 18 December 1970.
30. 'General Studies', *Cat-A-Mountain*, 1973.
31. Staff Meeting, 6 April 1966.
32. Ibid., 5 January 1970.
33. Ibid., 20 January 1971.
34. Staff Student Committee, 8 February 1971.
35. Newscuttings 1969-76, unidentified article, p. 1.
36. Ibid. 1955-69, *Shropshire Star*, 21 October 1965.
37. Ibid., *Newport Advertiser*, 20 October 1967.
38. *Cat-A-Mountain*, 1974.
39. 'Women at Harper', *Cat-A-Mountain*, 1968.
40. 'Ken Mathias', *Cat-A-Mountain*, 1988.
41. FGP, 22 April 1977.
42. Interview with Maurice Eddowes.
43. Ibid. Tom Fordham.
44. Ibid. Robin Harris.
45. FGP, 21 January 1977.
46. Farm Sub Committee, 20 December 1962.
47. Staff Meeting, 21 December 1962.
48. 'R. Kenney - Appreciation', *Cat-A-Mountain*, 1963.
49. Staff Meeting, 17 September 1962.
50. HAUC, An Outline of Present & Proposed Future Developments at the College [Autumn 1962?].
51. 'Principal's Notes', *Cat-A-Mountain*, 1965.
52. Staff Meeting, 12 July 1965.
53. Farm Sub Committee, 6 October 1969.
54. Ibid., 29 November 1968.
55. *Farm Guide*, 1971.
56. G.C. Baugh.
57. Newscuttings 1955-69, *Shropshire Journal*, January 1968.
58. 'SU President's Letter', *Cat-A-Mountain*, 1968.
59. Farm Sub Committee, 25 February 1977.
60. Ibid., 1 October 1969.
61. Ibid., 9 March 1973.
62. Newscuttings 1969-76, *Dairy Farmer*, February 1976; 'Harper Adams Club', *Cat-A-Mountain*, 1975.
63. Newscuttings 1969-76, unidentified article, p. 24.
64. 'Pig Husbandry Notes', *Cat-A-Mountain*, 1964.
65. HAUC, Progress Report on the Piggery Buildings, 28 February 1974; Report on Buildings Present and Future, 12 May 1975.
66. Farm Sub Committee, 12 March 1971.
67. 'Pig Technology Experiments', *Cat-A-Mountain*, 1972.
68. Newscuttings 1969-76, *Farmers Weekly*, October 1974.
69. FGP, 9 July 1965.

70. Ibid., 23 April 1971.
71. Ibid., 21 July 1972.
72. Ibid., 22 October 1976.
73. Ibid., 25 October 1974.
74. Farm Sub Committee, 13 March 1974.
75. FGP, 8 November 1974.
76. 'A Golden Jubilee', *Cat-A-Mountain*, 1977.
77. Telford, Holroyd and Wells.
78. 'Death of Cyril Block', *Cat-A-Mountain*, 1978.
79. Newscuttings 1977-94, *Newport Advertiser*, 22 August 1986.
80. Principal's Chronology.

Box 13

a. Information drawn from the Cat-A-Mountain and Harper Adams Club Files. I am grateful to Julie Brook for her research.

Chapter 14

The author is grateful to Tony Harris and Alan Costley for their considerable help with this chapter.

1. HAUC, Principal's Graduation Speech, 1988.
2. 'Tony Harris Principal 1977-94', *Cat-A-Mountain*, 1994.
3. Interview with Tony Harris.
4. Ibid. John Roberts.
5. Tony Harris personal papers, 1981.
6. Tony Harris, 'Is Agricultural Education on Course?', *Farming Issues*, 7 (Bristol, Lloyds Bank, 1992).
7. FGP, 17 January 1979, 13 July 1979.
8. Ibid., 20 January 1978, 26 October 1979.
9. Ibid., 21 April 1978.
10. 'Principal's Report', *Cat-A-Mountain*, 1978.
11. *Report of the Working Group on the Management of Higher Education in the Maintained Sector.* Cd 7130 (London: HMSO, 1978).
12. *The Development of Higher Education into the 1990s.* Cd 9524 (London: HMSO, 1985).
13. FGP, 18 October 1985.
14. *Cat-A-Mountain*, 1984.
15. *Ibid.*, 1992.
16. *Higher Education. Meeting the Challenge.* Cd 114 (London: HMSO, 1987).
17. *Cat-A-Mountain*, 1991.
18. Interview with Peter Watson Jones.
19. FGP, 7 October 1988.
20. HAUC, Strategic Plan, 1990-93.
21. *Cat-A-Mountain*, 1993, 1994.
22. Academic Board, 30 November 1988.
23. Strategic Plan, 1989-92.
24. Ibid., 1992-96.
25. Principal's Graduation Speech, 1989.
26. FGP, 23 January 1987.
27. HAUC, Rolling Plan, 1980-86.
28. FGP, 17 January 1992.
29. Academic Board, 10 June 1992.
30. Strategic Plan, 1993-97.
31. FGP, 22 January 1993.
32. Academic Board, 13 October 1977; FGP, 14 July 1978.
33. HAUC, Report of the Committee to consider Future Courses of Education at the College, 28 September 1966, 104/268.
34. Interview with Alan Costley.
35. Academic Board, 11 June 1980.
36. *Cat-A-Mountain*, 1982.
37. Tony Harris personal papers, 1980.
38. HAUC, CNAA Institutional Review, April 1989.
39. Newscuttings 1913, *Staffordshire Weekly Sentinel*, 19 July 1913.
40. Interview with Paula Large, nee Johnson.
41. FGP, 22 April 1983.
42. Ibid., 3 February 1989.
43. Interview with Ian Robson.
44. HAUC, Planning Proposals, Submission to NAB, 1987-88.
45. Academic Board, 29 March 1984.
46. 'PICKUP', *Cat-A-Mountain*, 1985.
47. Academic Board, 12 March 1986.
48. Tony Harris personal papers, 1989.
49. HAUC, Report prepared for the Visit of HMI 16-20 October 1978, 11/3.
50. HAUC, Aspects of Provision, 12-15 November 1990, a Report by HMI.
51. Rolling Plan, 1981-86.
52. FGP, 9 October 1987, 29 January 1988.
53. Ibid., 12 January 1990.
54. Newscuttings 1988, unidentified article.
55. Prospectus 1934, 6/22.
56. Newscuttings 1990-92.

Chapter 15

The author is grateful to Tony Harris for his considerable help with this chapter.

1. Interview with Peter Luscombe.
2. *Cat-A-Mountain*, 1978.
3. FGP, 21 October 1977.
4. Strategic Plan, 1989-92.
5. *The Development of Higher Education into the 1990s*. Cd 9524 (London: HMSO, 1985).
6. Interview with Tony Harris.
7. *Cat-A-Mountain*, 1984.
8. FGP, 1 July 1983.
9. Ibid., 4 May 1984.
10. 'Temperton Fund', *Cat-A-Mountain*, 1989.
11. Academic Board, 29 March 1984.
12. *Cat-A-Mountain*, 1985.
13. Interview with John Roberts.
14. Academic Board, 11 December 1991.
15. Principal's Graduation Speech, 1992.
16. Strategic Plan, 1993-97.
17. FGP, 5 October 1990.
18. Ibid., 8 June 1990, 25 January 1991.
19. Academic Board, 11 July 1990.
20. Information from Dick Wells.
21. Telford, Holroyd and Wells.
22. FGP, 10 July 1981.
23. Ibid., 31 May 1991.
24. Ibid., 28 June 1985.
25. FGP, 4 February 1994.
26. *Cat-A-Mountain*, 1987.
27. Report prepared for the Visit of HMI 1978, 11/3.
28. *Cat-A-Mountain*, 1986.
29. 'Principal's Notes', Ibid., 1983.
30. Tony Harris personal papers, 1983.
31. Rolling Plan, 1980-86.
32. 'Computers at Harper Adams', *Cat-A-Mountain*, 1992.
33. *Cat-A-Mountain*, 1993.
34. HAUC, The College Strategy, 8 May 1985.
35. FGP, 11 October 1991.
36. Ibid., 17 January 1992.
37. Ibid., 12 January 1990.
38. Interview with Ian Robson.
39. FGP, 20 January 1978.
40. Ibid., 20 April 1979.
41. Ibid., 19 June 1992.
42. *Cat-A-Mountain*, 1982.
43. Interview with Ed and Liz Pridgeon, nee Edgeworth.
44. 'Principal's Notes', *Cat-A-Mountain*, 1981.
45. Rolling Plan, 1981-86.
46. FGP, 21 April 1978.
47. Principal's Chronology.
48. HAUC, Julie Wickington and Andrew Wilcox, *Wildlife and Conservation at Harper Adams College* (Newport: Harper Adams Agricultural College, 1993), 8/6.
49. FGP, 9 October 1987.
50. Interview with Angela Billington, nee Holding.
51. Ibid. Ken Mathias.
52. *Cat-A-Mountain*, 1990.
53. Interview with Ian Baker.

Chapter 16

1. 'Principal's Appointment', *Cat-A-Mountain*, 1994.
2. FGP, 18 November 1994.
3. Ibid., 20 January 1995.
4. HAUC, Annual Review, 1998.
5. Interview with Wynne Jones.
6. *Cat-A-Mountain*, 1996.
7. 'SU President's Report', *Ibid.*, 1995.
8. Strategic Plan, 1996-2000.
9. HAUC, Tony Harris, Report on the Future of Agricultural Education presented to the Board of Governors, 13 December 1996.
10. Annual Review, 1997.
11. HAUC, Wynne Jones, Brief on the Dearing Report presented to the Board of Governors, December 1997.
12. FGP, 19 November 1995.

Index

Page numbers in italics refer to illustrations